141

Advances in Polymer Science

Springer-Verlag Berlin Heidelberg GmbH

Progress in Polyimide Chemistry II

Volume Editor: **H. R. Kricheldorf**

With contributions by
K. R. Carter, J. G. Dolden, C. J. Hawker,
J. L. Hedrick, H. R. Kricheldorf, J. W. Labadie,
K.-W. Lienert, R. D. Miller, T. P. Russell,
W. Volksen, D. Y. Yoon

 Springer

This series presents critical reviews of the present and future trends in polymer and biopolymer science including chemistry, physical chemistry, physics and materials science. It is addressed to all scientists at universities and in industry who wish to keep abreast of advances in the topics covered.

As a rule, contributions are specially commissioned. The editors and publishers will, however, always be pleased to receive suggestions and supplementary information. Papers are accepted for „Advances in Polymer Science" in English.

In references Advances in Polymer Science is abbreviated Adv. Polym. Sci. and is cited as a journal.

Springer WWW home page: http://www.springer.de

ISSN 0065-3195

ISBN 978-3-662-14718-4   ISBN 978-3-540-49814-8 (eBook)
DOI 10.1007/978-3-540-49814-8

Library of Congress Catalog Card Number 61642

© Springer-Verlag Berlin Heidelberg 1999
Originally published by Springer-Verlag Berlin Heidelberg New York in 1999
Softcover reprint of the hardcover 1st edition 1999
The use of registered names, trademarks, etc. in this publication does not imply, even in the absence of a specific statement, that such names are exempt from the relevant protective laws and regulations and therefore free for general use.

Typesetting: Data conversion by MEDIO, Berlin
Cover: E. Kirchner, Heidelberg
SPIN: 10648282   02/3020 - 5 4 3 2 1 0 – Printed on acid-free paper

Volume Editor

Professor Dr. H. R. Kricheldorf
Inst. für Technische und
Makromolekulare Chemie
Universität Hamburg
Bundesstraße 45

D-20146 Hamburg

E-mail: kricheld@chemie.uni-hamburg.de

Editorial Board

Preface

Over the past four decades polymers containing imide groups (usually as building blocks of the polymer backbone) have attracted increasing interest of scientists engaged in fundamental research as well as that of companies looking into their application and commercialization. This situation will apparently continue in the future and justifies that from time to time reviews be published which sum up the current state of knowledge in this field. Imide groups may impart a variety of useful properties to polymers, e. g., thermal stability chain stiffness, crystallinity, mesogenic properties, photoreactivity etc. These lead to a broad variety of potential applications. This broad and somewhat heterogeneous field is difficult to cover in one single review or monograph. A rather comprehensive monograph was edited four years ago by K. Mittal, mainly concentrating on procedures and properties of technical interest. Most reviews presented in this volume of Advances in Polymer Science focus on fundamental research and touch topics not intensively discussed in the monograph by K. Mittal. Therefore, the editor of this work hopes that the reader will appreciate finding complementary information.

Finally I wish to thank all the contributors who made this work possible and I would like to thank Dr. Gert Schwarz for the revision of the manuscripts of the contributions 3 and 4.

Hamburg, September 1998

Hans R. Kricheldorf

Contents

Contents of Volume 140

Progress in Polyimide Chemistry I

Volume Editor: H. R. Kricheldorf

Nanoporous Polyimides

J.L. Hedrick[1], K.R. Carter[1], J.W. Labadie[1], R.D. Miller[1], W. Volksen[1], C.J. Hawker[1], D.Y. Yoon[1], T.P. Russell[1], J.E. McGrath[2], R.M. Briber[3]

[1] NSF Center for Polymer Interfaces and Macromolecular Assemblies (CPIMA), IBM Research Division, Almaden Research Center, 650 Harry Road, San Jose, California 95120-6099, USA
[2] Virginia Polytechnic Institute and State University, 2111 Hahn Hall, Blacksburg, Virginia 24061-0344, USA
[3] University of Maryland, 2100 Marie Mount Hall, College Park, Maryland 20742-2115, USA

Foamed polyimides have been developed in order to obtain thin film dielectric layers with very low dielectric constants for use in microelectronic devices. In these systems the pore sizes are in the nanometer range, thus the term "nanofoam." The polyimide foams are prepared from block copolymers consisting of thermally stable and thermally labile blocks, the latter being the dispersed phase. Foam formation is effected by thermolysis of the thermally labile block, leaving pores of the size and shape corresponding to the initial copolymer morphology. Nanofoams prepared from a number of polyimides as matrix materials were investigated as well as from a number of thermally labile polymers. The foams were characterized by a variety of experiments including TEM, SAXS, WAXD, DMTA, density measurements, refractive index measurements and dielectric constant measurements. Thin film foams, with high thermal stability and low dielectric constants approaching 2.0, can be prepared using the copolymer/nanofoam approach.

Keywords: Nanofoams, Foamed polyimides, Dielectric constants, Microelectronic devices, Polyimides

Advances in Polymer Science, Vol.141
© Springer-Verlag Berlin Heidelberg 1999

1
Introduction

As minimum device dimensions continue to shrink and on-chip device densities increase, signal delays caused by capacitive coupling and crosstalk between closely spaced metal lines increase substantially. These effects are exacerbated as minimum feature sizes fall below 0.5 μm. Current production features have already reached 0.25 μm and are projected to decrease to 0.10 μm or less by early in the twenty first century. The chip manufacturing process is divided into the front-end-of-the-line (FEOL), which includes the active devices such as transistors, capacitors, resistors, etc., and the back-end-of-the-line (BEOL) which contains the on-chip wiring of the devices, establishment of signal and ground lines, insulator dielectrics, metal vias, bonding pads, etc. [1]. All of these interconnects are on the chip itself and their construction represents an important part of the semiconductor integration process. As dimensions decrease and device densities increase, the RC delays due to the interconnects begin to constitute a significant portion of the chip delay. Current advanced microprocessor chips contain over 5.5 million transistors and over 840 m of wiring (most Al(Cu)). Within 15 years, projections suggest that the transistor count will rise to 1 billion and the wiring lengths will exceed 10,000 m/chip!

Currently modern chips employ multilayer thin film wiring patterns to establish the interconnects, thereby minimizing wiring lengths. In this regard, modern logic chips often have 4–6 levels of metallization and dielectric insulators. The metal layers are electrically connected by conducting studs (usually tungsten) called vias. The BEOL capacitance may be separated into vertical capacitance (C_V) between metals in different layers and lateral (C_L), the intra-line capacitance between metal features in a common layer. As the metallization features decrease in size and become more densely packed, the lateral capacitance overwhelms the vertical capacitance. As the capacitance varies directly with the dielectric constant (ε) of the insulating medium and signal transmission velocity varies as the inverse square root, insulator dielectric constant becomes a critical feature in controlling signal delays. In addition, AC power consumption varies directly with the dielectric constant.

The current insulator for on-chip applications is silicon dioxide (ε=3.9–4.2). Current chip integration procedures have been optimized around the properties of this insulating material. Current integration demands for insulators for use with Al(Cu) wiring and tungsten vias require thermal stabilities in excess of 450 °C, good mechanical properties (strength, stability upon thermal cycling, etc.), resistance to crack generation and propagation, low defect densities, low water uptake and environmental stability, self-adhesion and adhesion to various substrates, ceramics and metals, chemical resistance to metallurgy at high temperatures, processability by photolithographic techniques and gas-phase etching procedures, capacity for planarization by etch-back, reflow or chemical mechanical polishing and others.

Among the many approaches to decreasing signal propagation delays in either the BEOL or module wiring, the reduction of the dielectric constant of the insulating media receiving the most attention [2–4]. It is in these critical wiring components that high performance organic polymers are finding increasing applications as dielectrics. The use of organic polymers in these devices potentially allows lower cost manufacturing because the materials can be spin-coated rather than vacuum processed for depositing. However, a major driving force for the implementation of polymeric insulators is their lower dielectric constants relative to the silicon dioxide. As mentioned, insulating materials for BEOL applications must be able to withstand the high temperatures necessary for the deposition of the metal lines and vias as well as for connecting chips to modules (i.e., C_4 soldering) [1]. As a minimum requirement, they must be able to withstand soldering temperatures without any degradation, outgassing, or dimensional change s (i.e., have a T_g significantly higher than the soldering temperature). Another major concern in the use of organic on-chip insulators is control of residual thermal stress, which is exacerbated with each additional level [5]. The stress results primarily from the mismatch in thermal expansion between the silicon chip and/or the ceramic substrate and the insulating material. In addition, the polymeric insulator must adhere well to the substrate and to itself (i.e., self-adhesion) to permit reliable fabrication of reliable multilevel structures [6]. The ability to process the polymer from a common organic solvent, and to planarize (both globally and locally) underlying topography are also critical features. Once deposited, the polymer should be solvent resistant to allow the fabrication of multilayer structures and permit lithographic imaging without dimensional changes.

Although a wide variety of polymers have been evaluated for use in microelectronics applications, polyimides have emerged as the favored class of materials [7–11]. Rigid and semi-rigid aromatic polyimides, in particular PMDA/ODA and BPDA/PDA, provide an excellent combination of thermal and mechanical properties which include a low thermal expansion coefficient, high modulus and tough ductile mechanical properties, the latter judged by high elongations to break. In addition, these properties are largely retained to 400 °C; above this temperature partial softening or flow is observed in some cases. These desirable mechanical properties reflect the high degree of molecular packing [12, 13]. For example, the polymer chains of PMDA/ODA polyimide assume locally extended conformations in a smectic-like layered order. Furthermore, imidization of thin PMDA/ODA polyimide films in contact with a substrate produces substantial orientation of the molecules parallel to the surface [14]. This orientation results in physical and optical properties that are anisotropic. This results in a larger thermal expansion coefficient and lower modulus out of the plane of the film as well as anisotropic solvent swelling behavior. These effects are even more pronounced for the more rigid BPDA/PDA polyimide. A further manifestation of the orientation, which is responsible for the excellent in-plane mechanical properties, is provided by a corresponding anisotropy in the dielectric constant [15]. In particular, the in-plane dielectric constant can be as much as 0.7–0.8 higher than the out-of-plane dielectric constant. An anisotropic insulator dielectric constant is clearly an unacceptable limitation for device design. Furthermore, the key advantage realized by the use of polymeric materials over inorganics is largely negated.

The most common approach to decreasing the dielectric constant of polyimides has been via the incorporation of comonomers containing perfluoroalkyl groups. Examples include the incorporation of fluorinated substituents either as hexafluoroisopropylidene linkages [16], main chain perfluoroalkyl groups [17] or pendent trifluoromethyl groups [18, 19]. DuPont [2] has reported the synthesis of polyimides based on 9,9-disubstituted xanthene dianhydrides, 6FXDA/6FDAM (9,9-bis(trifluoromethyl)xanthenetetracarboxylic dianhydride/2,2bis(4-aminophenyl)-1,1,1,3,3,3hexafluoropropane), as well as polyimides based on the TFMOB monomer (2,2'-bis(trifluoromethyl)benzidine) [3]. These polymers reportedly have dielectric constants as low as 2.5. The methodology to develop highly fluorinated polyimides is limited, to a certain extent, by synthetic difficulties associated with the incorporation of greater amounts of pendant perfluoroalkyl groups. Alternatively, the dielectric constant of polyimides can be lowered through the introduction of kinks and conjugation interrupting linkages in the polymer backbone to lower the molecular polarizability and reduce the chain-chain interactions [20]. Such structural modifications coupled with the judicious incorporation of fluorine-containing comonomers, leads to polyimides with dielectric constants in the range 2.4–2.8.

An alternative to structural modifications is the generation of polyimide foams which substantially reduces the dielectric constant while maintaining the desired thermal and mechanical properties of the aromatic polyimide. The reduction in the dielectric constant is simply achieved by incorporating voids which have a dielectric constant of 1. The advantage of a foam approach is readily apparent by examination of Fig. 1, which shows a Maxwell-Garnett modeling of composite structures based on a matrix polymer, with an initial dielectric constant of 2.8 [21]. Incorporation of a second phase of dielectric constant 1.0, as with the introduction of air-filled pores in a foam, causes a dramatic reduction in the dielectric constant. Moreover, it is important to note that the relationship between composition and overall dielectric constant is not linear; in particular, the most significant advantages are realized at modest levels of porosity. However, foams provide a unique set of problems for dielectric applications. It is obvious that the pore size must be much smaller than both the film thickness and any microelectronic device features. In addition, it is necessary that the pores be closed cell, i.e., the connectivity between the pores must be minimal to prevent the diffusion of reactive contaminants. Finally, the volume fraction of the voids must be as high as possible to achieve the lowest possible dielectric constant. All of these features can alter the mechanical properties of the film and affect the structural stability of the foam.

A new means of generating a polyimide foam with pore sizes in the nanometer regime has been developed [22–33]. This approach involves the use of block copolymers composed of a high temperature, high T_g polymer and a second component which can undergo clean thermal decomposition with the evolution of gaseous by-products to form a closed-cell, porous structure (Fig. 2).

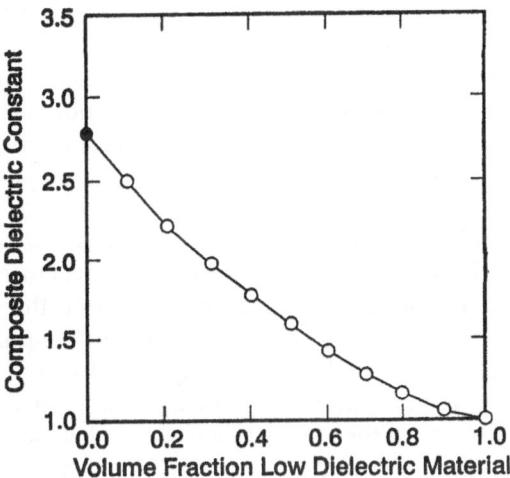

Fig. 1. Maxwell-Garnett theory used for the prediction of dielectric constant containing dispersed regions of air

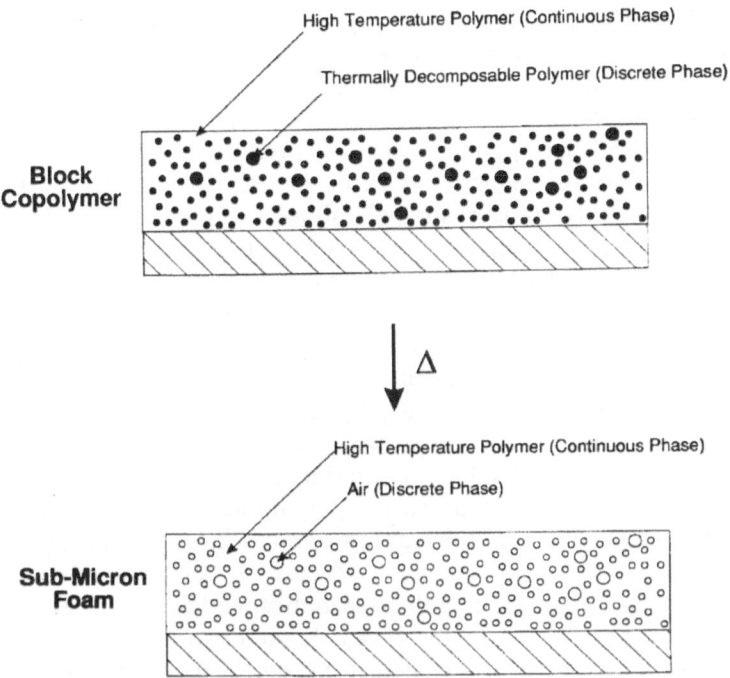

Fig. 2. Approach to the preparation of polyimide nanofoam using microphase separated block copolymers

This concept is similar to that developed by Patel [34] for the preparation of epoxy networks with well-defined microporosity. However, rather than using a phase separation process of two polymers which proceed via a nucleation and growth mechanism and leads to phase separation on the micrometer scale, the work described here takes advantage of the small size scale of microphase separation of block copolymers. These block copolymers can undergo thermodynamically controlled phase separation to provide a matrix with a dispersed phase that is spherical in morphology, monodisperse in size and discontinuous [35, 36]. Furthermore, the molecular structure and molecular weight of the segment allows precise control over both the size and volume fraction of the dispersed phase. By designing the block copolymers such that the matrix material is a thermally stable polymer of low dielectric constant and the dispersed phase is a labile polymer that undergoes thermolysis at a temperature below the T_g of the matrix to yield volatile reaction products, one can prepare foams with pores in the nanometer dimensional regime that have no percolation pathway; they are closed structures containing nanometer size spherical pores that contain air.

2
Criteria for High Temperature Polymer Continuous Phase

The successful adaptation of the block copolymer approach to polyimide nanofoams requires the judicious combination of polyimide with the thermally labile coblock. The material requirements for the polyimide block are stringent; (i) thermal and chemical stability to 450 °C and above, (ii) a high glass transition temperature (>375 °C), (iii) good mechanical properties, (iv) low water uptake, (v) a dielectric constant below 3.0, (vi) isotropic optical and electrical properties, (vi) processability either as the polyimide or as precursor polymers, (vii) ready availability of the monomeric precursors. Although many polyimide candidates were considered, including some semicrystalline materials, we focused our attention on the materials shown in Scheme 1. These materials seem to satisfy most of the criteria described above. The ODPA-FDA and PMDA-3F [37] were soluble in polar organic solvents such as NMP, etc., in imidized form. All were processable via the respective poly(amic acids), and PMDA-FDA and PMDA-3F [37] were available as polyamic ester derivatives. The anhydride 6FXDA was available from DuPont in experimental quantities and was processable via the polyamic acid precursor. The polyimides PMDA-FDA and 6FXDA-6F were insoluble in common solvents after processing. The films of PMDA-3F became insoluble after heating above 350 °C for 1 h. All cures were conducted under high purity nitrogen. Some thermal, optical and electronic properties of these materials are shown in Table 1. The poly(amic esters) of PMDA-FDA and PMDA-3F were prepared from diethyl pyromellitic diacid chloride (see Scheme 2). In these cases, we used starting materials enriched (>85%) in the meta isomer because of the improved solubility and the observation that the meta-polyamic esters (m-PAE) derivatives manifest an apparent T_g or softening

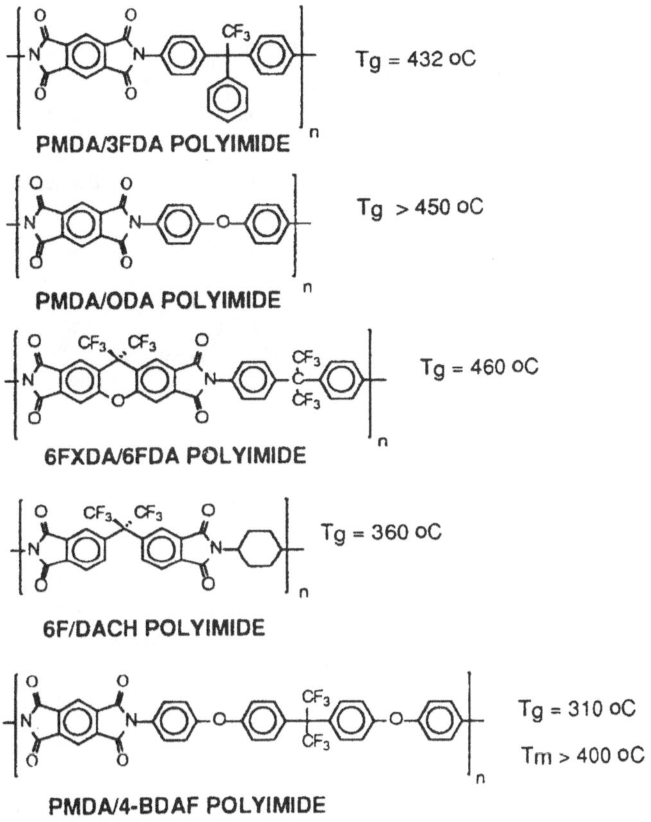

PMDA/3FDA POLYIMIDE   Tg = 432 oC

PMDA/ODA POLYIMIDE   Tg > 450 oC

6FXDA/6FDA POLYIMIDE   Tg = 460 oC

6F/DACH POLYIMIDE   Tg = 360 oC

PMDA/4-BDAF POLYIMIDE   Tg = 310 oC   Tm > 400 oC

Scheme 1

Table 1. Aromatic amine terminated thermally labile oligomers

Sample entry	Thermally labile block type	Polymerization method	Molecular weight (g/mol)	T_g (°C)
1a	poly(α-methylstyrene)	anionic	12,000	155
1b	poly(styrene)	anionic	14,000	100
1c	styrene/α-methylstyrene	anionic	14,000	100
1d	poly(styrene)	free radical	13,000	100
1e	poly(propylene oxide)	anionic	5,600	−65
1f	poly(methyl methacrylate)	group transfer	15,000	105

as evidenced by a large drop in modulus prior to imidization (Fig. 3). This characteristic is desirable for increased mobility to facilitate phase separation of the blocks prior to imidization.

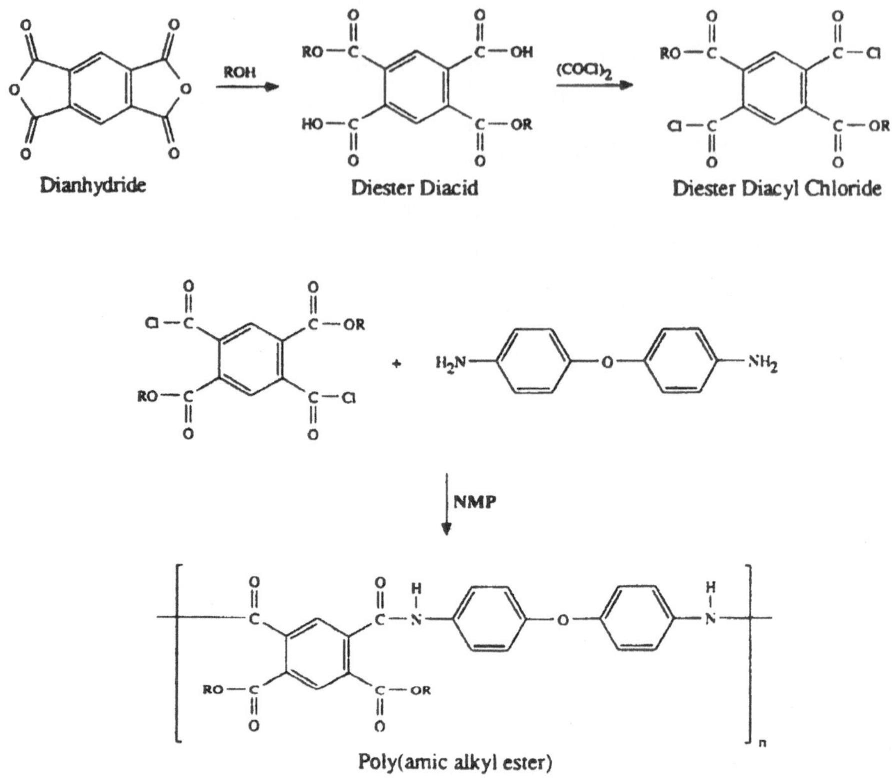

Dianhydride

Diester Diacid

Diester Diacyl Chloride

Poly(amic alkyl ester)

Scheme 2

3
Criteria for the Thermally Labile Coblock

The selection of suitable labile co-blockmaterials was based on certain criteria. A primary concern was the availability of a suitable synthetic route to well-defined functional oligomers suitable for incorporation into polyimides. This block must also decompose quantitatively into non-reactive species that can easily diffuse through a glassy polyimide matrix. The temperature at which decomposition occurs is also critical; it should be sufficiently high to permit standard film preparation and solvent removal yet below the T_g of the polyimide block to avoid foam collapse. The thermally labile coblocks investigated include poly(propylene oxide) [22–24, 27, 28], poly(methyl methacrylate) [22], poly(styrene) [23], poly(α-methylstyrene) [25], poly(lactides) [31, 33] and poly(lactones) [31, 33]. Each of these decomposes quantitatively into small molecules in the appropriate temperature regime. Although poly(propylene oxide) is stable in an inert atmosphere up to 300 °C, when exposed to oxygen, it decomposes rap-

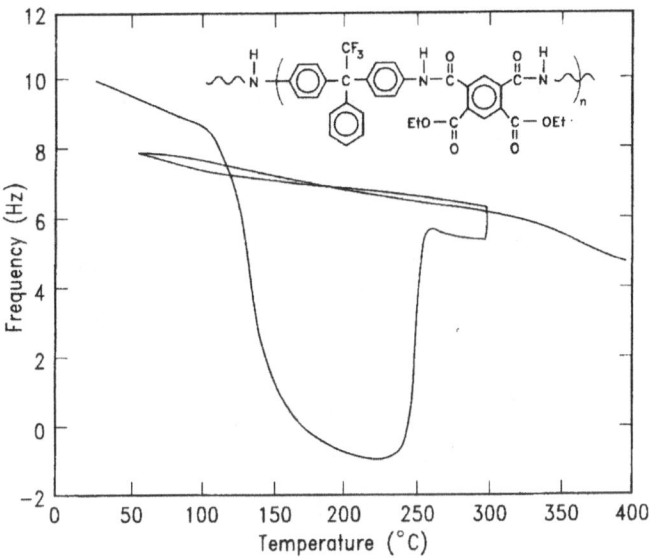

Fig. 3. Dynamic mechanical analysis of poly(amic alkyl ester). Sample heated to 300 °C and held (1 h), cooled and rerun to 450 °C

idly between 250 and 300 °C [22]. Figure 4 shows the thermogravimetric analysis (TGA) thermogram of poly(propylene oxide) heated isothermally at 275 °C in air. Within 20 min, quantitative decomposition is observed. The thermal decomposition temperature of poly(methyl methacrylate) is strongly dependent on the type of polymerization employed, since polymer regiostructure can influence the depolymerization process [38]. For example, poly(methyl methacrylate) prepared by free radical methods contains a substantial number of chain ends terminated via disproportionation. Typically, such poly(methyl methacrylate) degrades at relatively low temperatures. Conversely, poly(methyl methacrylate) prepared by anionic or group transfer methods has well-defined unreactive end groups and substantially higher decomposition temperatures. The poly(methyl methacrylate) used in this study had a decomposition temperature of 335 °C [22] and was synthesized according to Scheme 3. Poly(α-methylstyrene), polystyrene and styrene/α-methylstyrene copolymers are, in principle, ideally suited for use as thermally labile blocks, since well-defined functional oligomers can be prepared via anionic polymerization methods, and the polymers depolymerize by unzipping to monomer which can readily diffuse through the polyimide matrix (Fig. 5) [23]. The average number of monomer units generated per polymeric radical generated defines the zip length [39]. For poly(α-methylstyrene) the zip length is extremely high, ~1200, resulting in a nearly quantitative formation of monomer [39]. By comparison, poly(styrene) has a zip length of approximately 60 [39]. Thus, the polymer decomposition rate of the latter is significantly slower.

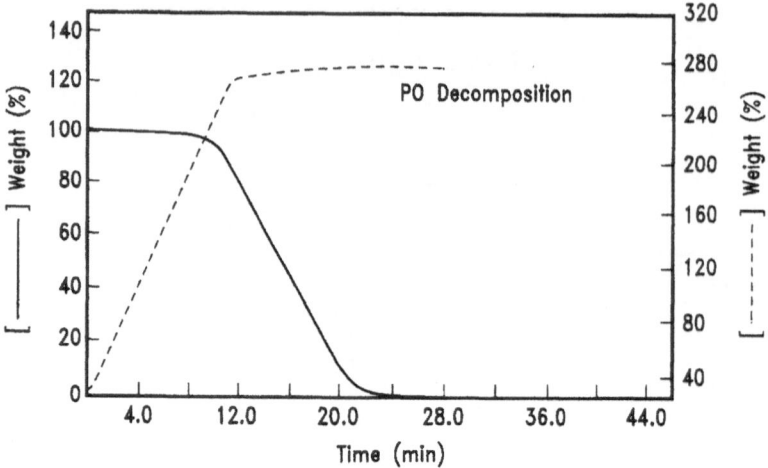

Fig. 4. Isothermal gravimetric analysis of poly(propylene oxide)

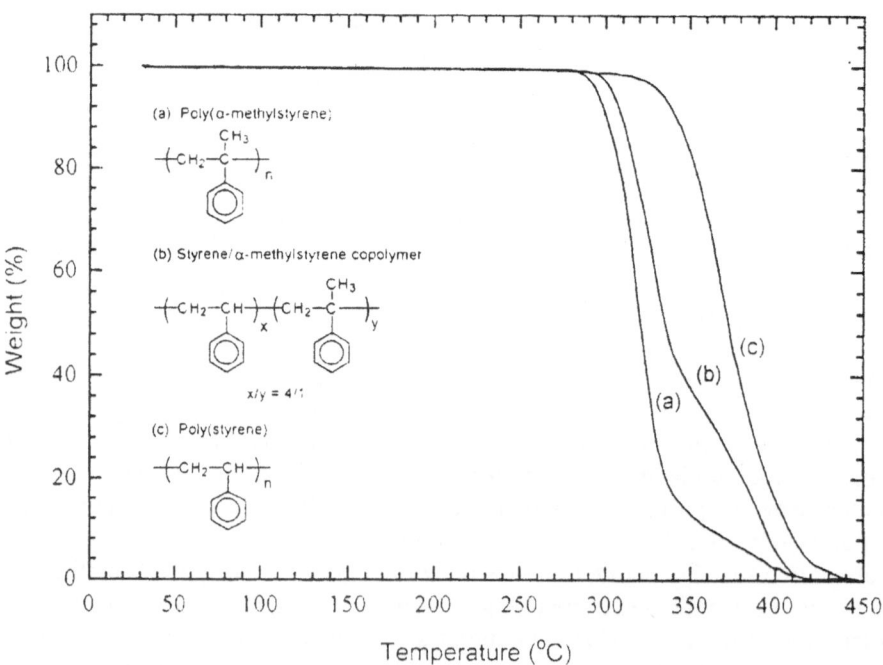

Fig. 5. Thermogravimetric analysis of poly(styrene), poly α-methylstyrene and copolymers

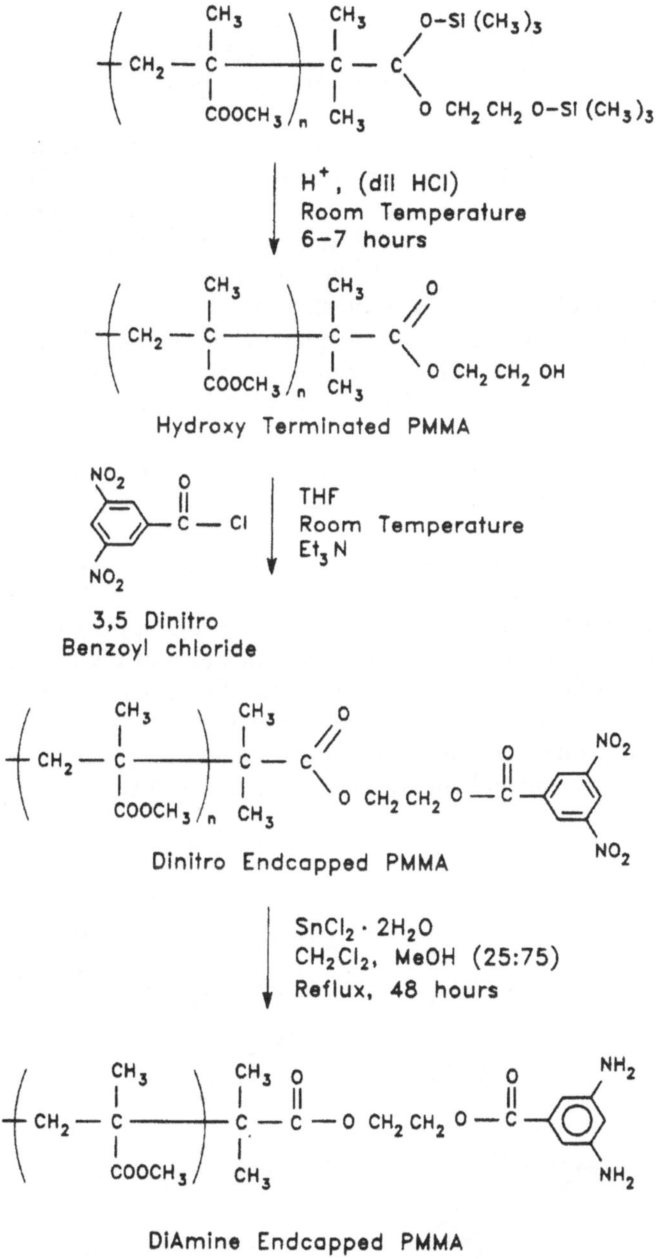

H⁺, (dil HCl)
Room Temperature
6–7 hours
Hydroxy Terminated PMMA
3,5 Dinitro
Benzoyl chloride
THF
Room Temperature
Et₃N
Dinitro Endcapped PMMA
SnCl₂ · 2H₂O
CH₂Cl₂, MeOH (25:75)
Reflux, 48 hours
DiAmine Endcapped PMMA

Scheme 3

Monohydroxyl terminated oligomers were prepared by anionic, group transfer polymerization methods and were converted to materials containing amino end groups, amenable towards polyimide copolymerization [22–33]. The aminophenyl carbonate end-capped propylene oxide oligomers were prepared by the reaction of the monohydroxyl terminated propylene oxide oligomers with 4-nitrophenyl chloroformate in methylene chloride containing pyridine (Scheme 4). The oligomers were then hydrogenated with Pearlman's catalyst (palladium hydroxide) to the desired amine. The molecular weights of the functionalized oligomers ranged from 5000 to 15,000 g/mol (Table 1).

Alternatively, functionality could be introduced in free radical processes through the use of "masked" or protected initiators. For example, an amino-functionalized polystyrene **1d** could be prepared by a novel "living" free-radical polymerization procedure using the appropriately functionalized AIBN initiator and 2,2,6,6-tetramethylpiperidinyloxy (TEMPO) [40, 41] (Scheme 5). Removal of the t-butoxycarbonyl protecting group leads to monoamino-terminated poly(styrene) with a polydispersity and functionality similar to those prepared by anionic methods (see Table 2). Likewise, a "masked" initiator approach was used for the "living" ring opening polymerization (ROP) of poly(lactones), poly(lactides) and poly(carbonates). Jérôme et al. [42] and Kricheldorf et al. [43a,b] have reported the synthesis of end-functional aliphatic polyesters by ring-opening polymerization (ROP) of ε-caprolactone (ε-CL) and lactides (LA). Polymerization initiated with functional aluminum alkoxides of the general structure $Et_{(3-p)}Al(OR)_p$, where p=1 and 3, proceeds via a living "coordination insertion" mechanism, under suitable conditions (temperature, concentration, and solvent), with the formation of α-functional chains of predictable molecular weights. The use of nitrophenethyl alcohol as the co-initiator allows the direct

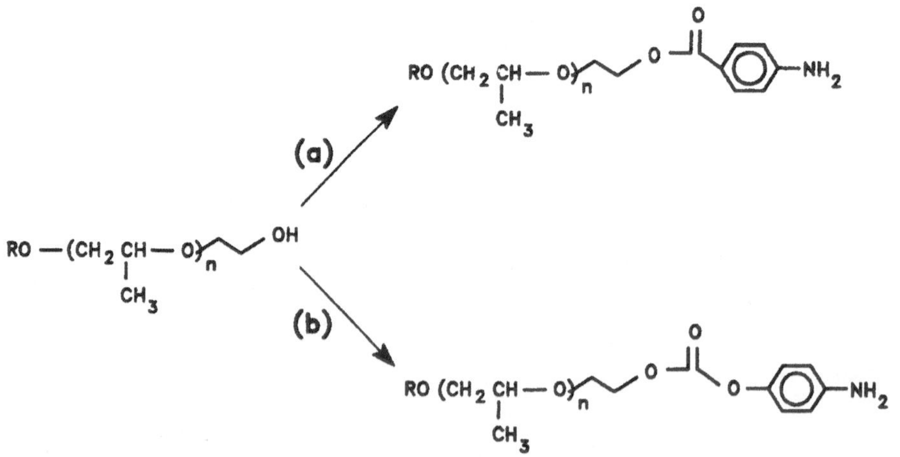

Scheme 4

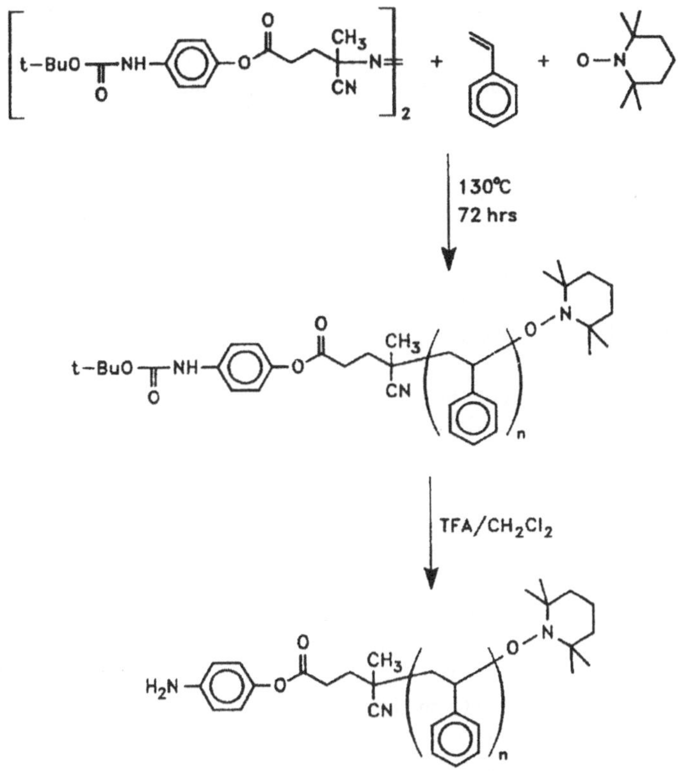

Scheme 5

introduction of a single nitrophenyl end functionality (Scheme 6). However, when using nitrophenyl ethyl alcohol as the co-initiator, the alcohol/Al(Et)$_3$ratio must be 2:1 in order to achieve the desired molecular weight control [43c–e]. The nitro group of the polymer chains could be reduced by standard methods (Pd/C and H$_2$) to yield the desired amino-terminated polymers. The characteristics of the oligomers prepared in this fashion are shown in Table 2.

An alternative to the above approach which leads to graft polyimides would be the use of reactive oligomers having diamino functionality at one end of the chain, which lead to graft copolyimides. The monohydroxy functionalized poly(propylene oxide) oligomers were derivatized to the 3,5-diamino benzoate functionality by first reacting the monohydroxy terminated poly(propylene oxide) oligomer with 3,5-dinitrobenzoyl chloride in dry tetrahydrofuran solvent containing pyridine as the catalyst and acid acceptor (Scheme 7). The solution was filtered to remove pyridine hydrochloride salt and concentrated to a clear viscous liquid. The reaction was followed using ^{1}H-NMR to confirm that complete conversion of the hydroxy group to the dinitrobenzoate functionalized poly(propylene oxide) was achieved. The hydrogenation of the dinitrobenzoate

Table 2. Characteristics of block copolymers

Copolymer entry	Polyimide type and form	Thermally labile block type	Thermally labile block composition, wt%			Volume Fraction of labile block, %
			Charge	Incorporated		
				^{1}H NMR	TGA	
2a	3FDA/PMDA (alkyl ester)	Poly(propylene oxide)	15	9.9	9	11
2b	3FDA/PMDA (alkyl ester)	Poly(propylene oxide)	25	23	22	27
2c	ODPA/FDA (imide)	Poly(propylene oxide)	15	13.1	13	–
2d	ODPA/FDA (imide)	Poly(propylene oxide)	25	22.5	22.5	–
2e	PMDA/FDA (alkyl ester)	Poly(propylene oxide)	15	14	9.2	12
2f	PMDA/FDA (alkyl ester)	Poly(propylene oxide)	25	22	18.4	–
2g	ODPA/FDA (imide)	Poly(α-methyl-styrene)	14	13	14	16
2h	ODPA/FDA (imide)	Poly(α-methyl-styrene)	25	24	24	29
2i	3F/PMDA (alkyl ester)	Poly(α-methyl-styrene)	15	14	15	27
2j	3F/PMDA (alkyl ester)	Poly(α-methyl-styrene)	25	–	24	27
2k	3F/PMDA (alkyl ester)	Poly(styrene)	20	18	19	2
2l	3F/PMDA (alkyl ester)	α-methylsty-rene/styrene copolymer	20	15	14	18
2m	PMDA/ODA (amic ester)	Poly(propylene oxide)	25	22	23	28
2n	PMDA/ODA (amic ester)	Poly(methyl methacrylate)	25	20	21	23

terminated poly(propylene oxide) with Pearlman's catalyst (palladium hydroxide on carbon) produced the aromatic diamine. The molecular weights of the amino benzoate functionalized poly(propylene oxide) were determined using both ^{1}H-NMR and standard potentiometric titration of the amine end groups with a standardized solution of HBr in acetic acid. These results are shown in Table 1.

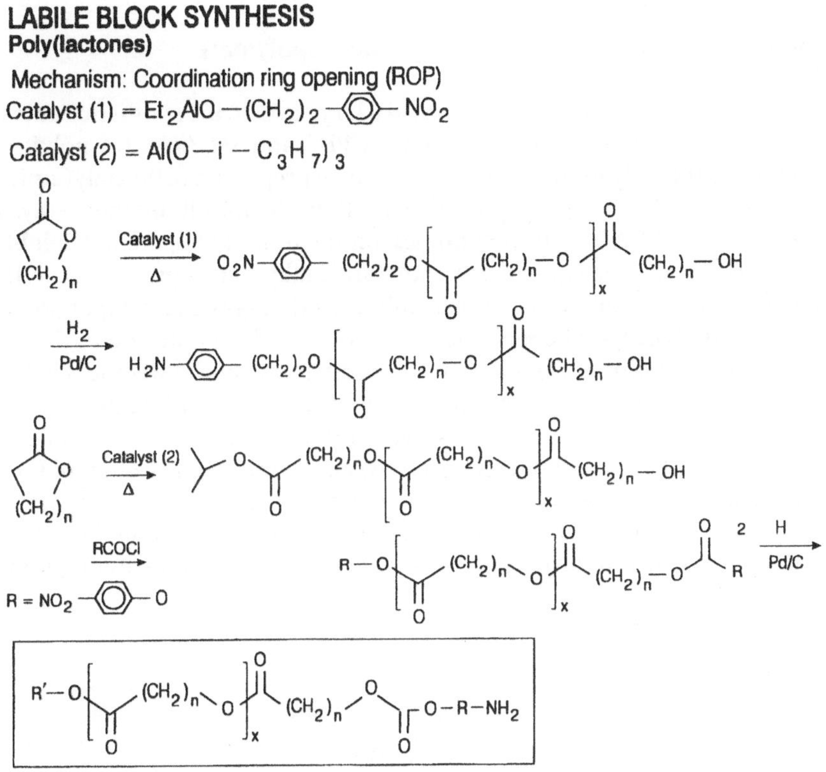

LABILE BLOCK SYNTHESIS
Poly(lactones)

Mechanism: Coordination ring opening (ROP)

Catalyst (1) = $Et_2AlO-(CH_2)_2-\langle\bigcirc\rangle-NO_2$

Catalyst (2) = $Al(O-i-C_3H_7)_3$

Scheme 6

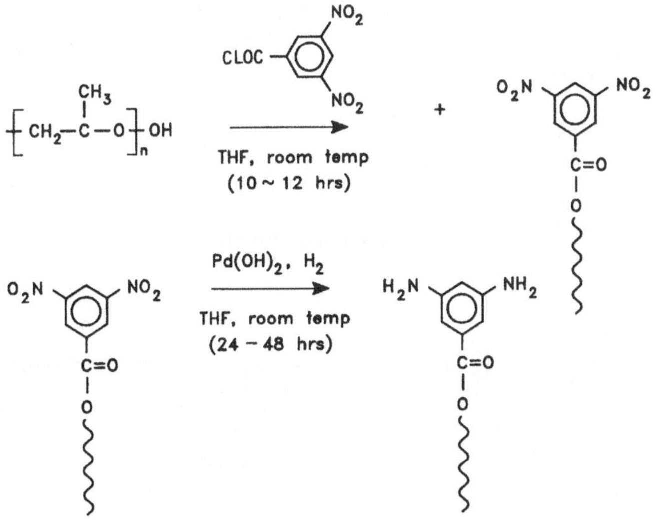

Scheme 7

4
Synthesis and Characterization of Polyimide Copolymers

The solubility of the polyimide dictates, to a large extent, the synthetic route employed for the copolymerization. The ODPA/FDA and 3FDA/PMDA polyimides are soluble in the fully imidized form and can be prepared via the poly(amic-acid) precursor and subsequently imidized either chemically or thermally. The PMDA/ODA and FDA/PMDA polyimides, on the other hand, are not soluble in the imidized form. Consequently, the poly(amic alkyl ester) precursors to these polymers were used followed by thermal imidization [44]. For comparison purposes, 3FDA/PMDA-based copolymers were prepared via both routes. The synthesis of the poly(amic acid) involved the addition of solid PMDA to a solution of the styrene oligomer and diamine to yield the corresponding poly(amic acids) (Scheme 8). The polymerizations were performed in NMP at room temperature for 24 h at a solids content of ~10% (w/v). Chemical imidization of the poly(amic-acid) solutions was carried out in situ by reaction with excess acetic anhydride and pyridine at 100 °C for 6–8 h. The copolymers were subjected to repeated toluene rinses in order to remove any unreacted styrene homopolymer.

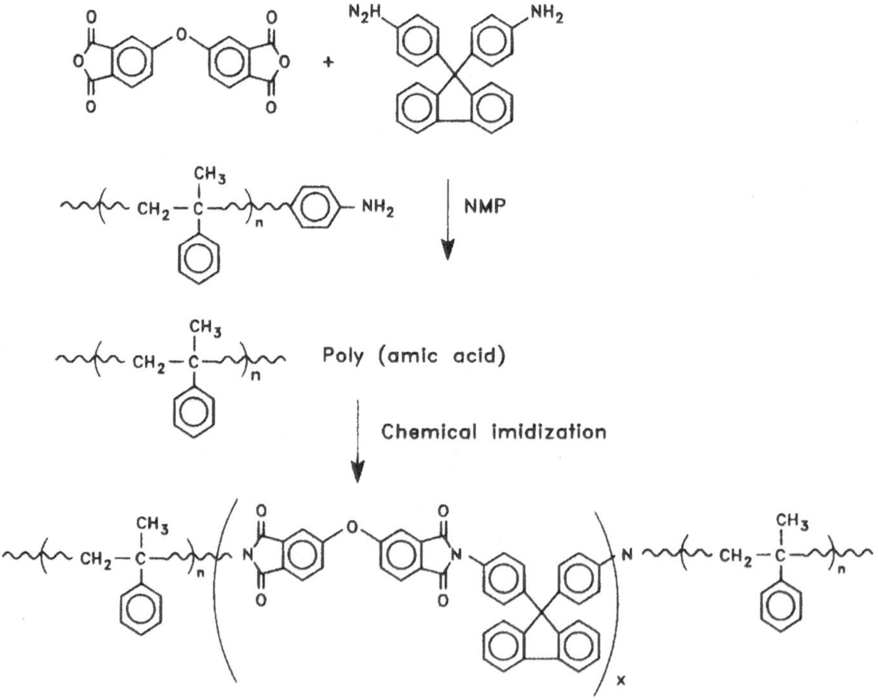

Scheme 8

The synthesis of the corresponding poly(amic alkyl ester), on the other hand, involved the incremental addition of PMDA diethyl ester diacyl chloride in methylene chloride to a solution of the oligomer and 3FDA in NMP containing pyridine as the acid acceptor (Scheme 9). Alternatively, the diamino functional oligomers could be utilized in an analogous fashion to yield the graft copolymers (Scheme 10). In these experiments, the meta isomer of PMDA diethyl ester diacyl chloride was used primarily due to its enhanced solubility and lower soften-

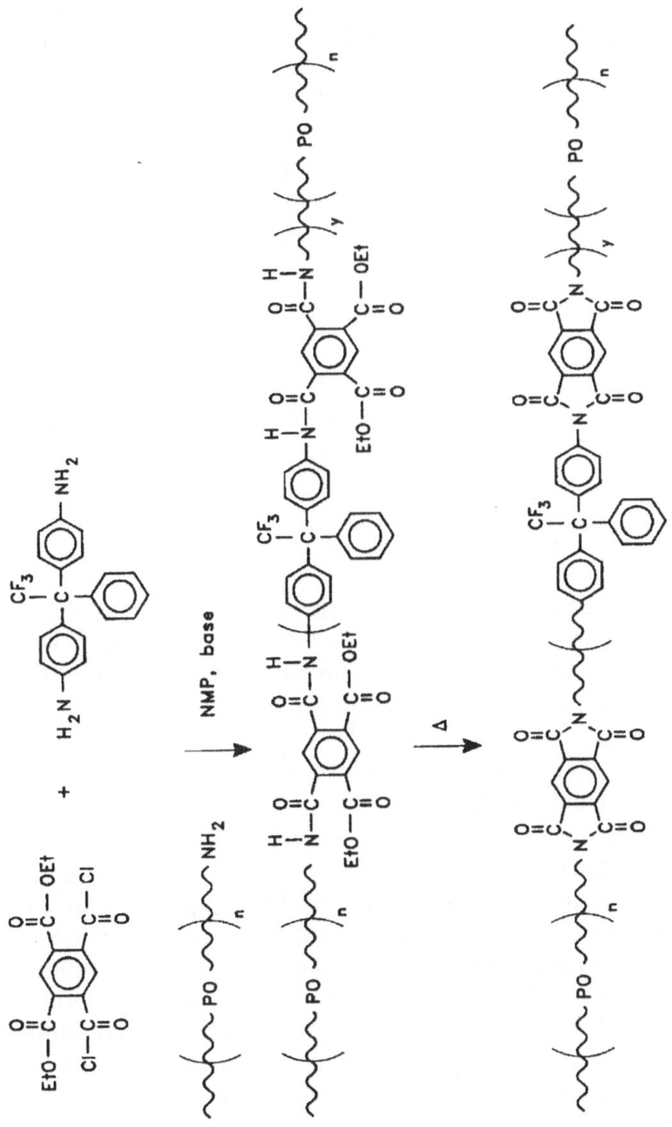

Scheme 9

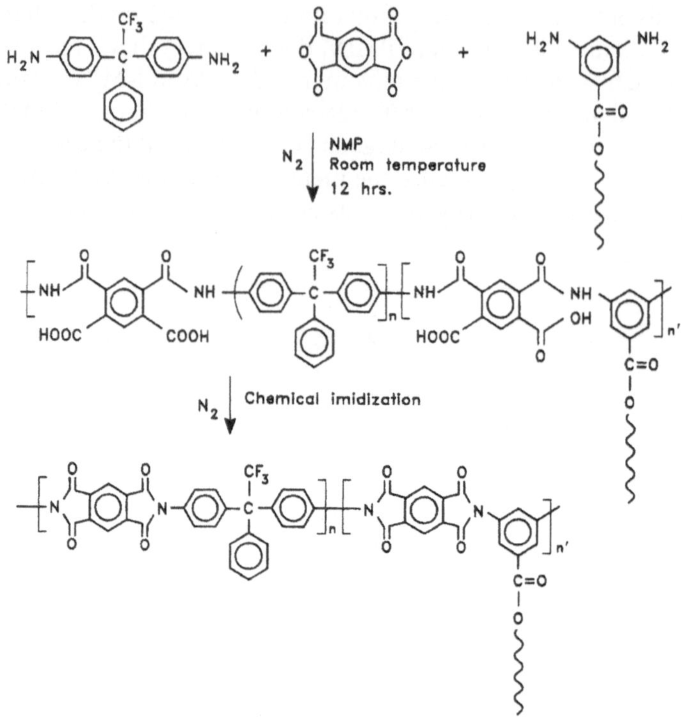

Scheme 10

ing temperature, and to facilitate comparison with previous studies [23–33]. The solids composition was maintained at ~15% for each of the polymerizations. The copolymers were isolated by precipitation from methanol/water, washed with water to remove remaining salts, and rinsed with methanol and toluene followed by drying.

The "softening" or T_g characteristic of the poly(amic acid ethyl ester) precursors allows sufficient mobility for a crosslinking reaction to occur at temperatures well below the T_g of the polyimide [30]. In this respect, the ethynyl functionality has been shown to crosslink thermally in the temperature range 180–300 °C and the resulting networks possess the requisite thermal stability [30]. Poly(amic ethyl ester)s have been prepared with several ethynyl compositions using 1,1-bis(4-aminophenyl)-1-(4-ethynlphenyl)-2,2,2-trifluoroethane as a co-diamine [45], and swelling measurements of the imidized networks showed substantially less solvent uptake than the corresponding non-crosslinked material [30]. The decreased interaction of the labile block degradation products with the crosslinked polyimide matrix is anticipated to result in nanofoams with controlled pore size and minimal interconnectivity.

Table 3. Characteristics of polyimide/PPO graft copolymers

Sample entry	Polymer sample	PPO $(M_w \times 10^{-3})$	Target (wt%)	Measured	
				^{1}H NMR	TGA
3a	PMDA-3F (Polyimide)	3.5	25.0	24.9	23.6
3b	PMDA-3F (Polyimide)	3.5	15.0	15.0	14.3
3c	PMDA-3F (Alkyl ester)	3.5	25.0	19.8	19.1
3d	PMDA-3F (Polyimide)	7.9	25.0	21.9	22.1
3e	PMDA-3F (Polyimide)	7.9	15.0	15.7	16.1
3f	ODPA-FDA (Polyimide)	3.5	25.0	24.0	–
3g	ODPA-FDA (Polyimide)	7.9	25.0	16.0	–
3h	PMDA-FDA (PAA)	3.5	15.0	12.5	–
3i	PMDA-FDA (PAA)	3.5	25.0	20.0	–
3j	PMDA-FDA (PAA)	7.9	15.0	7.5	–
3k	PMDA-FDA (PAA)	7.9	25.0	13.0	–

The characteristics of selected copolymers, prepared both in the fully imidized form and as the poly(amic alkyl ester) precursor to the polyimide, are shown in Tables 2 and 3. The weight percentage, or loading levels, of the labile blocks in the copolymers was intentionally maintained low (~20 wt%) in order to produce discrete spherical domains of the block embedded in the polyimide matrix. At higher loadings, phase separation could, in principle, produce cylindrical or more interconnected morphologies which are undesirable. The incorporation of the labile block in the copolymer was assessed by thermal gravimetric analysis (TGA) and by ^{1}H-NMR. For most of the copolymers, incorporation levels of the labile block agreed closely with that theoretically expected from the feed ratios. The use of monofunctional oligomers in the polyimide syntheses described above affords an ABA triblock copolymers architecture, where the thermally labile component comprised the terminal A blocks and the stable polyimide is the B block. It should be noted with either ABA or graft copolymers that the molecular weight of the polyimide block remains the same upon thermal decomposition of the labile coblock. This is important for mechanical and physical property considerations. Conversely, the use of the difunctional oligomers produced a graft molecular architecture and, as in the previous case, upon degradation of the labile coblock, the polyimide molecular weight remains unchanged (Table 3).

4.1
Thin Film Processing Conditions

The processing window for film and foam formation was established primarily with the 3FDA/PMDA imide-based copolymers, since these copolymers are soluble and can be isolated and characterized at various stages of the processing or imidization. It is critical that the decomposition of the labile block should occur substantially below the T_g of the polymer matrix. Furthermore, the casting solvent must be effectively removed, prior to labile block degradation, to minimize plasticization effects on the polyimide matrix, which would further narrow the processing window (i.e., temperature difference between the polyimide T_g and decomposition temperature of labile block). Samples were cast from NMP, cured, and the processing window for film and foam formation was established by ^{1}H-NMR, TGA and dynamic mechanical analysis (DMA). Since most of the labile blocks in this study were stable to 300 °C in nitrogen, samples cured to this temperature to complete imidization and/or to remove the casting solvent should retain their labile block composition. ^{1}H-NMR showed complete solvent removal and the full T_g of the polyimide phase (PMDA-3FDA) was achieved. Volksen and co-workers [44] found that the temperature range over which imidization occurred for the poly(amic ethyl ester) derived from PMDA/ODA was 240–355 °C, with a maximum in the rate at 255 °C. I.R., ^{1}H-NMR and calorimetry measurements indicated that the imidization was essentially quantitative under these conditions, with minimal loss of the labile block composition as measured by TGA. However, the processing temperature of the α-methylstyrene based copolymers was limited to 265 °C to minimize poly(α-methylstyrene) decomposition during solvent removal and curing. Under these conditions, ~1–3% solvent remained and complete imidization was not accomplished, factors which decreased the T_g of the polyimide.

4.2
Block Copolymer Morphology

Dynamic mechanical analysis was one technique used to access the extant of microphase separation. It is essential that separation occur with high phase purity in order to obtain a nanofoam while minimizing pore collapse. Selected dynamic mechanical spectra of the polyimide copolymers are shown in Figs. 6 and 7. Two transitions were observed in each case, indicative of the presence of microphase separated morphologies. For the imide α-methylstyrene copolymer, the transition occurring near 160 °C is similar to that seen for the α-methylstyrene oligomer used in the synthesis and the damping peaks associated with the α-methylstyrene are sharp, indicating that the phases are pure and have discrete boundaries. For the α-methylstyrene-based copolymers and other copolymers containing labile coblocks which rapidly degrade, the transitions of the imide block show a strong dependence on the fraction of the α-labile block in the copolymer. The copolymer containing ~15 wt% α-methylstyrene (copolymer 2g) shows an imide transition which is nearly identical to that of the polyimide homopolymer. However, the copolymer with more α-methylstyrene shows an imide transition which is substantially depressed (2h). Furthermore, in this case, the transition appears at nearly the same temperature (~320 °C) as the decomposition temperature of poly(α-methylstyrene).

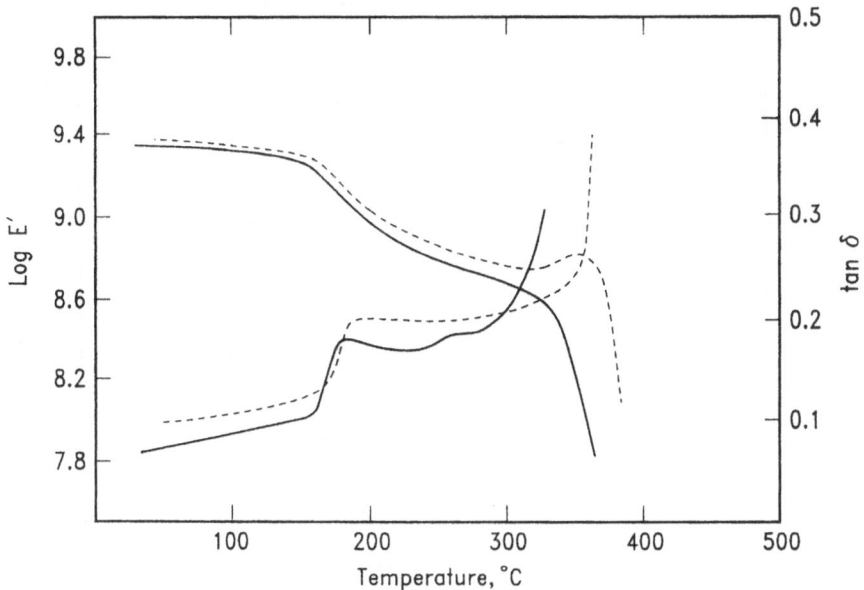

Fig. 6. Dynamic mechanical spectra of copolymers **2g** (---) and **2h** (—)

Small angle X-ray scattering has proved to be an effective method for examining the block copolymer morphology, as well as for studying the in situ foaming process. In the case of the imide/propylene oxide copolymers, the initial scattering profile showed a scattering peak at a scattering vector, g, of 0.017 $Å^{-1}$, corresponding to a Bragg spacing of ~367 Å. From the volume fraction of propylene oxide in the copolymer (~0.12), this yields an average size of the propylene oxide domains of ~44 Å. It should be noted that in this case only a single reflection is observed, with no indication of higher order reflections. If the propylene oxide domains were arranged periodically on a lattice, higher order reflections would be expected. Consequently, while the microphase-separated domains are periodic, lattice distortions cause a rapid loss of interferences at higher g. These distortions arise from two sources. First, the microphase-separated morphology is far from an equilibrium state. This is not surprising when one considers the conditions under which the sample has been prepared. It is evident that the copolymer has been trapped in a highly nonequilibrium state. There is no question, however, that the copolymer has microphase separated. Second, the size of the propylene oxide domains is polydisperse. The persistence of scattering at much smaller values of g also suggests this. Similar data was obtained for the imide-caprolactone and imide-α-methylstyrene copolymers, and in each case microphase separated morphologies were achieved with domain sizes of 100 Å or less.

Small angle neutron scattering (SANS) and neutron reflectivity (NR) have also been used to characterize the morphology and scattering from PDMA-

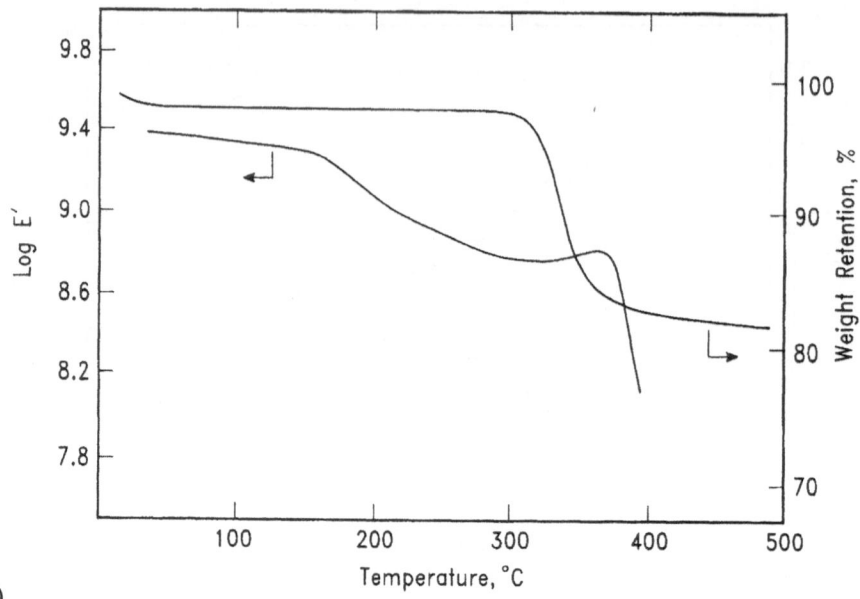

a)

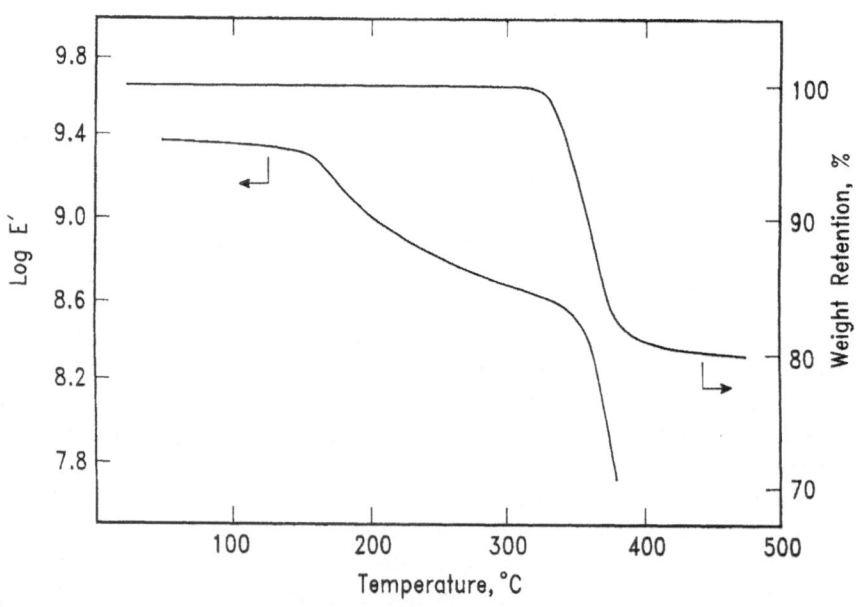

b)

Fig. 7. a Dynamic mechanical spectra and thermogravimetric spectra for copolymer **2g**.
b Dynamic mechanical spectra and thermogravimetric spectra for copolymer **6h**

3F/polyimide propylene oxide based nanofoam materials [46, 47]. SANS has the advantage over SAXS in that the scattering experiments can be done with the films directly on the Si substrate due to the high transmission of Si for neutrons and the fact that the wafers do not contribute any significant scattering in the small q range. In addition, there is sufficient neutron contrast for the PMDA-3F/polyimide propylene oxide materials without deuterium labeling due to the low hydrogen atom content of the PMDA-3F polymer relative to the poly(propylene oxide).

Figure 8a shows a set of SANS curves for a PMDA-3F/polyimide propylene oxide block copolymer containing 25% PPO on a Si substrate for the poly(amic ethyl ester) (prior to imidization), the cured polyimide and during the foaming process. The starting (unimidized) film thickness is 9650 Å. The most interesting aspects of the data is that the small angle scattering is almost flat for the poly(amic ethyl ester) film and upon imidization a pronounced peak develops with a spacing of about 260 Å. This peak remains at approximately the same q value and intensity after foaming. The data implies that the poly(amic ethyl ester) film as spun from solution is not microphase separated and that microphase separation occurs upon imidization. Calculation of the neutron scattering length densities (SLD) for the poly(amic ethyl ester), polyimide and poly(propylene oxide), indicate that contrast between the poly(propylene oxide) and poly(amic ethyl ester) is only slightly less than for poly(propylene oxide)/polyimide and if microphase separation was present in the initial poly(amic ethyl ester) films, it should be readily observed by SANS. The relatively small change in the scattering upon foaming is due to the fact that the SLD for PPO is close to zero (i.e., almost the same as a void) and hence the change neutron contrast with foaming is negligible. This is in direct contrast to the SAXS results where the voids dominate the scattering. Shown in Fig. 8b is SANS taken data during imidization. A 5100 Å-thick film of poly(amic ethyl ester) was heated in an Argon atmosphere from 25 to 300 °C. Data was collected for half-hour intervals as the temperature was increased. The scattering was constant at temperatures up to 250 °C. At 250 °C a peak develops during the half-hour (indicating the onset of microphase separation) but stays constant in position and intensity for the second half-hour at 250 °C. At 300 °C the peak increases in intensity and moves to lower q. The dependence of the scattered intensity for the film on temperature indicates that the microphase separation occurs during the imidization process and does not begin at temperatures below 250 °C. The microphase separation process is clearly coupled to the imidization reaction and the morphology is not fully developed until the curing process is largely complete at 300 °C.

5
Foam Formation

The balance of these different factors is found in the dependence of nanofoam formation on the labile block content in the copolymer. For copolymers with a low composition of labile block, the degradation is rapid, forming a nanoscopic

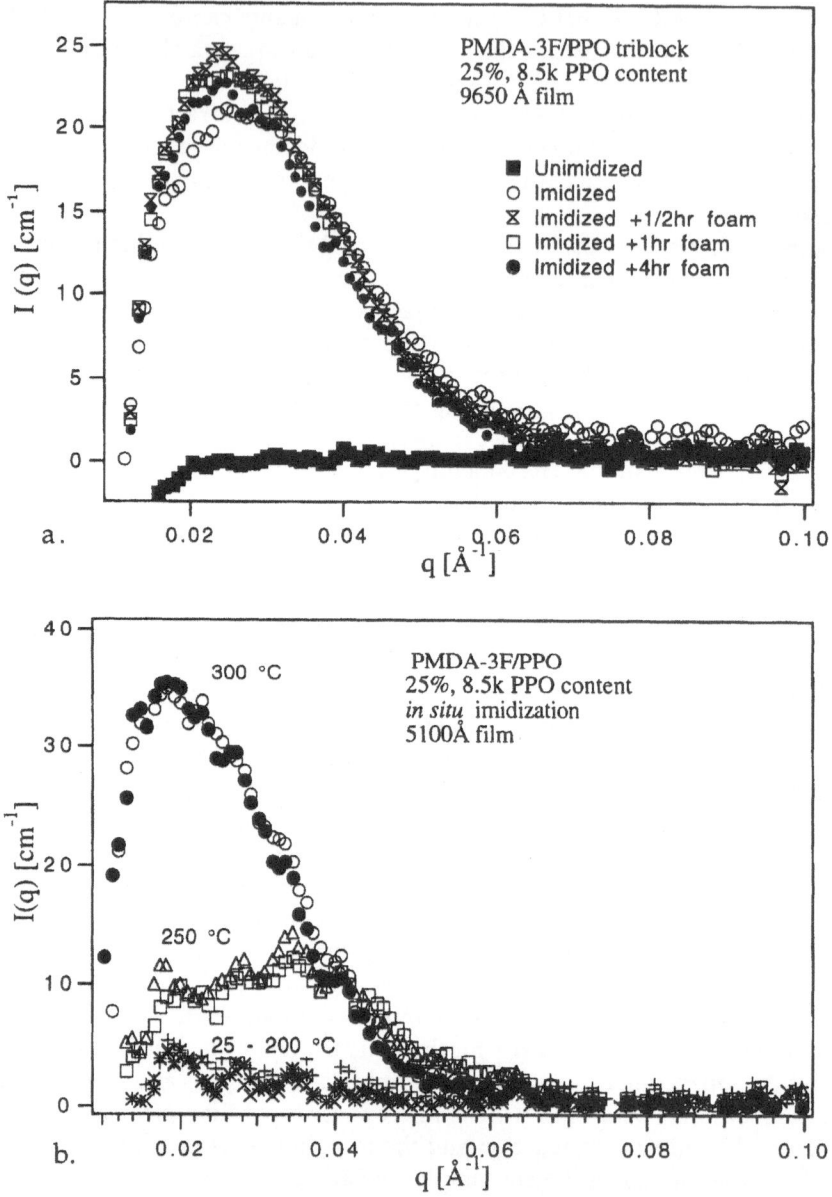

Fig. 8. a SANS curves for a 9650-Å-thick film of a PMDA-3F/PPO triblock copolymer on a Si wafer substrate as the poly(amic ethyl ester) and as the cured polyimide during foaming. The poly(amic ethyl ester) shows almost no small angle scattering indicating that the films are homogeneous as spun from solution. **b** A 5100 Å-thick film during the imidization process showing that the microphase separated morphology develops during the imidization curing reaction

reservoir of degradation products that must be removed in order to form voids. Since the permeability, which is a function of both the solubility and diffusion coefficient, is high, these reservoirs can be depleted effectively. This, as observed above, will cause an initial depression (i.e., plasticization) of the modulus followed by a gradual recovery of the modulus as the concentration of degradation products diminishes (Fig. 6). However, for copolymers containing higher fractions of the labile blocks, a saturation concentration of the degradation products in the imide matrix must be reached (Fig. 7). Although the rate at which the degradation products are removed is presumably rapid, within the time scale defined by the relaxation of the plasticized imide matrix, the removal is not rapid enough to prevent a collapse of the nanofoam structure. Consequently, the retention of the foam structure depends upon a delicate balance between the rate of decompositon of the labile coblock, the solubility of the degradation products in the imide matrix and the rate at which the degradation products diffuse out of the matrix. For each labile coblock surveyed, mild decomposition temperatures were employed to control the decomposition rate of the labile block in such a way as to minimize plasticization.

The generation of the nanofoam was accomplished by subjecting the copolymer film after curing and solvent removal to a subsequent thermal treatment to decompose the labile coblock. The temperature range and degradation atmosphere varied depending on the labile block type. For instance, the propylene oxide-based copolymers were heated to 240 °C in air for 6 h (Fig. 9), followed by a post-treatment at 300 °C for 2 h to effect the complete degradation of the propyl-

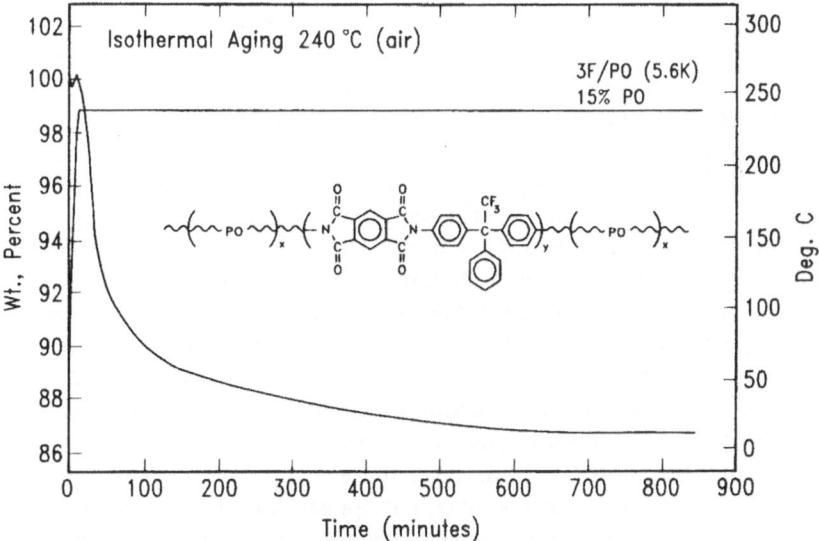

Fig. 9. Isothermal gravimetric analysis ITGA at 240 °C for copolymer **6a**

ene oxide component [22]. The degradation process was followed by thermo-gravimetric analysis (TGA) and ^{1}H NMR, and, under these conditions, quanti-tative degradation was observed with no evidence of residual by-products nor chemical modification of the polyimide. Conversely, the degradation of the sty-renic-based copolymers required a step-wise process so as to control the degra-dation rate of these blocks (Figs. 6 and 7) [23, 25].

5.1
Porosity Determination

The porosity of the resulting foams was determined using a number of tech-niques. Of these, film density determinations in comparison with the respective homopolymer is conceptually the simplest technique, since the resulting poros-ity is inversely proportional to the sample density. Film densities were deter-mined using a density gradient column $Ca(NO_3)_2$. This technique requires films which can be removed from the substrates. Also, any connectivity of the pores can lead to ambiguous results, since the films progressively drift down the col-umn (higher densities) as the fluid percolates through the pores. IR spectrosco-py can also be used to determine porosity in comparison with the pure polyim-ides, assuming that the sample thicknesses can be measured and normalized [48]. This technique, however, leads to significantly higher values for the poros-ity when the pore sizes become large enough to scatter significant amounts of the infrared light. Scattering techniques such as small angle X-ray and neutron scattering (SAXS, SANS) can be used to determine porosity since the scattering density tracks the sample density [22, 23, 46, 47]. The former technique requires free standing thick films, while the latter can be used for films on silicon sub-strates. Similarly, X-ray and neutron reflectivity (XR, NR) can also be used to de-termine the porosity of samples on substrates. Finally, transmission electron mi-croscopy may be used to determine porosity, providing significant contrast is achievable [49]. The results obtained by TEM were comparable to those deter-mined by density and IR techniques for samples with small pore sizes. TEM analysis permits much higher spatial resolution than IR, and is better suited to samples which scatter strongly at IR wavelengths.

The density measurements on thermally processed polymers clearly show the formation of a foamed polymer. The initial density values for selected foams to-gether with the respective polyimide homopolymers are shown in Tables 4 and 5. The density values for the ODPA/FDA and PMDA/FDA homopolyimides were both 1.28 g cm^{-3} and 3FDA/PMDA is 1.34. Most of the propylene oxide-based copolymers derived from these copolymers ranged from 1.09 to 1.27 g cm^{-3}, which is ~85–99% of that of the polyimide homopolymers, irrespective of the ar-chitecture of the copolymer (i.e., triblock vs graft). This is consistent with 1–15% of the film being occupied by voids. From these data (i.e., comparison of Tables 2 and 3 with Tables 4 and 5, respectively), it appears that the volume fraction of propylene oxide in the copolymer (i.e., ~80% or less). Thus, the efficiency of foam formation is poor, irrespective of the copolymer architecture. Conversely,

Table 4. Characteristic of polyimide foams

Sample entry	Initial labile block composition, vol%	Density (g cm^{-3})	Volume fraction of voids (porosity[a]) %
3FDA/PMDA polyimide	–	1.35	–
4a	11	1.17	13
4b	27	1.1	18
ODPA/FDA polyimide	–	1.28	–
4c	–	1.2	6
4d	–	–	12
PMDA/FDA polyimide	–	1.28	–
4e	–	1.17	7
4f	–	1.11	12
4g	16	1.23	3.1
4h	29	1.18	7.5
4i	27	1.13	16
4j	27	–	30
4k	22	1.17	14
4l	18	1.18	19
PMDA/ODA polyimide	–	1.41	–
4m	20	1.41	0
4n	25	1.41	0

[a]Porosity determined by IR technique

Table 5. Foams derived from polyimide/PPO graft copolymers

Sample entry	Sample	PPO ($M_w \times 10^{-3}$)	wt%	vol%	Porosity (%)
5a	PMDA-3F (Polyimide)	3.5	25.0	30.0	18.7
5b	PMDA-3F (Polyimide)	3.5	15.0	19.0	13.4
5c	PMDA-3F (PAE)	3.5	22.0	24.0	15.9
5d	ODPA-3F (Polyimide)	3.5	24.0	27.0	0.4
5	ODPA-3F (Polyimide)	7.9	16.0	19.0	9.0
5f	PMDA-FDA (PAA)	3.5	20.0	23.5	0.8
5g	PMDA-FDA (PAA)	7.9	7.5	9.0	5.5
5h	PMDA-FDA (PAA)	7.9	13.0	15.5	8.7

the propylene oxide-based copolymers with PMDA/ODA as the imide component did not show the expected density drop, and the values were essentially identical to that of the homopolymer. In PMDA/ODA-based systems, molecular ordering and orientation were found to be critical in controlling the stability of the foam structure. Where the characteristic in-plane molecular orientation and molecular ordering of homogeneous PMDA/ODA films were maintained, relaxation rates were clearly enhanced in the presence of the pores, leading to collapse of the foam structure well below the matrix T_g [22]. Moreover, there is some suggestion that the presence of pores may enhance stress relaxation rates in step strained imide foams. Thus, the recreation of these effects in a suitably designed block copolymer (where the presence of a second phase should maintain the structural integrity) may be a means of promoting rapid stress relaxation in PMDA/ODA imide films, yielding the sought-after low stress materials [5].

The porosity values, as determined by the density column measurements, for the foamed copolymers derived from the styrene and α-methylstyrene coblock were unreliable [25]. In all cases, the density gradient method yielded porosity values which were consistently lower than those described above. During the course of the measurement, the film which settled initially to a specific height in the column drifted downwards in the column to higher densities. This, apparently, results from the fluid penetrating into the porous film, which eventually leads to density values comparable to the 3FDA/PMDA polyimide homopolymer. Consequently, an alternative means of measuring porosity was required in these cases. It has been shown in some cases that infrared spectroscopy provides a spectral means of determining the void content in polymeric materials. By independently measuring the film thickness, the IR absorbance, when calibrated against the bulk polymer, coupled with the refractive index determined from the presence of interference fringes, yield results that are in quantitative agreement with density gradient methods. The porosities of the foamed copolymers measured by IR are shown in Table 4. The foams derived from the block copolymers comprised of the shorter block lengths show foaming efficiencies comparable to those observed for foams derived from the imidized copolymers. Conversely, the foam derived from the copolymer containing the high molecular weight α-methylstyrene coblock (copolymer **6j**) showed porosities which reach and even exceed the initial volume fraction of α-methylstyrene coblock. However, many of the foams prepared from copolymers containing α-methylstyrene and styrene labile blocks were somewhat cloudy or even opaque in some cases. This effect leads to significant scatter in the IR results. For these samples, transmission electron microscopy (TEM) was used to assess the porosity [49] and the void sizes were larger than expected based on the copolymer morphologies.

5.2
Polyimide Nanofoam Morphology

Convincing evidence that nanofoams are generated via the block copolymer process is shown in Fig. 10a. Here, an electron micrograph of a foam is shown derived from an PMDA/3FDA imide-propylene oxide copolymer. The porous structure of a foam is clearly evident where the white areas represent the voids from the degraded labile phase [49]. The average size of the pores by TEM is ~60 Å which is slightly larger than that calculated by small-angle X-ray scattering. One important feature from these data is that the pores appear not to be sig-

Fig. 10. a TEM micrograph of polyimide foam derived from imide/propylene oxide triblocl copolymer. **b** TEM micrograph of polyimide foam derived from imide/propylene oxide graf copolymer

nificantly interconnected. This feature is critical for the end use of porous materials. In addition, there is no TEM evidence of very small pores, which is consistent with the small-angle X-ray scattering data (minimal scattering at large q values). Furthermore, the periodic nature of the pores is clearly evident. Shown in Fig. 10b is a foam derived from FDA/PMDA polyimide propylene oxide graft copolymer. Non-interconnected discrete pores are formed with a pore size of ~50 Å, commensurate with the SAXS data.

Small-angle X-ray scattering provides a useful means of evaluating the formation of the nanofoam structure due to the length scales probed and the large X-ray contrast between the void and the polymer matrix. Shown in Fig. 11 is the X-ray scattering as a function of the scattering vector, g, obtained in heating the ODPA/FDA copolymer with propylene oxide from 60 to 250 °C at a rate of 5 °C min^{-1}. The initial scattering profile showed a peak in the scattering at

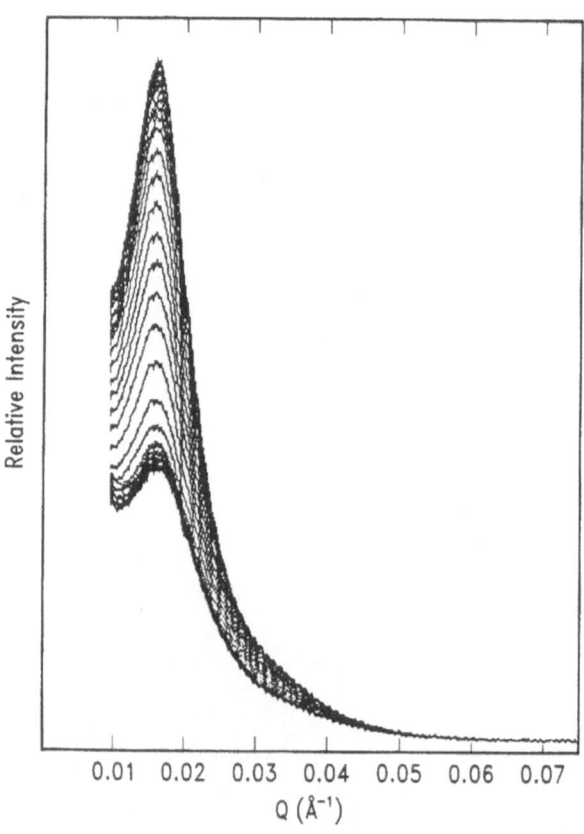

Fig. 11. Small-angle X-ray scattering of copolymer **2f** as a function of the scattering vector, Q, during heating from 60 to 250 °C at 5 °C min^{-1}. The scattered intensity is seen to increase with increasing temperature as the propylene oxide block decomposes

g=0.017 Å^{-1}, corresponding to a Bragg spacing of ~367 Å. From the volume fraction of propylene oxide in the copolymer (~0.12), this yields an average size of the propylene oxide domains of ~44 Å. It should be noted that only a single reflection is observed, with no indication of higher order reflections. If the propylene oxide domains were arranged periodically on a lattice, higher order reflections would be expected. Consequently, while the microphase-separated domains are periodic, lattice distortions cause a rapid loss of interference at higher Q. These distortions arise from two sources. First, the microphase-separated morphology is far from an equilibrium state. Second, the size of the propylene oxide domains is polydisperse. The persistence of scattering at much smaller Q also suggests this.

As the sample is heated, the intensity of the scattering increases dramatically and at ~220 °C a nearly five-fold increase in the scattering is evident. It should be noted that the scattering increases over the entire scattering vector range but with larger increases being evident at the lower scattering vectors. In addition, the position of the reflection does not change as a function of temperature (i.e., the foam structure generated is identical to the initial block copolymer morphology).

Neutron reflectivity has been used to quantify the scattering length density (SLD) normal to the film surface for thin films on Si wafer substrates. Figure 12a shows the neutron reflectivity curve for a 860 Å-thick PMDA-3F/PPO block copolymer which has been foamed for 4 h at 300 °C. The smooth line is the fit of the reflectivity data using the SLD profile shown in Fig. 12b. Also shown in Fig. 13b is the SLD profile measured for the film prior to foaming the film becomes about 5% thinner on foaming (the thickness decreases from 910 to 860 Å). As can be seen from the data, there is a significant skin-core effect even for the unfoamed polyimide. Data for the poly(amic ethyl ester) (not shown) shows a uniform SLD through the film indicating that this effect develops during the imidization reaction. The skin-core effect is increased upon foaming with an approximately 100 Å-thick skin forming next to the air interface having an SLD which is equivalent to pure polyimide (3.1_10^{-6} Å^{-2}. There is a similar skin next to the Si substrate but the SLD does not quite reach that of the pure polyimide. The decrease in density in the center of the film corresponds to a void content of about 15%. The TEM results indicate that the morphology of the voids is somewhat tortuous and the presence of the skin may help to isolate the voids from the outside environment creating at type of closed cell foam.

The SLD profiles obtained from neutron reflectivity were confirmed using dynamic scanning mass spectroscopy (DSIMS) to measure the fluorine concentration as a function of depth. As can be seen from Fig. 12c, the DSIMS data shows the same qualitative skin-core features as observed by neutron reflectivity [47].

In contrast, the foams derived from the α-methylstyrene and styrene-based copolymers showed pore sizes ranging from 200 to ~1800 Å, values which are considerably larger than the size of the microdomains of the initial copolymer (Fig. 13). Furthermore, the pores in these samples appear to be more interconnected than those obtained from the fully imidized copolymers. Finally, the

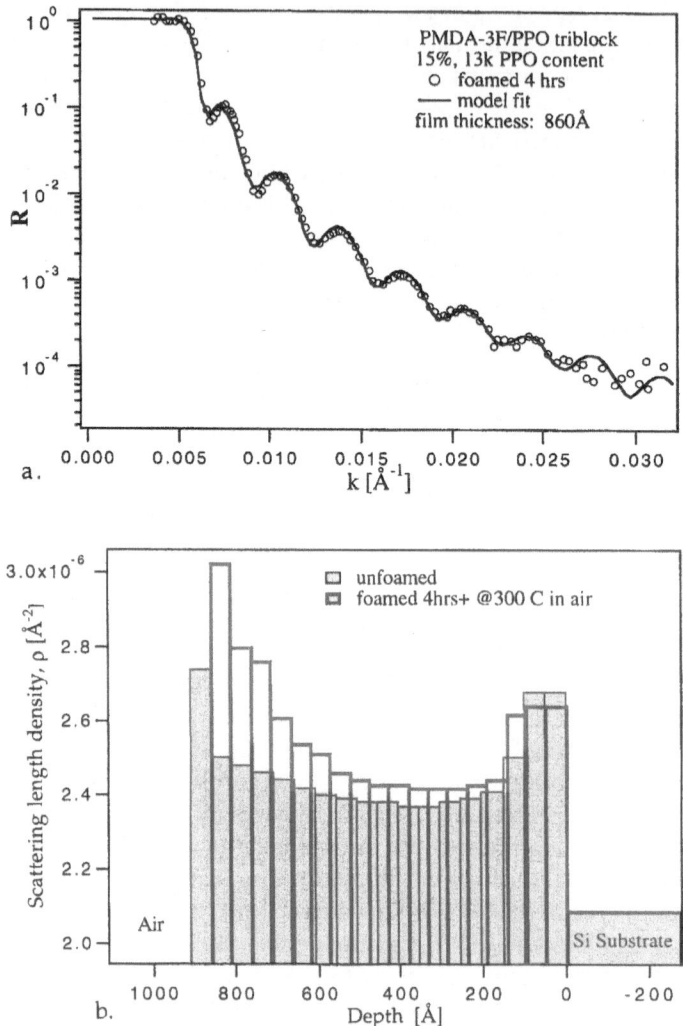

Fig. 12. a Neutron reflectivity data for a 860 Å-thick PMDA-3F/PPO triblock film which has been foamed for 4 h. The line through the data is the fit using the scattering density profile shown in b. **b** Scattering length density profiles for a unfoamed and foamed PMDA-3F/PPO triblock. The foaming results in a decrease in the overall film thickness and an increase in the density of the skin layer, particularly at the air interface. The density in the center of the film corresponds to about 15% voids. density skin at the top and bottom surfaces of the film

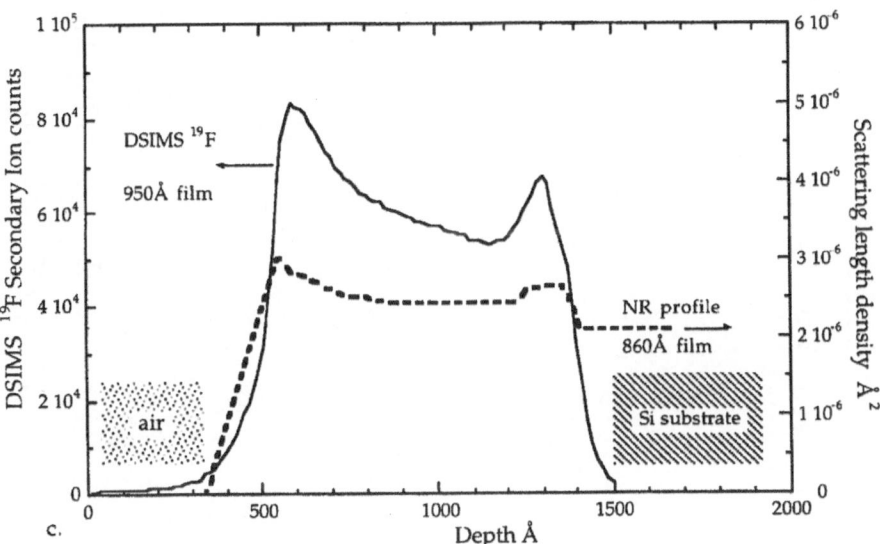

Fig. 12 c Comparison of DSIMS and neutron reflectivity data for similar films. The data is consistent with both techniques showing the presence of a high

pores are anisotropic in shape rather than spherical. The block copolymers derived from intrinsically rigid or semi-rigid polyimides with either poly(methyl methacrylate) or poly(propylene oxide) labile blocks did not show the expected nanofoam formation upon thermolysis. The rapid collapse of the foam well below the nominal glass transition temperature, T_g, was attributed to the anisotropic mechanical properties, characteristic of thin films of these ordered polyimides. In contrast, in the case of the α-methylstyrene-based copolymers with the rigid polyimides, highly porous materials were prepared as a result of the plasticization of the matrix and subsequent blowing.

Although nanofoam formation was observed in the amorphous high T_g polyimides containing PPO blocks, the volume fraction of voids does not directly correspond to the volume fraction of propylene oxide in the initial copolymer. A decrease in the volume fraction of voids present in the matrix in comparison to the initial volume fraction of the propylene oxide in the copolymer can be rationalized by considering the distribution of sizes in the propylene oxide phase. Since the copolymer blocks are not monodisperse, a distribution of propylene oxide microdomain sizes is expected. The pressure exerted on a pore will vary as γ/r, where γ is the surface tension and r is the pore radius. Consequently, the higher pressure on the smaller pores will tend to cause them to collapse. In addition, there will always be a small amount of propylene oxide co-mixed within the imide phase, which will also be removed upon decomposition. However, this

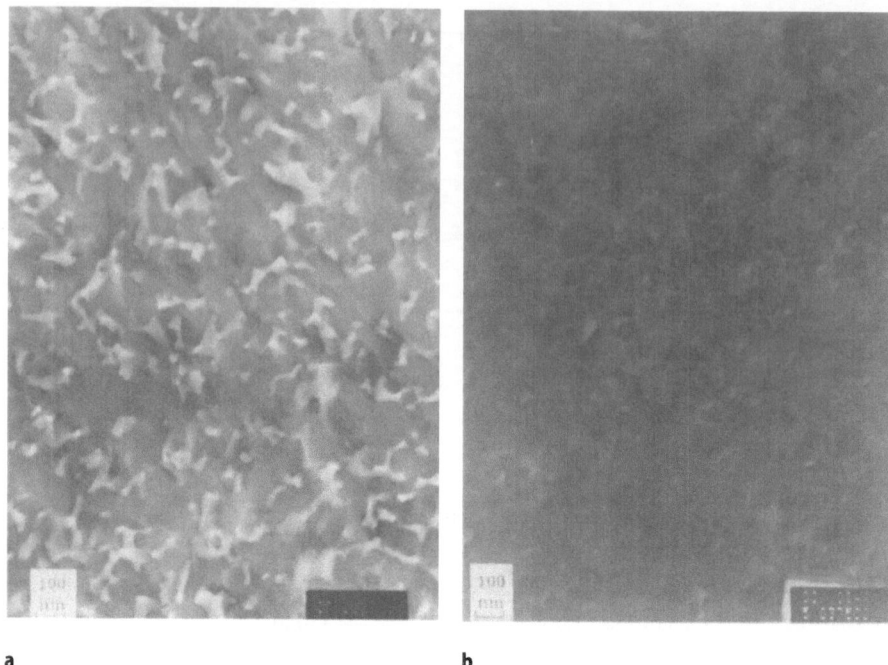

a   b

Fig. 13. a TEM micrograph of a polyimide foam derived from a imide/α-methylstyrene triblock copolymer. **b** TEM micrograph of a polyimide foam derived from a imide/styrene copolymer

feature alone cannot account for the significant discrepancies, particularly when
the initial propylene oxide content was high. In this case, the large volume fraction of the propylene oxide will necessarily result in smaller volumes of the imide phase between the propylene oxide domains. During the degradation of the
propylene oxide, the imide can be plasticized by the decomposition products
and cause a collapse of the void structure. However, the extent of interaction of
the poly(propylene oxide) degradation products with the polyimide matrix is
not clear. Although there are approximately 11 major degradation products
upon the thermolysis of poly(propylene oxide) [26], acetaldehyde and acetone
comprise nearly 80% of these products. These polyimides are polar materials
with solubility parameters close to those of the major degradation products of
poly(propylene oxide). In fact, ODPA/FDA and PMDA/FDA films show a significant level of uptake of acetone and acetaldehyde, validating the hypothesis of favorable interactions and the increasing possibility of plasticization. Although
the duration of the plasticization may be minimal due to the elevated temperature, this transient mobility coupled with any residual thermal and solvent loss
stresses may be sufficient to cause a partial collapse of the foam structure. The
foaming efficiency studies suggest that there is probably an optimal void size

and volume fraction that can be incorporated into the imide matrix. These parameters will vary for each polyimide and labile block combination.

For the copolymers with styrene and α-methylstyrene blocks, the TEM results demonstrate that the size of the pores generated are much larger than the size of the initial copolymer microdomains (e.g., the copolymers are transparent while the foams are translucent). This suggests that the degradation products, either α-methylstyrene or styrene, act as blowing agents. As the blocks decompose by unzipping, some of the monomer diffuses into the polyimide matrix, plasticizing the polyimide. The decomposition of these blocks, particularly α-methylstyrene, is quite rapid, however, because of the unzipping process probably more rapid than the monomer diffusion rate. The respective monomers, at these elevated temperatures, exert a substantial pressure on the surrounding matrix. Since the matrix is at least partially plasticized, significant blowing occurs (Fig. 13). Some of the larger pores can coalesce with time resulting in a partially interconnected pore structure.

5.3
Properties of Polyimide Nanofoams

The modulus was measured for two foamed samples derived from PMDA-3F/PPO [52]. The PPO block molecular weights were 3.5 kg/mol and the incorporation levels were 15 and 25 wt%. These copolymer samples produced foams with porosities of 14 and 19% respectively, and the measured moduli were 1.74 and 1.65 GPa. These numbers should be compared with that of the homopolymer PMDA-3F, which is 2.7 GPa. As expected, the modulus of the polymers decreases with increasing porosity (i.e., decreasing density). Residual thermal stress measurements recorded using a Flexus apparatus for foams generated on silicon wafers show that the film stresses decrease substantially (10–45%) in going from the cured polyimide to the respective foam (Fig. 14), consistent with drop in the modulus of the foams relative to the homopolymer (Table 6).

Table 6. Nanofoam properties from polyimide/PPO, ABA triblock copolymers

Polymer	Labile block (wt%)	Stress (MPa)	n(TE)	n(TM)	Void volume (%)	Dielectric constant	%Water uptake QCM
PMDA-3F	0	46	1.62	1.60	–	2.85	3.04
PMDA-3F (Foam)	24	25	1.46	1.42	18	2.35	2.80
PMDA-FDA	0	50	1.66	1.65	–	2.95	–
PMDA-FDA (Foam)	13	27	–	–	–	–	–
PMDA-4BDAF	0	49	1.63	1.56	–	2.85[a]	–
PMDA-4BDAF (Foam)	20	21	1.51	1.46	16	2.30[a] 2.70[b]	–
6FXDA-6F	0	31	1.56	1.50	–	2.55	2.90
6FXDA-6F (Foam)	16	28	1.47	1.44	14	2.25	2.70

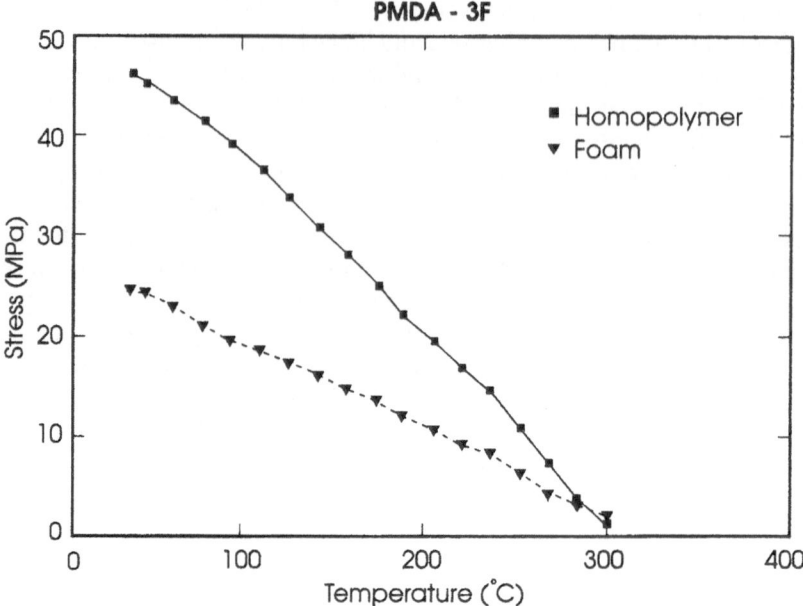

Fig. 14. Stress vs temperature plots for polyimide and polyimide foam

Measurements of the dielectric constant of PMDA-3F polyimide homopolymer from 1 Hz to 1 GHz show practically no frequency dependence [51]. This
suggests that there is no concern of any high frequency AC losses in PI films due
to water or ionic impurities. This should also be the case for the polyimide nanofoams, although this has not yet been demonstrated. Figure 15 shows a plot of
the dielectric constant (1 MHz) of a foam derived from PMDA-3F/PPO prepared
from the poly(amic ester) precursor vs the void volume fraction as determined
by density measurements [51]. The data is also included in Table 6 for this and
other samples. Also included in the figure is a plot of the expected dielectric constant as calculated from Maxwell-Garnett theory [21] based on spherical voids.
The agreement between experiment and theory is reasonably good, although
there are limited experimental points. The measured dielectric constants are actually somewhat lower than those predicted by theory. The thermal stability of
the foam (11% porosity) derived from PMDA-3F/PPO (PAE precursor) is shown
in Fig. 16, which plots both dielectric constant (2.45) and film thickness as a
function of temperature. Both quantities are stable to 350 °C.

The equilibrium water uptake of the homopolymers and the corresponding
foams were measured at 22 °C and a relative humidity of 90%, using a quartz
crystal microbalance (Table 6). The rate of water uptake was linear with RH, suggesting a Fickean diffusion process for both the homopolymers and the porous
materials. The water uptake for the PMDA/4-BDAF samples was not measured,

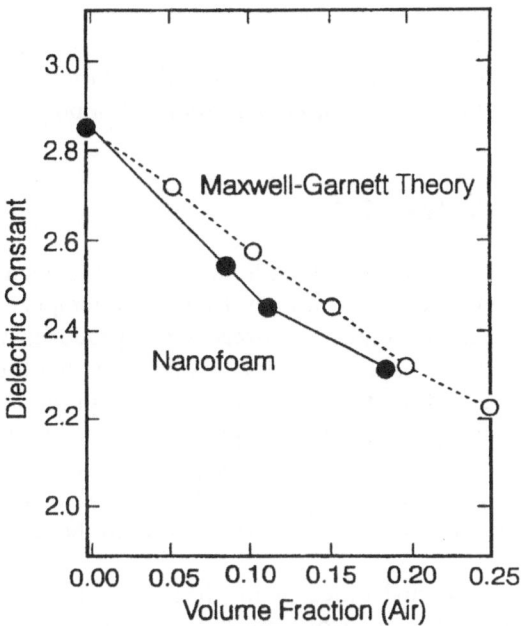

Fig. 15. Dielectric constant vs void volume fraction for foams derived from PMDA/3FDA/propylene oxide copolymers

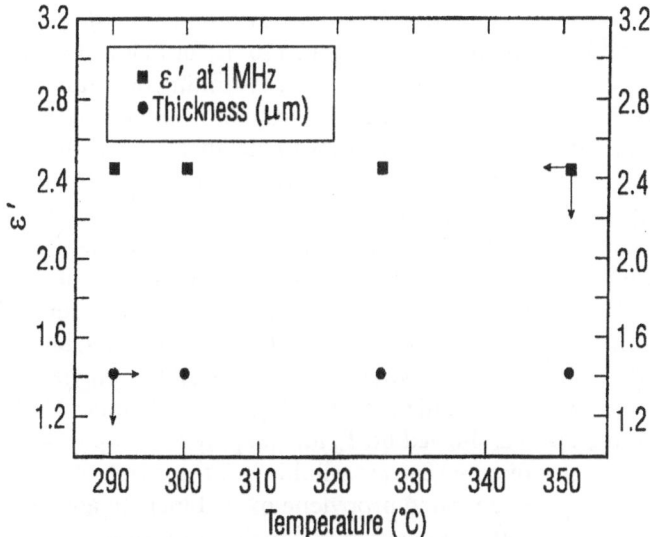

Fig. 16. Thermal stability of foams from PMDA/3FDA polyimide/propylene oxide copolymers

because the glass transition temperature of the polymer (T_g 305–325 °C) was lower than desired for devise integration. The key observation is that the extent of water uptake for the foams is not excessive and, in the examples studied, is actually slightly lower than for the cured polyimide homopolymers themselves.

6
Manipulation of Porosity Content

With PMDA-3F/PPO, porosities as high as ~18% have been achieved, while maintaining the nanoporous morphology. However, attempts to increase porosity by incorporating higher volume fractions of PPO produced highly porous but opaque films. A plausible explanation for this result is that the decomposition rate of the PPO is faster than the diffusion rate through the polyimide and the degradation products, particularly at high volume fractions, plasticized the polyimide acting as blowing agents, increasing pore size and connectivity. One possible solution is to crosslink the polymer, thus providing internal "solvent resistance" to the degradation products produced upon thermolysis of the labile block [30].

Pendant aromatic ethynyl functionality has been shown to react between 200–300 °C and the resulting networks produced from ethynyl substituted polymers possess the requisite thermal stability. However, this reaction temperature range is 150–200 °C below the T_g of fully cured PMDA-3F (T_g~440 °C), making crosslinking in the functionalized but preimidized form unlikely. To be effective, the crosslinking reaction must also be complete prior to the decomposition of the labile block [30]! This dilemma is avoided by using the poly(amic ester) (PAE) precursor of PMDA-3F.

As mentioned, in the precursor PAE derived from the meta isomer of diethyl pyromellitic acid, a softening occurs about 120 °C, far below the onset of imidization at 250 °C, offering the possibility of crosslinking of pendant ethynyl substituents prior to imidization. With this system, it should be possible to the crosslink and imidize in N_2 at 300 °C prior to the final thermal decomposition of the PPO block in air. For this purpose, the comonomer 1,1-bis(4-aminophenyl)-1-(4-ethynylphenyl)-2,2,2- trifluoroethane (3FET) was prepared [45]. Triblock copolymers containing this comonomer and either propylene oxide or poly(α-methylstyrene) were prepared, as described in Scheme 11. The labile coblock compositions ranged from 15–40 wt% (17–44 vol.%), values somewhat higher than in previous studies. The mole percent of 3FET in the PAE block was varied from 5–40 wt%. Samples of the ethynyl containing PAE copolymer were simultaneously crosslinked and imidized by heating to 300 °C under N_2. DMA analysis of the cured materials showed no T_g for the polyimide up to 475 °C, consistent with the presence of extensive crosslinking. A transition observed at –60 °C was seen for the phase-separated propylene oxide block. In air, a slight drop in modulus around 150–250 °C was attributed to decomposition of the labile block and partial plasticization of the polyimide matrix. Foaming was initiated after curing and crosslinking by heating in air at 240 °C. The samples cured in N_2 to

diethyl ester diacyl chloride of PMDM   3FDA

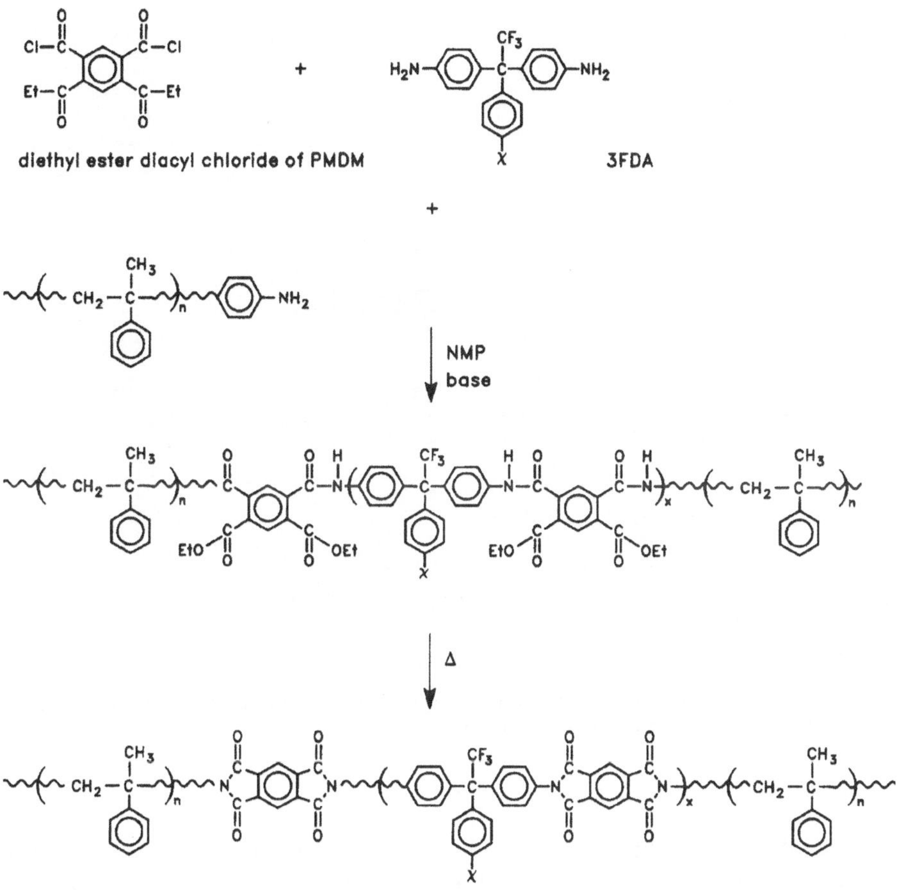

NMP
base

Δ

	X
Copolymer 1	75 mol%–H, 25 mol%–C≡CH
Copolymer 2	60 mol%–H, 40 mol%–C≡CH

Scheme 11

300 °C were transparent, although microphase separated. However, the foamed samples were all opaque, suggesting a pore size much larger than the initial co-polymer morphology. Although density measurements of porosity by flotation were irreproducible and IR estimates were inaccurate due to scattering, TEM studies confirmed the porous nature of the foam. High contrast TEM studies showed that the pores extend through the sample (800 Å) (Fig. 17). The morphology of the foams was not consistent with that of the imidized copolymer, confirming that plasticization and blowing had occurred, leading to enlarged and interconnected pores, in spite of the extensive crosslinking. Subsequent

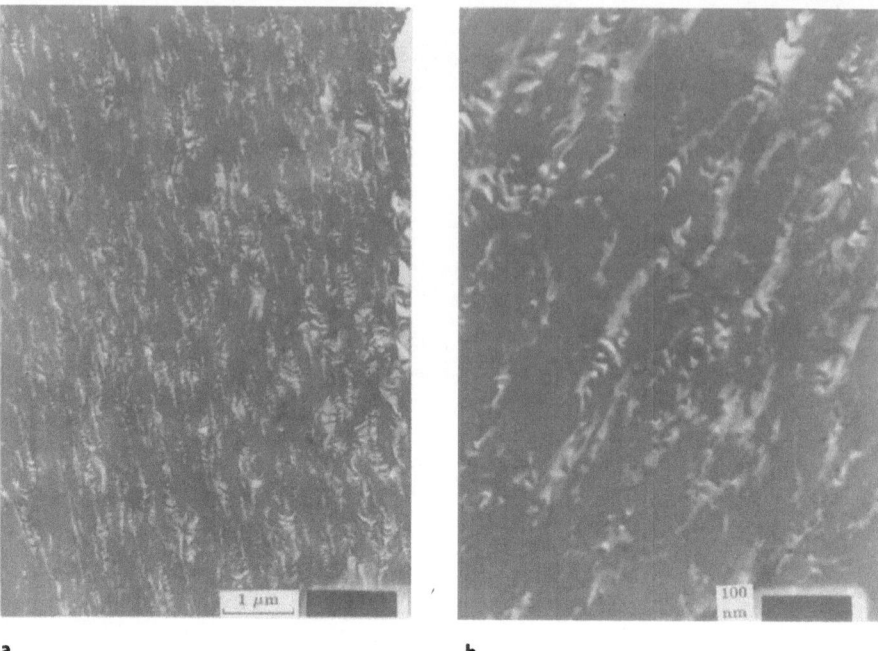

a   b

Fig. 17. TEM micrograph of crosslinked polyimide foam derived from imide/propylene oxide copolymer bearing ethynyl functionality

studies of the crosslinked polyimide containing no PPO showed that the crosslinking had practically no effect on the uptake of either acetone or acetaldehyde, major decomposition products of PPO. In this system, it was demonstrated that polymer crosslinking could not be used to stabilize nanofoams with increased porosities produced from copolymers containing higher loading levels of the labile block [30a]. Similar morphologies were generated with the poly(α-methylstyrene) based block copolymers [30b].

7
Conclusions

The generation of nanofoam materials by the self-assembled block copolymer approach has been demonstrated. The nanofoam approach has been shown to work for a variety of different polyimide matrices, in combination with a variety of different thermally labile coblocks. Numerous synthetic approaches were surveyed as a means to appropriately functionalized thermally labile coblocks, including group transfer, anionic, ROP "living" free radical and others. (The block copolymers were prepared either by the poly(amic acid) or poly(amic alkyl ester) precursor to the polyimide.) In each case, microphase separated morphologies were observed by DMTA, SAXS and SANS measurements. Upon degradation of the thermally unstable coblock, discrete voids in the nanometer range have been observed by TEM and confirmed by small angle X-ray scattering measurements. The mechanical and thermal behavior of the nanofoams is not significantly degraded from those of the unfoamed polyimides. Some properties are significantly enhanced, such as residual thin film stress reduction and decreased anisotropy. Most importantly, low dielectric polyimide foams have been generated showing that the introduction of nanoscale voids in the polymers can reduce the dielectric constant of thin films leading to measured dielectric constants as low as 2.3 while maintaining the thermal stability and other desired properties of polyimide films.

Acknowledgements. The authors gratefully acknowledge partial support of the NSF-funded MRSEC Center for Polymer Interfaces and Macromolecular Assemblies (CPIMA) (NSF DMR-9400354) and the partial support of the Advanced Technology Program of NIST under cooperative agreement 70NANB3H1365 and IBM.

8
References

1. Tummala RR, Keyer RW, Grobman WD, Kapen S (1989) In: Tummala RR, Rymaszewski EJ (eds) Microelectronic packaging handbook, chap 9. Van Nostrand Reinhold, New York
2. Muraka SP (1996) Solid State Tech Mar:83
3. Jang S-P, Havemann RH, Chang MC (1994) Mater Res Soc Symp Proc 337:25
4. Singer P (1996) Semiconductor Intl May:88
5. Hedrick JL, Brown HR, Volksen W, Plummer CJG, Hilborn JG Polymer (to appear)
6. (a) Brown HR, Yang ACM, Russell TP, Volksen W, Kramer EJ (1988) Polymer 29:1807; (b) Hedrick JL, Labadie JW, Russell TP, Palmer T (1991) Polymer 32:950; (c) Hedrick JL, Russell TP, Labadie JW (1991) Macromolecules 24:4559; (d) Hedrick JL, Hilborn J, Labadie J, Volksen W (1990) J Polym Sci Polym Chem Ed 28:2255
7. Numata S, Fujisaki K, Makino D, Kinjo N (1985) Proceedings of the 2nd Technical Conference on Polyimides: Society of Plastic Engineers, Inc. Ellenville, New York, p 164
8. Pfeiffer J, Rhode O (1985) Proceedings of the 2nd Technical Conference on Polyimides: Society of Plastic Engineers, Inc, Ellenville, New York, p 336
9. Cassidy PE (1980) Thermally stable polymers: synthesis and properties. Marcel Dekker, New York
10. Mittal KL (ed) (1984) Polyimides. Plenum Press, New York
11. Lupinski JG, Moore RS (eds) (1989) Polymeric materials for electronic packaging and interconnection. ACS Symposium Series 407
12. Takahashi N, Yoon DY, Parrish W (1984) Macromolecules 17:2583
13. Russell TP (1986) J Polym Sci Polym Phys Ed 22:1105
14. Gattiglin E, Russell TP (1985) J Polym Sci Part B Polym Phys Ed 27:2131
15. Boese D, Lee H, Yoon DY, Swalen JD, Rabolt JF (1992) J Polym Sci Part B Polym Phys Ed 30:1321
16. Haidar M, Chenevey E, Vora RH, Cooper W, Glick A, Jaffe M (1991) Mater Res Soc Symp Proc 227:35
17. Critchlen JS, Gratan PA, White MA, Pippett J (1972) J Polym Sci A-1 10:1789
18. Harris FW, Hsu SLC, Lee CJ, Lee BS, Arnold F, Cheng SZD (1991) Mater Res Soc Symp Proc 227:3

19. (a) Sasaki S, Matuora T, Nishi S, Ando S (1991) Mater Res Soc Symp 227:49; (b) Auman BC, Trofimenko S (1992) Polym Prep Am Chem Soc Div Polym Chem 34:244; (c) Auman BC, Trofimenko S (1993) Macromolecules 26:2779

20. St Clair AK, St Clair TL, Winfree WP (1988) Proc Am Chem Soc Div Polym Mater Sci Eng 59:28

21. (a) Maxwell-Garnett JC (1982) Philos Trans R Soc 15:2033; (b) Kantor Y, Bergman DJ (1982) J Phys C 15:2033

22. Hedrick J, Labadie J, Russell TP, Hofer D, Wakharker V (1993) Polymer 34:4717

23. Hedrick JL, Hawker CJ, Di Pietro R, Jérôme R, Charlier Y (1995) Polymer 36:4855

24. (a) Hedrick JL, Russell TP, Hawker C, Sanchez M, Carter K, Di Pietro R, Jérôme R (1995) ACS Symposium Series 614, American Chemical Society, Washington, DC, p 425; (b) Hedrick JL, Carter K, Cha H, Hawker C, Di Pietro R, Labadie J, Miller RD, Russell T, Sanchez M, Volksen W, Yoon D, Meccerreyes D, Jérôme R, McGrath JE (1996) Reactive and Functional Polymers 30:43; (c) Hedrick JL, Di Pietro R, Plummer CJG, Hilborn JG, Jérôme R (1996) Polymer 37:5229

25. Hedrick JL, Di Pietro R, Charlier Y, Jérôme R (1995) High Perform Polym 7:133

26. Charlier Y, Hedrick JL, Russell TP (1995) Polymer 36:4529

27. Charlier Y, Hedrick JL, Russell TP, Jones A, Volksen W (1995) Polymer 36:987

28. Hedrick JL, Russell TP, Labadie J, Lucas M, Swanson S (1995) Polymer 36:2685

29. Carter KR, Di Pietro R, Sanchez MI, Russell TP, Lakshmanan P, McGrath JE (1997) Chem Mater 9:195

30. (a) Hedrick JL, Carter KR, Sanchez M, Di Pietro R, Swanson S, Jayaraman S, McGrath JE (1997) Macromol Chem Phys 198 (in press); (b) Charlier Y, Hedrick JL, Russell TP, Swanson S, Sanchez M, Jérôme R (1995) Polymer 36:1315

31. Hedrick JL, Russell TP, Sanchez M, Di Pietro R, Swanson S, Meccerreyes D, Jérôme R (1996) Macromolecules 29:3642

32. Labadie JW, Hedrick JL, Wakharkar V, Hofer DC, Russell TP (1992) 15:925

33. Hedrick JL, Russell TP, Sanchez M, Di Pietro R, Swanson S, Mecerreyes D, Dubois Ph, Jérôme R (1997) Chem of Mat (in press)

34. Patel N (1990) Multicomponent network and linear polymer system: thermal and morphological characterization. PhD Thesis, UPI and SU (under McGrath JE)

35. Bates FS, Fredrickson GH (1990) Annu Rev Phys Chem 41:525

36. Leibler L, Fredrickson GH (1995) Chem Brit 31:42

37. Rogers ME, Moy TM, Kim YS, McGrath JE (1992) Mat Plas Soc Symp 13:264

38. (a) Manring LE (1989) Macromolecules 22:2673; (b) Manring LE (1988) Macromolecules 21:528; (c) Manring, LE (1988) Macromolecules 21:528; (d) Inaba A, Kashiwagi T, Brown JE (1988) Polym Degrad Stab 21:1

39. (a) Bywater S, Black PE (1965) J Phys Chem 69:2967; (b) Sinaha R, Wall LA, Bram T (1958) J Chem Phys 29:894

40. (a) Hawker CJ (1994) J Am Chem Soc 116:11,185; (b) Hawker CJ, Elce E, Dao J, Volksen W, Russell TP, Barclay GG (1996) Macromolecules 29:2686

41. Hawker CJ, Hedrick JL (1995) Macromolecules 28:2993

42. (a) Dubois Ph, Ropson N, Jérôme R, Teyssié Ph Macromolecules (accepted); (b) Stassen S, Archambeau S, Dubois Ph, Jérôme R, Teyssié Ph (1994) J Polym Sci Polym Chem 32:2445; (c) Barakat I, Dubois Ph, Jérôme R, Teyssié Ph (1994) J Polym Sci Polym Chem 32:2099; (d) Barakat I, Dubois Ph, Jérôme R, Teyssié Ph (1993) J Polym Sci Polym Chem 31:505; (e) Degée Ph, Dubois Ph, Jérôme R, Teyssié Ph (1992) Macromolecules 25:4242

43. (a) Kricheldorf HR, Kreiser-Saunders I, Scharnagl N (1990) Makromol Chem Macromol Symp 32:285; (b) Kricheldorf HR, Boettcher C (1993) Makromol Chem 194:1653; (c) Carter KR, Richter R, Kricheldorf HR, Hedrick JL (1997) Macromolecules 30; (d) Carter KR, Hedrick JL, Richter R, Furuta PT, Mecerreyes D, Jérôme R (1996) Mater Res Soc Symp Proc 431:487; (e) Carter KR, Richter R, Hedrick JL, McGrath JE, Mecerreyes D, Jérôme R (1996) Polym Prep 37:607

44. Volksen W, Yoon DY, Hedrick JL, Hofer D (1991) Mater Res Soc Symp Proc 227:23
45. Jensen BJ, Hegenrother PM, Ninokogu G (1943) Polymer 34:639
46. Briber RM, Fodor JS, Russell TP, Miller RD, Carter KR, Hedrick JL (1996) Proceedings from the Materials Research Society Fall Meeting (accepted)
47. Fodor JS, Briber RM, Russell TP, Carter KR, Hedrick JL, Miller RD (1996) Polymer (submitted)
48. Sanchez MI, Hedrick JL, Russell TP (1995) J Polym Sci Part B Polym Phys 33:253
49. Plummer CJG, Hilborn JG, Hedrick JL (1995) Polymer 36:2485
50. (a) Fodor JS, Briber RM, Russell TP, Carter KR, Hedrick JL, Miller RD (1997) J Poly Sci Poly Phys Ed 35:1067; (b) Fodor JS, Briber RM, Russell TP, Carter KR, Hedrick JL, Miller RD (1997) Macromolecules (in press)
51. Cha HJ, Hedrick JL, Di Pietro R, Blume T, Beyers R, Yoon DY (1996) Appl Phys Lett 68:1930
52. (a) Leterrier Y, Manson J-AE, Hilborn JG, Plummer CJG, Hedrick JL (1995) Proc MRS 371:487; (b) Hilborn JG, Plummer CJG, Leterrier Y, Hedrick JL (1995) MRS Proc

Received: March 1998

Poly(ester-imide)s for Industrial Use

Klaus-W. Lienert

Beck Elektroisolier-Systeme, Postfach 280 180, D-20514 Hamburg, Germany

Poly(ester-imide)s are a class of polymers known for more than 35 years. They are used to-day in large tonnage as electrical insulating materials. The patent literature reviewed shows that predominant research activities in the past were focused on improving the electrical, thermal and mechanical properties. In recent times new applications for these polymers have been found, such as engineering thermoplastics, adhesives, printed circuit boards and membranes. Excellent properties and easy processing will probably lead to a continuous growth of poly(ester-imide) business.

Keywords. Saturated poly(ester-imide)s, Unsaturated poly(ester-imide)s, Imide modified alkyds, Imide modified coatings, Thermoplastic poly(ester-imide)s, Wire enamels, Impregnating materials, Engineering plastics, Other applications

1
Introduction

Poly(ester-imide)s are a class of thermally stable polymers having a wide use mainly in the electrical industry. In 1997 the demand of the electrical industry for saturated poly(ester-imide)s based wire enamels was approximately 40,000 tons and for unsaturated poly(ester-imide)s for coil impregnation it was also several thousand tons world-wide.

After the successful synthesis of a poly(ester-imide) and its use as wire enamel [1], a development of this class of polymers started which is not yet finished today. The progress in chemistry has been reviewed in several papers [2, 3] where general principles were described. The economical importance of poly(ester-imide)s made companies "uncommunicative" and publication work was directed to patent applications. A large number of patents are known on poly(ester-imide)s. They cover various areas ranging from raw materials, resin compositions, synthesis and varnish formulations, to uses and processing.

Patent applications are written to protect know-how. Patents contain examples to prove the inventions. Tables existing in patents show the advantages of the invented products or processes in comparison to the state of the art. The data published in patents over the years cannot be compared with each other, because the test methods (some standardized, some not) have been changed, and misleading conclusions could be made. On the other hand, the products marketed by different companies have technical data sheets made in accordance with the valid standards, e.g. IEC (International Electrical Commission). It is difficult, if not impossible, to get a link between the data from a patent applications and data sheets. Therefore no tables from patents have been included in this review. The tables listing properties are summaries of data sheets from our company products.

The information presented here and selected from patents is that which, in the opinion of the author, is the basic work done, showing the progress in technology. The review is not and cannot be complete. The reason is that know-how concerning poly(ester-imide)s is filed in patents having different classifications, and is difficult to retrieve (e.g. if it is filed in a patent having as the main claim the processing machine). Another reason is that particular company know-how never was patent protected because patent violation could not be proved, and the know-how was better kept secret.

The examples cited here refer to the field of electrical insulating materials if not otherwise indicated.

2
History

In the 1950s poly(ethylene terephthalate) based polyesters branched with glycerine or trimethylolpropane were used for coating copper winding wires [4, 5]. The excellent thermal stability of the polyimides was known [6] and also the difficulties in handling these materials. To get the good flexibility, processability and storage stability of the polyesters and the high thermal properties of the polyimides, the terephthalic polyesters were modified with imide moieties and the poly(ester-imide)s were created (**1**). The imide building blocks were responsible for the improved heat shock (combined mechanical and thermal test where a copper wire is wound around a mandrel and the specimen obtained is stored at elevated temperature) and thermal endurance (temperature index, IEC 172). Also the dielectric losses at elevated temperatures were minimized. On the other hand, the imide units cause a reduction of the adhesion of the films on the copper surface and, depending on the imide content, a reduction of the "cut-through" (softening temperature of the polymer film measured as described in the IEC 851 document).

In the same period it was found that using tris-(2-hydroxyethyl)-isocyanurate (THEIC) for polymer branching increases the cut through of the polyester film substantially [7]. The poly(ester-imide)s were modified with THEIC, and improved materials were obtained (Fig. 1) [8].

Table 1 shows the average properties of commercial wire enamels, glycerine and THEIC branched polyesters (PE) and glycerine and THEIC branched poly(ester-imide)s (PEI).

In the mid-1960s the unsaturated polyesters used for impregnating magnet wire coils [9] were modified to improve the thermal property level, following the experience with the wire enamels [10]. In order to obtain unsaturated poly(ester-imide)s soluble in styrene and having low viscosities, imide structures are used as endcapping groups. Figure 2 illustrates structures of unsaturated poly(ester-imide) resins, one branched by trimellitic anhydride (**2**) and the other by THEIC (**3**).

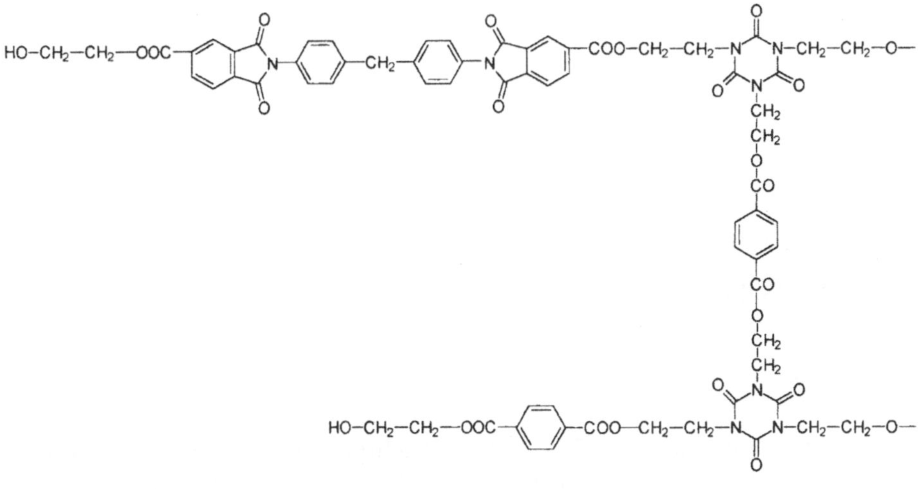

(1)

Fig. 1. THEIC – poly(ester-imide)

Table 1. Properties of wire enamels coated on copper wire

		Glycerine PE	THEIC PE	Glycerine PEI	THEIC PEI
Wire diameter	mm	0.71	0.71	0.71	0.71
Diameter increase due to enamel	mm	0.052	0.065	0.062	0.065
Mandrel winding test 1*d, pass at elongation of	%	20	25	25	25
Heat shock 1*d, pass after 30 min at	°C	130	155	200	200[a]
Cut trough	°C	280	420	280	400
Tangent delta steep rise	°C	120	150	140	185
Temperature index		135	155	180	220

a Heat shock 10% elongation plus 1*d

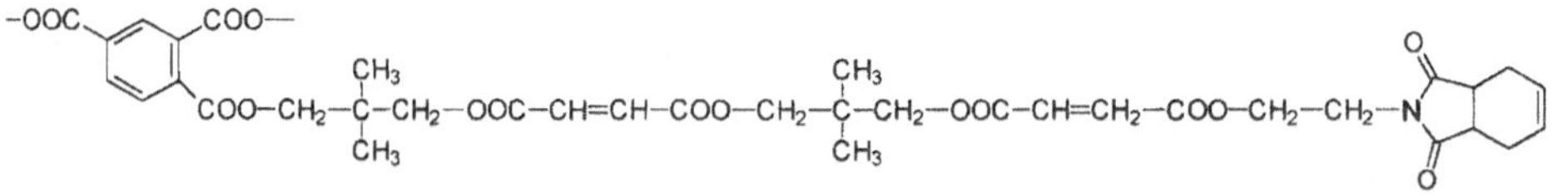

(2)

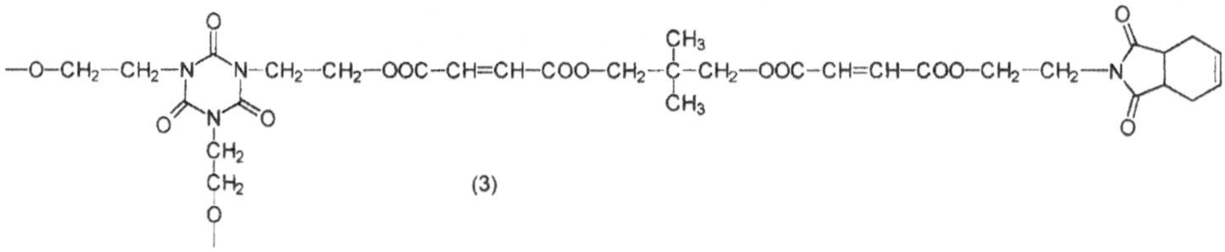

(3)

Fig. 2. Unsaturated poly(ester-imide)s

3
Saturated Poly(ester-imide)s for Wire Enamels

3.1
Monomers

A saturated poly(ester-imide) is made by modifying linear or branched polyesters with imide-containing structures (Fig. 3). The synthesis, which is described below in more detail, starts from polyester components like diols and triols (e.g. glycol and THEIC), diacids or reactive diacid derivatives, like terephthalic acid

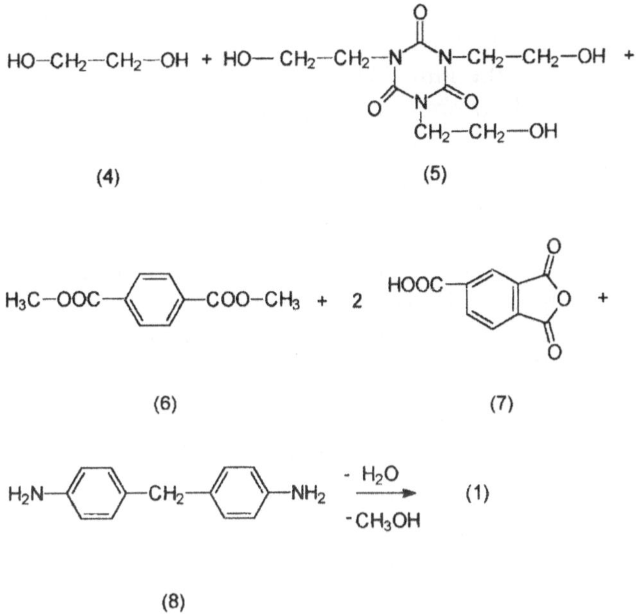

Fig. 3. Reaction scheme poly(ester-imide) formation

or dimethylterephthalate. The imide moieties are formed by reactions between a primary amine having another reactive group, like 4,4'-diaminodiphenylmethane or aminoethanol, and diacids able to form a five-membered imide having another reactive group, like trimellitic anhydride or pyromellitic anhydride.

Most of the commercially available poly(ester-imide)s are products made from a large excess of glycol, thus having an OH-number between 100 and 300 mg KOH/g resin, depending on the product, and a molecular weight which is usually adjusted to be below 5000.

The basics of poly(ester-imide) structures and synthesis were patented in the 1960s [1, 8, 11–13], but also in later years companies have patented poly(ester-imide)s of various compositions and for different applications [14–24]. There are also patents where a polyester is blended with a low molecular weight imide [25], or a poly(ester-imide) is modified with amide structures [26]. These particular polymers are not reviewed here.

The largest amount of poly(ester-imide)s manufactured world-wide is used for wire enameling. Therefore the raw materials used in the manufacturing process have to be available world-wide at acceptable prices.

The vast majority of recipes existing in the patent literature contains ethylene glycol (1). The reason is that poly(ester-imide)s used as thermosets are curing by a transesterification, so that ethylene glycol is formed and eliminated from the film by diffusion and evaporation. The boiling points of other diols are too high for this kind of application. When other diols are used additionally, they remain in the cured film, imparting special properties to the film. The use of 1,4-butanediol flexibilises the film [27], and for 1,4-cyclohexanedimethanol containing resins the authors are claiming good heat resistance and resistance to breakdown [28].

Work on improving the thermal resistance and particularly the resistance to carbonization (short circuiting of layers of enameled wires under the influence of temperature) via special glycols led to diphenols [29, 30]. Diphenols are not reactive under the conditions of a normal poly(ester-imide) synthesis. In synthesis the lower aliphatic diesters of diphenols were used [29–32]. The use of acid chlorides in the polyester reaction with aromatic OH-groups was also protected by patents [33–35] but it seems unlikely that this reaction was performed on the production scale.

2,2-bis(4-Hydroxyphenyl)propane is claimed in most of the "diphenol patents". In one case [36] bisphenol-A is used in combination with ethylene carbonate to prepare, together with other monomers, a wire enamel resin, having a good flexibility of 1*d and 30% elongation and a heat shock of 2*d at 200 °C.

A patent was filed on a poly(ester-imide) where octachlorodihydroxybiphenyl is used [37].

For nonelectrical insulating uses a rigid poly(ester-imide) with optical anisotropy is known, containing a diphenol with an imide structure, made from the reaction of 4-hydroxyphthalic acid with p-aminophenol [38]. The diacetylated derivate is used in the polymer forming reaction.

A novel poly(ester-imide) is claimed [39] with good heat resistance and mouldability containing 4,4'-dihydroxydiphenylsulfone, along with its symmetrical di- and tetramethylderivates.

Ethylene glycol is one major OH-component in poly(ester-imide)s used in wire enamels. The other OH-component is a polyol used as branching agent. It was known from other types of wire enamels that a high degree of branching increases the thermal stability of the cured film. The first poly(ester-imide) made in this way was based on pentaerythritol [1], but common for the products in the market is the cheaper glycerine. These glycerine-based poly(ester-imide)s were modified in various ways (e.g. with fatty acids [41]) to improve the property level. Today they are rarely used, and only for special applications, but patenting is continuing [42].

The use of tris-(2-hydroxyethyl)-isocyanurate (THEIC) [8] (5) improves the thermal properties of poly(ester-imide) based wire enamels dramatically (Table 1). The cut through increases over 100 °C when glycerine was replaced by THEIC. A dramatic increase of the tangent delta steep rise (a glass transition temperature measured via a dielectric loss method, IEC 317) was also obtained. The temperature index, giving in degrees Centigrade the maximum temperature for the 20,000 h use of the material (IEC 172), was passing the 200 °C limit.

THEIC was for many years scarce on the market. Wire manufacturers tried therefore to prepare it themselves by the reaction between isocyanuric acid and ethylene oxide. Another synthetic procedure starts from isocyanuric acid and the commercial ethylene carbonate. The in situ synthesis of THEIC in the presence of the other monomers has been described [42]. Because a purification step of THEIC is not available to eliminate the byproducts formed (e.g. the incomplete reactants or the higher alkylene oxylated derivatives), the resulting wire enamels do not have standard qualities.

Higher alkylene oxylated derivatives of isocyanuric acid were claimed to be useful for poly(ester-imide) preparations [43], and also 1,3,5-tris(2-hydroxyethyl)-hexahydro-1,3,5-triazine-2,4-dione, prepared by reacting ethylene oxide and hexahydro-1,3,5-triazine-2,4-dione [44]. No wire enamels made from these polyols have been seen on the market.

For an injection and extrusion molding application a poly(ester-imide) was patent protected, where the polyol is a novolac resin [45]. The product was claimed to have having improved char forming properties.

The diacid mainly used in the production process for poly(ester-imide) based wire enamel is terephthalic acid (or its dimethylester) (6). Resins are known where isophthalic [46] and phthalic acid [47] are used. The large tonnage products all contain the terephthalic unit. The cured films have better thermal and mechanical properties (e.g., a higher hardness), important when the coated wires are processed in high speed winding machines.

Nevertheless, the efforts made to produce wire enamels starting out from other acid components must be mentioned here. Reacting p-aminobenzoic acid with trimellitic anhydride yields an imide having two free COOH-groups for polyester chain formation. The use of N-(4-carboxyphenyl)trimellitimide in a poly(ester-imide) is claimed [48]. In an other patent [49], glycine, aminonaphthoic and aminobenzoic acids are described for the same type of chemistry. It seems that these intermediates have not been used for regular production, probably being economically uncompetitive.

Another diacid which is found frequently in the poly(ester-imide) literature is ithaconic acid. Ithaconic acid reacts with amines forming N-substituted 3-carboxy-2-pyrolidones. When one mole of 4,4'-diaminodiphenylmethane is reacted with two moles of ithaconic acid, a diacid is obtained, which is useful for making a poly(ester-imide) resin [50–52].

Aliphatic hydroxy acid esters, e.g., dihexyl ester of malic acid and triethyl citrate, are claimed as being useful as starting materials for resins as insulating electric wires [53].

Patents on poly(ester-imide) resins where one of the components is ethylene carbonate [54, 55] or an imide modified linear poly(ethylene carbonate) [56] also exist. In another case the carbonate group is introduced into the molecule using diphenylcarbonate via transesterification reaction [57].

Serious doubts may be cast on the marketability of wire enamels based on poly(ester-imide)s made from other acids than terephthalic acid. This consideration is also valid for resin made from the exotic tris-omega-carboxyalkyl isocyanurate [58].

The thermal improvement of the polyesters was achieved in an easy and cheap way by incorporating imide structures into the chain. For the synthesis of the five-membered nitrogen-containing imide ring, two molecules have to react: one with a minimum single amine group and a second functional group suited for further reactions, and one anhydride with another reactive group. The imide structure is also obtained when the amine is replaced by an isocyanate.

The reaction of trimellitic anhydride (7) with ethanolamine (9) giving the hydroxy acid (10) and with 4,4'-diaminodiphenylmethane (8) giving the diacid (11) has been published in the first poly(ester-imide) patent [1]. The second one is nowadays the predominant reaction for making poly(ester-imide)s. Trimellitic anhydride is the basic dianhydride for introducing the imide structure into the polyesters. Nearly every example in patents is based on trimellitic anhydride alone or mixtures with other anhydrides, e.g., tetrahydrophthalic anhydride [59]. The imides made from aromatic anhydrides are thermally more stable than the ones resulting from aliphatic structures (Fig. 4). Both types have been protected by patents, and products made from them are on the market.

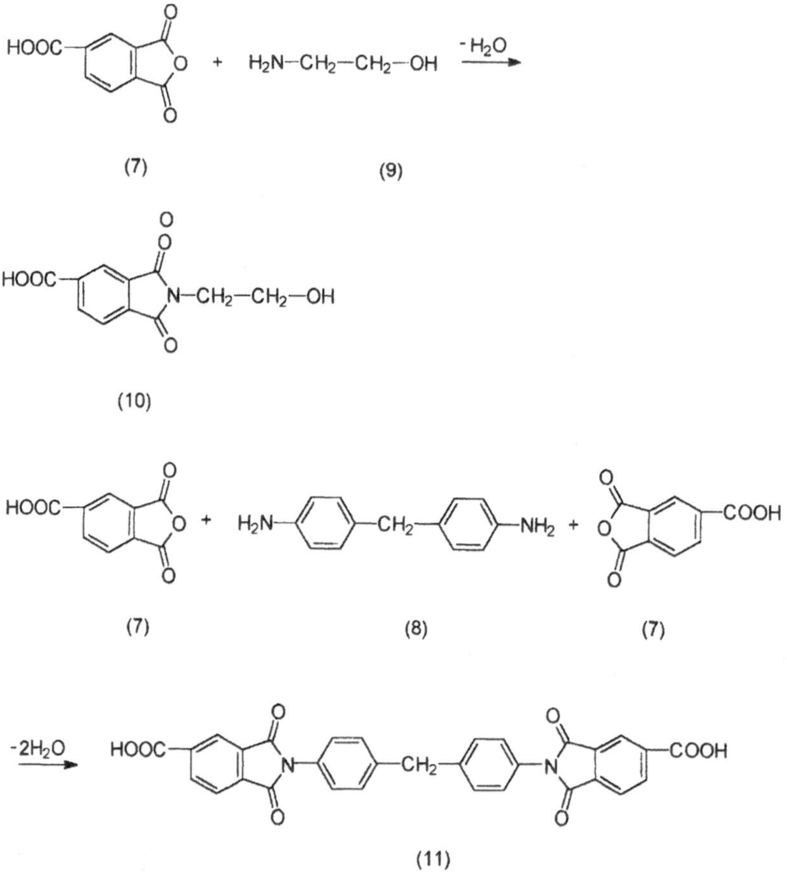

Fig. 4. Imide forming reactions

Also known are poly(ester-imide)s prepared from 3,3',4,4'-benzophenonetetracarboxylic dianhydride (**12**) (Fig. 5). They were claimed to have excellent thermal properties [60], excellent refrigeration medium resistance [61] or to be useful for making high performance printed circuit boards [62].

For a new kind of application, the use in an electrophotographical element, the 1,4,5,8,-naphthalentetracarboxylic dianhydride (**13**) is claimed [63]. The

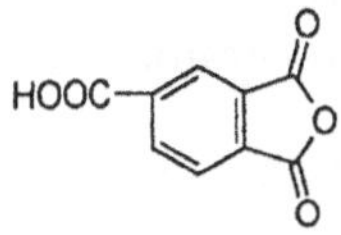

(7)

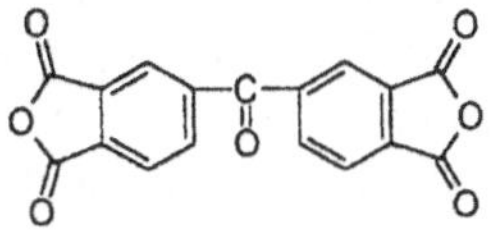
(12)

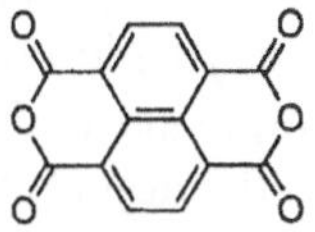
(13)

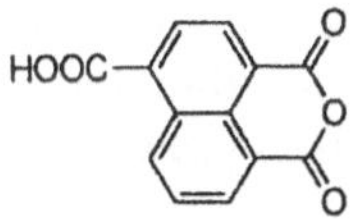

(14)

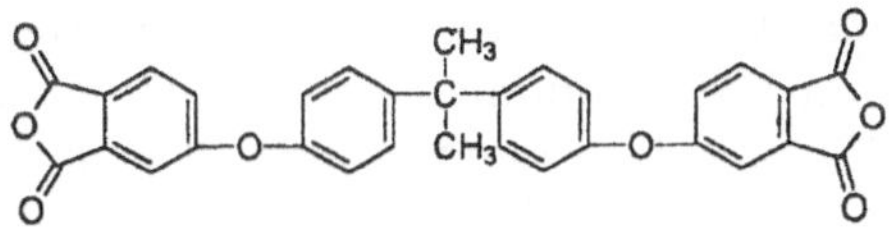

(15)

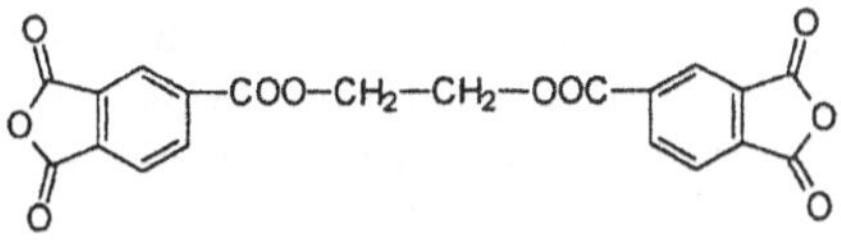

(16)

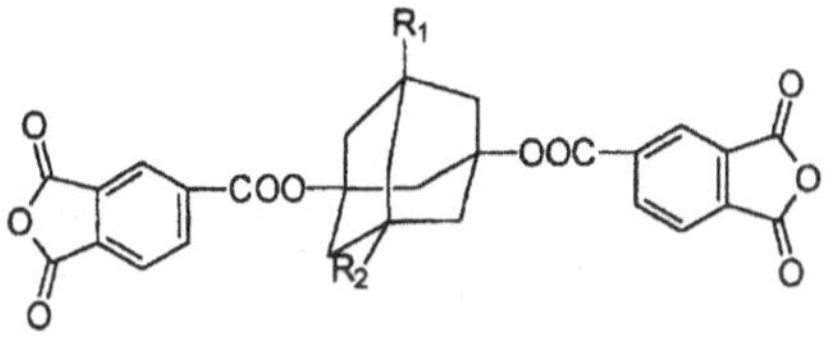

(17)

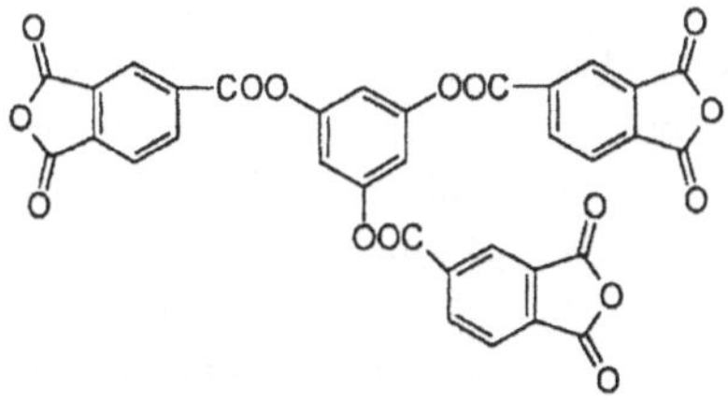

(18)

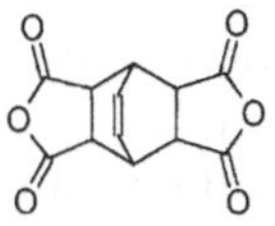
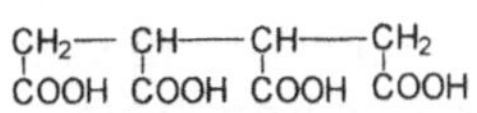

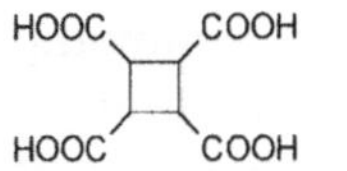

(19) (20) (21)

Fig. 5. Anhydrides used for imide synthesis

1,4,5,-naphthalenetricarboxylic anhydride (14) is also claimed for making poly(ester-imide)s [64].

Efforts made to get the trimellitic anhydride structure into bigger monomeric molecules were leading to novel poly(ester-imide)s. The bisphenol-A dianhydride (15) is used from making a wire enamel showing at 220 °C a dissipation factor of 3.2 [65]. The addition of 2 mol of trimellitic anhydride to 1 mol of ethylenglycol followed by a rearrangement gives the dianhydride (16) claimed in a patent [66]. The dianhydride (17) reacted with diaminodiphenylether is giving after curing a poly(ester-imide) having a Tg of 235 °C [67].

The trianhydride (18) is used for making poly(ester-imide) resins for advanced composites and adhesives [68].

Between the nonaromatic acids or acid derivatives claimed for poly(ester-imide) resins are bicyclo(2,2,2)-oct-2-ene-2,3,5,6-tetracarboxylic dianhydride (19) for wire enamels [69], and 1,2,3,4-butanetetracarboxylic acid (20) for wire enamels, lamination materials and adhesive sheets [70]. The butanetricarboxylic acid anhydride monomethylester [71] and the cyclobutanetetracarboxylic acid (21) [72] are also claimed as poly(ester-imide) forming raw materials.

Despite the large number of amines known to be suited for poly(ester-imide) formation, only a few have been used for industrial production. Ethanolamine (9) described in the first patent [1] was used and it remained interesting for a period of time [73, 74]. However it was soon replaced by the 4,4'-diaminodiphenylmethane (8) [1]. The reason was that the aromatic diamine imparts into the polyesterimide a higher Tg, a higher resistance to oxidative degradation of the wire enamels, and an improved hardness of the cured film. The disadvantage of the diamine is that, being a solid, it leads to more handling in production. Also, being rated as having carcenogenic potential, the necessary safety precautions are more sophisticated. The reaction product of the diamine with two moles of trimellitic anhydride is the diimidodicarboxylic acid (11) called DID acid. Being insoluble in all the usual solvents, stirring of this precipitate formed in the reaction vessel during the synthesis requires highly powered tailor-made stirring equipment. The DID acid is made soluble when reacted with the diol and triol present in the reaction vessel. The imide modified polyester wire enamels made from 4,4'-diaminodiphenylmethane are today the standard products for single coat magnet wires world-wide. The diamine is a bulk chemical, used mainly for 4,4'-diphenylmethanediisocyanate synthesis and is available world-wide in a sufficient amount, of good quality, and at reasonable prices.

In the patents where poly(ester-imide) syntheses are claimed as suitable for imide forming, 4,4'-diphenylmethanediisocyanate is also mentioned as starting material. There are several patents known dedicated to this subject [75, 76]. It can replace 4,4'-diaminodiphenylmethane in the formation of the DID-acid. There are some practical problems rendering this replacement somewhat difficult. In the manufacturing process the diisocyanate reacts at first with the cresol, normally used as solvent, and the resulting urethane then reacts with the anhydride. In this second reaction step, cresol is formed along with carbon dioxide. Since the volume of the evolved carbon dioxide is high, the reaction equipment has to be designed properly to handle it. Because 4,4'-diphenylmethanediisocyanate is sensitive to temperature and moisture during storage, and because it requires careful processing, it is probably not much used for poly(ester-imide) production. From the analysis of a commercial wire enamel resin it is impossible to find out what raw material was used, the diamine or the diisocyanate.

Nowadays ethanolamine (9) is of no importance in the production of poly(ester-imide)s for wire enamels, but it is still a key monomer for the manufacturing of unsaturated poly(ester-imide)s.

Aromatic monoamines, having a minimum one additional functional group, and being found in the patent literature of poly(ester-imide)s, are the aforementioned p-aminobenzoic acid [48, 77], the p-aminophenol [78], the aminoterephthalic acid [79] and the p-aminobenzenesulfonic acid [80], but probably none of these monomers were ever used for large scale production. Also the diaminobenzenes had no economical success as wire enamel raw materials, probably because they were too expensive, too toxic and because they have, in bulk, insufficient storage stability. 3,5-Diaminotriazole-(1,2,4) was claimed as raw material for wire enamel resin yielding films with improved hardness [81].

Polyesters modified with diimidodicarboxylic acids derived from aliphatic diamines and trimellitic anhydride have been patented, like methylene-N,N'-bis-trimellitimide, made from hexamethylenetriamine [82], or the 1,4-dimethylaminocyclohexane derivative [83]. The diacid obtained from trimellitimide, made from the anhydride and dry ammonia, complexed with a divalent metal, e.g., Zn, Co and others, is used in the preparation of a poly(ester-imide), described as giving a wire enamel with good flexibility on copper wire [84].

Other special aliphatic imide forming compounds claimed in patents, like 2-oxazolidone [85] or epsilon-caprolactame, lead to polymers having aliphatic chains. These chains are responsible for the lower T_g of the cured films in comparison with the polymers having aromatic diamines. No commercial poly(ester-imide) wire enamel in western Europe contains them.

A triamine specially claimed for poly(ester-imide)-based wire enamels melamine [86]. Melamine is not too reactive. To obtain the poly(ester-imide) a polyester precondensate is made followed by the addition of melamine and trimellitic anhydride. At 245 °C the reaction is completed. The resulting resin dissolved in N-methylpyrrolidon gives a hard film on wire with a good heat shock resistance. A high molecular weight diamine, the polypropyleneglycol diamine with a molecular weight of 2000, is used for poly(ester-imide) synthesis, useful for molding and extrusion purposes [87].

3.2
Synthesis

The synthesis of poly(ester-imide) resins used in wire enamels is performed world-wide in a high tonnage. The resins are produced in reaction vessels, with volumes between 1 m^3 and 25 m^3, from ethylene glycol, THEIC, dimethylterephthalate, trimellitic anhydride and 4,4'-diaminodiphenylmethane. The process involves esterification reactions between the diols, the triols and the acids and a cycle forming reaction between the diamine and the anhydride.

The esterification reactions are catalysed usually by titanates, tetra-*n*-butyl titanate being the most preferred. However large number of other catalysts are known, and as examples zinc acetate [88], cerium and vanadium octoate [89] and litharge [90] are mentioned here. The imide forming reaction needs no catalyst.

Of only historical importance are the two-step preparations of the resins. One process starts with the synthesis of the polyester and afterwards the diamine and the anhydride are added [91–93]. In another process a polyester is made from all the components, excepting the diamine, and in a second stage the diamine is added [94]. Also known is a process where first the DID acid is made and then the polyester forming compounds are added [95, 96] (Fig. 6).

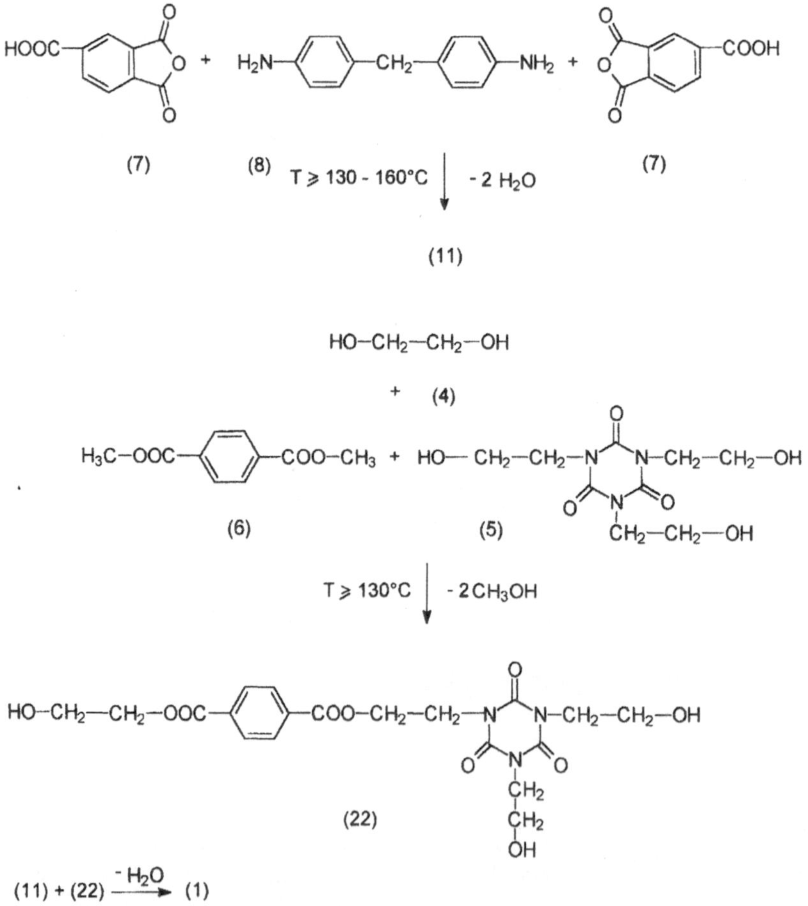

Fig. 6. Reactions during poly(ester-imide) resin formation

All these processes have only one purpose, to avoid the blocking of the stirrer in the reaction vessel due to the precipitation of the DID acid. When all the monomers are charged into the reaction vessel and the temperature of the mixture is reaching 130–160 °C, the trimellitic anhydride reacts with the 4,4'-diaminodiphenylmethane and the totally insoluble DID acid is formed. Water is generated and is distilled off. Around 130 °C the transesterification between dimethylterephthalate and ethylenglycol and THEIC starts and methanol is formed, which distills off from the reaction mixture. The insoluble DID acid reacts slowly between 160 and 190 °C with the excess of ethylene glycol and with the THEIC polyester giving the final poly(ester-imide). Depending on the imide content of the resin, this reaction is finished sooner or later. All the DID acid has to be reacted, because the tiny needles of the acid left in the varnish cause surface defects on the copper wire during the enameling process.

To get better stirring and heat transfer in the process, technical grade cresol is used to dilute the reaction mixture. Cresol is also used later on to formulate the ready-to-use wire enamels. The disadvantage of adding cresol is the capacity decrease of the vessel. Also the distillate contains cresol and has to be incinerated. An excess of ethylene glycol can be used as diluent. The glycol must be removed by vacuum distillation at the end of the process to achieve the desired molecular weight. The glycol recovered and purified can be reused. In this case the distillates are cresol free. Instead of dimethylterephthalate terephthalic acid is used by some manufacturer. The advantage is that there is no dimethylterephthalate which can sublime, and the distillate is free of methanol. The disadvantage is that the terephthalic acid is insoluble in the reaction mixture, even when cresol is used, and it reacts slower than the dimethylester. An improperly produced resin contains not only free DID acid but also free terephthalic acid. Raw material prices, reaction vessel design, imide content of the resin, production costs and costs of the distillate treatment are the deciding factors for the process used in modern poly(ester-imide) production.

The experience that the manufacturing equipment is less stressed when the polyester is formed first, and the availability of cheap poly(ethylene terephthalate) scrap, led to patents where all kind of processes were claimed to make poly(ester-imide) wire enamels from this polyester [97–101]. The problem is that clean, unpigmented and granulated poly(ethylene terephthalate) is needed for profitable production. There are some indications that this process is used today for industrial productions of wire enamels. Poly(ethylene terephthalate) is not only used as a raw material for the synthesis, but it can also be blended with a poly(ester-imide) to give a useful wire enamel [102].

Synthesis in other solvents like n-dodecanol [103] and diethylenglycolmonomethylether [104] is known and will be discussed under the topic noncresylic wire enamels. N-Methylpyrrolidon is an excellent solvent for the synthesis of poly(ester-imide) resins, e.g., as described in [105], but it is too expensive to be competitive. The preparation of a liquid crystalline poly(ester-imide) claimed for Si wafer coating is performed in pyridine [106].

3.3
Wire Enamel Formulations

The poly(ester-imide) resins prepared world-wide in large tonnage are used for formulating wire enamels, which are coated using continuous working enameling machines and round and rectangular copper wires by repeated application. Commercially available poly(ester-imide) based wire enamels use cresol, also known also as cresylic acid, as solvent. Other solvents have no economical importance nowadays (see discussion below).

The reason is that cresol is an excellent and cheap solvent, it can be adjusted to an optimum boiling range, and it is its own flow additive. Cresol can be blended with hydrocarbons, e.g., xylene, technical solvents like Solvesso® or Shellsol®, without losing much solvency power, and varnishes are less sensitive to drawing agent residues from bare copper wire manufacturing. Cresol has a high heat of combustion of 34220 KJ/kg. Modern enameling machines burn catalytically the evaporated solvents from wire enamels. The exhaust gases are used to preheat the air entering the machine. The disadvantages of cresol are its toxicity, having a low MAK-value of 5 ppm and a very low smell detection limit between 0.8 and 0.0005 ppm. This all requires a very careful handling of products containing cresol.

Technical grade cresol is available on the market from coal tar and coal liquefaction distillation processes. The main components are m- and p-cresol and smaller amounts of methylated and ethylated isomers. Large amounts of phenol and o-cresol are undesirable in wire enamel formulations. A recent paper gives a good overview of isomers and their influence on viscosity and solvency power [107].

In addition to the poly(ester-imide) resin and solvents a wire enamel contains catalysts, additives for improving flow and in some cases other resins for improving properties.

The curing of the poly(ester-imide) resins in the ovens of the enameling machines occurs via a transesterification reaction. Ethylene glycol is formed and evaporated. This reaction is catalysed by titanium esters like butyl or polycresyl titanates. Combinations are known of butyl titanate with toluensulfonic acid [108], of toluensulfonic acid with Zn-, Co- or Mn-octylates [109], and of titanates in combination with carboxylates of Ce, Co, Zn, Fe, Pb, and Zr [110]. More exotic is stearyltitanate [111]. The most effective catalysts and catalysts concentrations for a given wire enamel formulations is company know-how which is not disclosed.

Additives used in formulations have the purpose of improving the flow of the enamel, to improve the thermo-mechanical properties like heat shock, or to give better adhesion of the cured film to the copper surface. Some of the additives have multiple effects. Because these effects are very specific to a given poly(ester-imide) resin and varnish formulation, the product classes are here only enumerated. Phenolics [112, 113], epoxies [114, 115], and silicones [116] are well known and some of them were claimed. Better understood is the effect of the phenol blocked polyisocyanates on thermal and mechanical properties, like Desmo-dur® CT stabil. This is an isocyanurate with three phenol blocked isocyanate groups. During curing the blocking agent is released and the isocyanate reacts with the OH-groups from the poly(ester-imide) resin, giving additional crosslinking. Additives containing sulphur heterocycles improve adhesion by copper complex formation [117, 118]. The specific additives effective with their products are part of the know-how of the companies and are kept secret.

Open secrets are the effects of other polymers blended with poly(ester-imide) enamels. Polyhydantoins, are high temperature resistant polymers [119]. Wire enamels containing only polyhydantoins have heat shocks of 1*d/30 min at 300 °C and a tan delta steep rise of about 240 °C, that means 100 °C respectively, 60 °C higher than a poly(ester-imide). Being expensive materials and having a thermal endurance of only 170 °C, polyhydantoins have been blended with po-ly(ester-imide)s. These polyhydantoin modified varnishes have been used in the past frequently for coating heavy round and rectangular wires [120, 121]. The production of polyhydantoins for the wire enamel industry was stopped and no source of this materials is known today. Blends of poly(ester-imide) with polya-mides [122], with polyamides and epoxy resins [123], with poly(amide-imide)s [124], thermoplastic polysulfones [125], and PTFE [126] are known, all of them fulfilling a special technical requirement.

The toxicity of cresol led to solvents being screened to identify a replacement. The substitute had to have similar properties to cresol in respect of solvency power, boiling range and heat of combustion, to make the noncresylic varnishes processable on the existing enameling machines. Claimed as solvents have been diethylene glycol methyl or ethyl ethers [127–130], di- and triethylene glycol [131], monophenyl glycol [132, 133], propylene glycol monophenyl ether [134], a mixture of diacetone alcohol and N-methylpyrrolidone [135], acetophenone [136], dimethylphthalate [137], a mixture of the dimethylesters of adipic, glutar-ic and succinic acids marketed as DBE [138], alkylene carbonates [139], dimeth-ylformamide [140], and a mixture of ethylene glycol with glycol monoethers [141]. Solvesso® is claimed as solvent, without any other cosolvents, for a resin solubilised by using benzylic alcohol for partial endcapping [142].

A comparison of potential cresol replacements is given in Table 2 [143]. None of the solvents shown there is a full equivalent to cresol. Because of the reduced solvency power for poly(ester-imide)s the varnishes contain less hydrocarbons blended in and the heat of combustion is lower than that of a cresylic varnish, undesired for the existing installed enameling machines.

The property levels of cresol free solvent based poly(ester-imide) enamels are inferior to the cresylic products. In Table 3 the cresylic poly(ester-imide) var-nish, solved in a mixture of cresol and Solvesso® and the cresol free varnish 1, solved in a mixture of diethyleneglycol monomethyl and monoethyl ether plus some small amounts of Solvesso® have the same resin composition. The cresylic product is superior. The reason is that the preparation of the cresol free resin was made using diethyleneglycol monomethyl ether as solvent in the synthesis, causing a lower average molecular weight of the polymer. The cresol free po-

Table 2. Noncresylic solvents for wire enamels [143]

Name	Cresol	Methyldi-glycol	Propylen-carbonate	Butyro-lactone	DBE	Monophenyl-glycol	PG-Diacetate
Formula	C_7H_8O	$C_5H_{12}O_3$	$C_4H_6O_3$	$C_4H_6O_2$		$C_8H_{10}O_2$	$C_7H_{12}O_4$
CAS-No.	1319-77-3	111-77-3	108-32-7	96-48-0		122-99-6	623-84-7
MW	approx. 108	120	102	86	approx.158	138	160
Mp °C	–18	–65	–50	–43.5	–20	12	–75
Bp °C	210	194	242	206	198°–225	245	190
Flash point °C	91	71	123	104	102	121	95
Density (g/cm^3)	1.035	1.030	1.208	1.130	1.10	1.11	1.06
Heat of combustion (KJ/kg)	34,220	25,100	17,600	23,400	22,400	30,650	22,000
MAK-value (ppm)	5	none	none	none	none	none	none
Price index	100	130	175	363	118	183	198

(PG-Diacetate=Propylenglycol diacetate)

Table 3. Properties of cresol free wire enamels coated on copper wire

		Cresylic PEI	Cresol-free PEI 1	Cresol-free PEI 2	Water-based PEI	Hot melt PEI
Viscosity mPas/23 °C		700–900	360–460	400–600	110–140	200–400/180
Solids (1 g/1 h/180 °C)	%	38–40	44-46	44-46	38–40	90–92
Wire diameter	mm	0.71	0.71	0.71	0.71	0.50
Diameter increase due to enamel	mm	0.065	0.068	0.0682	0.065	0.040
Mandrel winding test 1*d, pass at elongation of	%	25	20	20	15	20
Heat shock 1*d, pass after 30 min at	°C	200[a]	200	200	200	200
Cut through	°C	400	370	385	335	350
Tangent delta steep rise	°C	190	190	190	185	190

a Heat shock 10% elongation plus 1*d

ly(ester-imide) resin 2 is prepared without any solvent and has, on a molar basis, 80% more THEIC and 100% more DID acid, but the varnish made from it is still inferior to the cresylic product. The reason is the interference of the diethylene glycol ethers used as solvents with the transesterification reaction of the curing process. Similar effects can be expected from the ester based solvents.

Noncresylic solvent based poly(ester-imide) wire enamels have been widely used in the past in some east European countries because the enameling plants were not equipped with catalytic burning units for the exhausted gases.

Water based poly(ester-imide) wire enamels were developed in the 1960s and 1970s. The resins were made water soluble in different ways. Resins with a defined acid number were neutralized with alkanolamines [144–148] or ammonia [149]. In another process the poly(ester-imide) resins were submitted to an aminolysis with alkanolamines [150–153] or ammonia [154–156], when the resin network is more or less degraded. Solvents for this poly(ester-imide) is water and usually a small amount of high boiling solvents like N-methylpyrrolidon or diethylene glycol monomethylether. Titanium catalysts stable to hydrolysis, like titanium-ammonium lactate and titanium lactate, were used [157]. To improve thermal and mechanical properties, phenol blocked isocyanates can be added to the water based poly(ester-imide)s. The blocked isocyanates are dispersed by means of an ethoxylated nonylphenol and are added to the water based wire enamel. Improvement of the property level is claimed [158].

The water based poly(ester-imide) wire enamel from Table 3 has the same resin composition as the cresol free solvent based resin 2. It can be seen that the mechanical and thermal properties of the water based varnish are inferior to the noncresylic product. It was also found that minor amounts of drawing agent residues from copper wire manufacturing were highly detrimental to the surface quality of the enameled wires.

In 1970 the next step was made in the development of poly(ester-imide) wire enamels by developing the solvent free hot melt resins [159]. The principle is the production of a resin containing mainly oligomers, being solid at ambient temperature. For processing of these prepolymers a heated applicator (140°–180 °C) with closed dies is required and the wires have to be passed through the enameling machine with a very low degree of vibration. The resin is coated on the wire as a melt with only four passes, because higher film thicknesses could be obtained with this technology. For many years a small amount of wire production has existed in western Europe using a hot melt poly(ester-imide) resin. Many patents are known dealing with this technology [160–168] and covering the whole field of poly(ester-imide)s, from glycerine to THEIC containing resins, from low to high imide content, covering catalysts, additives, etc. Table 3 lists the typical property level of a hot melt resin. The weak point of the material is the low cut through of only 350 °C and the necessity of having a special applicator for the enameling machines.

When using noncresylic based wire enamels the old enameling machines must compensate for the energy deficit of the coatings. New enameling machines are thermodynamically more efficient, and have a modified design, permitting manufacture of enameled wire without large additional energy costs. From all the alternatives for cresol based poly(ester-imide) wire enamels, the hot melt technology is the most environmentally friendly, the safest regarding transportation and storage of the varnish, and also gives good productivity.

3.4
Conventional Enameling of Poly(ester-imide) Wire Enamels

From the approximately 40,000 tons of poly(ester-imide) wire enamels, 400,000 tons of coated magnet wire are made world-wide by a continuous enameling process. First an excess of enamel is applied onto the copper wire, which is then adjusted for thickness with wiping devices, dies or felts. After this treatment the solvents contained in the enamel are evaporated and the resins are cured in an oven at temperatures between 300 and 600 °C. As the solvent content of the enamels is reduced to 60%, evaporation of this solvent and curing overlap in time. Blister-free wires can only be produced if the wet films applied are very thin. This means that the film thicknesses required for insulation can only be achieved by repeated application. It is usual to coat the wire six to ten times. After each coat the wire passes through the curing oven and is then cooled in a current of air, coated again, and so on. Also, large-scale statistical surveys have shown that good concentricity of the enamel film and thus better overall characteristics of the enameled wires can only be ensured by applying numerous individual coats.

Smooth running of the wire is one of the essential requirements for producing high-grade enameled wires and preventing cracks.

A modern enameling plant permits economical wire production and at the same time low solvent emissions by utilizing the heat of combustion of the solvents contained in the enamel. The solvents are usually burned catalytically using noble metal catalysts that have a permanent degree of efficiency of >95% at working temperatures of around 600 °C. The waste gas emission levels produced are below the limits currently specified by the laws.

In enameling plant the air entering the system is preheated by the exhaust gas in heat exchangers only. State-of-the-art enameling ovens achieve a high degree of thermal efficiency with direct or indirect air recirculation systems. The circulating air is returned to the enameling furnace (Fig. 7).

The suitability of an enameled wire for a particular application depends to a very large extent on the thermal class to which it belongs. Classification of a wire according to the usual IEC standards is carried out primarily on the basis of its temperature index, breakdown voltage at nominal temperature, and the heat shock characteristics of the insulation. The IEC specifications are internationally binding. In many countries the national standards are completely IEC-harmonized.

In western Europe, especially in Germany, the poly(ester-imide)s have penetrated the market to a degree above the average. The increasing trend towards poly(ester-imide)s on the world market is reducing the proportion of polyviny-

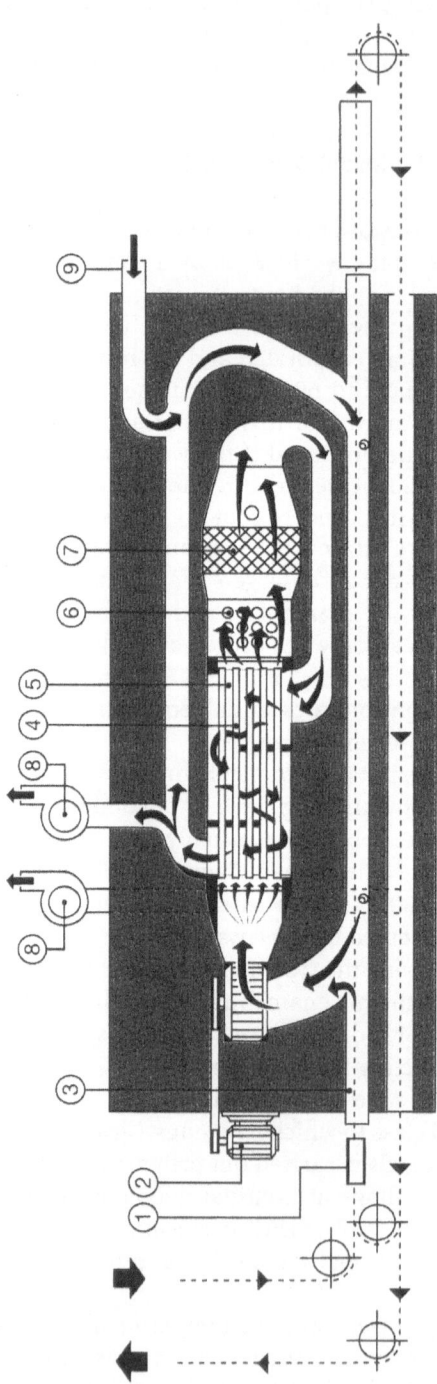

1. Enamel applicator
2. Blower motor for air circulation
3. Oven
4. Heat-exchanger pipes
5. Heat exchanger
6. Heater for catalyst
7. Catalyst
8. Waste air blower
9. Fresh air intake

Fig. 7. Scheme of a modern enameling machine (SICME NEM 400)

lacetals and conventional polyesters. There are many reasons for the popularity of the poly(ester-imide)s.

In transformer and motor construction, including hermetic motors, poly(ester-imide) wires have conquered a wide field of applications due to their ease of use and good compatibility with impregnating and refrigerating agents. Experience over a long period has now shown applications in distribution transformers to be safe, as the enamel film is highly resistant to hydrolysis. The products' thermal properties allow them to be allocated to classes 155–200, so that there is no need for dual coat wires in some cases. A production speed some 30% higher than that for polyester enamels reduces the production costs.

3.5
New Enameling Technologies

Besides the manufacturing of enameled wires in a conventionally way, starting from copper wire and a solvent-based varnish, other technologies have been developed.

In 1973 and the years thereafter, aqueous poly(ester-imide) dispersions were patented [169]. The conventionally produced resins were ground, dispersed in water, and flow control agents like 1,4-butanediol or butyl glycol, and non-hydrolyzsble titanium catalyst were added. Technically high grade wire insulations comparable with conventional enamels were produced with this systems. Special applicators were required for satisfactory processing of the dispersion. Concentric, uniform films could only be produced with a motor-driven applicator roller and wiping felts, and a complex pump circulation system was necessary to prevent sedimentation of the binder. The investments for this new applicators was not accepted by the market.

Poly(ester-imide)s aminolysed with ammonia, or neutralized, when they had an acid number, form stable dispersions in the presence of a surfactant (e.g. sodium laurylsufate), which can be deposited electrophoretically [170, 171]. Formulations were claimed giving improved mechanical and thermal properties with blocked isocyanates [172], or giving better processability when amine modified epoxies were used together with the poly(ester-imide) [173]. Mica added to the dispersion increases the film thickness, and improves the insulation [174]. The advantages of electrophoretic coating lie in low emission levels and good edge coating in the case of rectangular wires. This is where the process was most widely tested. No large scale industrial production is known based on this technology. Another use of the electrodeposition of poly(ester-imide) was claimed in the manufacturing of printed base plates [175] in the printing industry.

A further alternative for cutting down on solvents is to work with radiation curing of wire enamel systems [176–178]. A characteristic property of UV-curing systems is their very low energy requirement for the curing process, but the thermal stability of the obtained films requires improvement. The handicap is the double bonds, acrylic or olefinic, necessary for the curing reaction, which polymerize to produce a thermally weak aliphatic structure. On the other hand adhesion of such enamel films to the copper conductor is still insufficient.

Solder resist materials, usable in electrical and electronic field, based on poly(ester-imide)s made from 1,2,3,4-butanetetracarboxylic acid and p-aminobezoic acid or ethanolamine have been patent protected. UV sensitivity was obtained by using fumaric or p-phenylendiacrylic acid [179]. In another patent cinnamic acid [180] is used for a temperature resistant solder resist formulation. UV-curable foils made from poly(ester-imide)s containing a benzophenonetetracarboxylic diimide are known [181].

Powder coating is also an energy-saving, low-emission process with many industrial applications. Powders were applied by the fluidized-bed method or electrostatically. In the enameled wire industry powders are rarely used for rectangular conductors requiring good edge coverage. Because of the causal relationship between layer thickness and particle size, powder coating is restricted to heavy round and rectangular wires and fine grained powders, which are very expensive to produce. Genuine poly(ester-imide) powders are known for coating electrical wires [182], including amide [183] and epoxy modification [184]. For the latter system a heat shock of 175 °C was claimed.

3.6
Other Poly(ester-imide) Wire Enamels

Poly(ester-imide) resins have excellent thermal and mechanical properties, and wire enamels made from them are used in stressed electrical appliances. Other uses for imide modified polyesters are also known, where a balance between two thermal properties, cut through and soldering temperature, is required, e.g., in solderable poly(ester-imide)s and poly(ester-imide)s used in polyurethane wire enamels. A third application is in selfbonding wire enamels, where a softening of the film in a given temperature range is desired.

Solderable wires are used in telecommunications and the construction of analytical instruments, but they are becoming more and more common in small motors and dry-type transformers. The special characteristic of solderable wire enamels is the direct soldering. When the enameled wire is dipped in a solder bath at temperatures above 350 °C the coating melts and leaves the bare copper wire, avoiding the need for elaborate mechanical or chemical removal of the enamel. This characteristic achieved in solderable poly(ester-imide) by a special polymer composition and in polyurethane based wire enamels is due to the thermally reversible splitting of the polyurethane group. Nevertheless, good thermal stability is necessary and product temperature indexes >155 are required. For temperature indexes of 130, simply polyurethane wire enamels were used.

Essential for solderability is the absence of THEIC from the molecule of poly(ester-imide). To get acceptable thermal properties, branching of the polymer is necessary. There are known solderable poly(ester-imide)s containing glycerine, ethylene glycol, terephthalate and DID acid [185–187]. Another resin contains only glycerine, ethylene glycol and DID acid [188]. Glycerine can be re-

Table 4. Properties of solderable wire enamels

		Polyurethane	Polyurethane imide modified	Solderable PEI	Standard THEIC-PEI
Viscosity (mPas/23 °C)		70–80	140–160	750–850	700–900
Solids (1 g/1 h/180 °C)	%	25–28	27–29	35–37	38–40
Wire diameter	mm	0.50	0.50	0.71	0.71
Diameter increase due to enamel	mm	0.050	0.045	0.060	0.065
Mandrel winding test 1*d, pass at elongation of	%	25	20	20	25
Heat shock 1*d, pass after 30 min at	°C	155	200		
2*d, pass after 30 min at	°C			185	200a
Cut through	°C	230	250	350	400
Tangent delta steep rise	°C	110	140	170	190
Solder test 375 °C	s	0.9			
420 °C	s		0.8		
460 °C	s			6.0	
520 °C	s			1.0	not possible

a Heat shock 10% elongation plus 1*d

placed as branching molecule in this type of formulation by trimellitic anhydride [189] or a mixture of glycerine with trimellitic anhydride [190]. A similar resin additionally containing terephthalate was also claimed [191]. The replacement of THEIC with glycerine in a poly(ester-imide) resin containing 1,4-cyclohexane dimethanol [192] leads to a good solderability [193].

In the case of the solderable poly(ester-imide)s, a decomposition of the enamel film occurs at temperatures of 460 °C or higher, leaving no residues on the copper surface. In the case of imide modified polyurethanes, the film is built up by the reaction of a blocked isocyanate made, e.g., from trimethylol propane and toluene diisocyanate blocked by phenol, with an OH-poly(ester-imide). A catalyst, like dibutyltin dilaureate promotes during curing the reaction of the imidized polyester with the blocked isocyanate, when a polyurethane is formed and the blocking agent is released. Since thermal stability of the urethane bond is weak, decomposition of the polymer occurs at lower temperatures. Imide free polyurethanes can be soldered at temperatures of 375 °C. Imide modification makes soldering possible at temperatures around 420 °C. Imide modified polyurethanes have linear OH-poly(ester-imide) [194, 195], branched by glycerine [196] or trimethylolpropane [197]. Even a THEIC branched material has been claimed [198].

Table 4 summarizes wire enamels such as a standard polyurethane, an imide modified polyurethane, a solderable poly(ester-imide), and a standard poly(ester-imide) for comparison of properties and soldering characteristics.

With selfbonding enamels, applied as a top coat over the base insulation, the layers of a winding are bonded into a compact unit without the use of an impregnating agent or a coil support after the individual wires have been fixed together by heat or solvents. For this purpose thermoplastic bond coats are used, such as polyamides, polyvinylbutyrals, or phenoxy enamels, where only the solvent is evaporated during enameling and crosslinked duroplastic selfbonding enamels, where, during the manufacture of the wire, the curing reaction is taken as far as the so-called B stage. The reaction is completed during the bonding process, when the polyamide is reacted with the hardener. The polyamide-based selfbonding wire enamels have the disadvantage of getting tacky in high humidity environments (e.g. in south east Asia), making unwinding of the wire from the spools laborious. Self bonding wire enamels developed on the basis of poly(ester-imide)s do not have this disadvantage. For this application the cured resin has to have a softening temperature around 175 °C, achieved by using linear resins [199] (e.g., from trimellitic anhydride, ethylene glycol and hexamethylene diamine). Blends of linear poly(ester-imide)s with polysulfones were also claimed [200]. Probably many other formulations had been tested in that region but, never being patent protected, they are unknown in the literature.

4
Unsaturated Poly(ester-imide)s

The other class of poly(ester-imide)s used in the electrical industry is the unsaturated imide modified polyesters. These products have been in the patent literature since 1965 [10]. Unsaturated polyesters and the imide modified unsaturated polyesters are used for impregnating electrical coils. They bond the windings, protecting the coil against water and chemicals, have a better thermal conductivity than air, and are electrically insulating. The unsaturated polyesters have a temperature index above 130 and the unsaturated poly(ester-imide) above 155. The imide modification of the polyester structure improves the thermal properties and the mechanical properties at elevated temperatures. Using an unsaturated poly(ester-imide) resin for impregnating, the size of an electrical motor can be reduced considerably without losing power. But in this case, the higher service temperature of the smaller motor requires superior impregnating resin.

Unsaturated poly(ester-imide)s are prepared, for example, by reacting tetrahydrophthalic anhydride (23) with aminoethanol (9) giving the imidoalcohol (24). The alcohol is further reacted with maleic anhydride (25), neopentyl glycol (26) and trimellitic anhydride (7) to give the unsaturated poly(esterimide) (2) (Fig. 8). In patent examples, ethylene or propylene glycol were found instead of neopentyl glycol. Some maleic anhydride may be replaced by phthalic, isophthalic, or terephthalic acids. There are known linear and branched resins, e.g., by glycerine or trimethylolpropane. The ready-to-use product is a solution of the unsaturated poly(ester-imide) in a monomer (in most cases styrene or vinyl toluene), but diallylphthalate and acrylates were also claimed. Viscosity is adjusted to a given value with the monomer and the reactivity with a peroxide. A satisfactory storage stability of the materials is obtained by adding a suitable stabilizer, mostly a radical scavenger, such as a quinone or phenol. The electrical coil is impregnated by dipping, dip rolling, or trickling and is then cured at temperatures between 130 and 160 °C, where the resin copolymerises radically with the monomer.

Unsaturated poly(ester-imide) resins are known having imide structures at the end of the molecule or in the backbone. Chain termination by imide is frequently done with the tetrahydrophthalic anhydride (23) aminoethanol (9) reaction product [10, 201–204]. In the synthesis of these resins, maleic anhydride is used. It is cheap and a world-wide available raw material, having the advantage of generating per mol only one mol of distillate. Patents are known where fumaric acid is claimed, e.g., in [205], but it is probably not frequently used in industrial processes. Anyway, under the conditions of the esterification reaction, a part of the maleate is isomerised into the fumarate [206]. To get materials with good heat resistance the use of THEIC is claimed for branching the unsaturated poly(ester-imide) resins [202]. Flexibility and heat resistance are improved when fatty acids, e.g., castor oil fatty acid [205, 207], are build into the macromolecule.

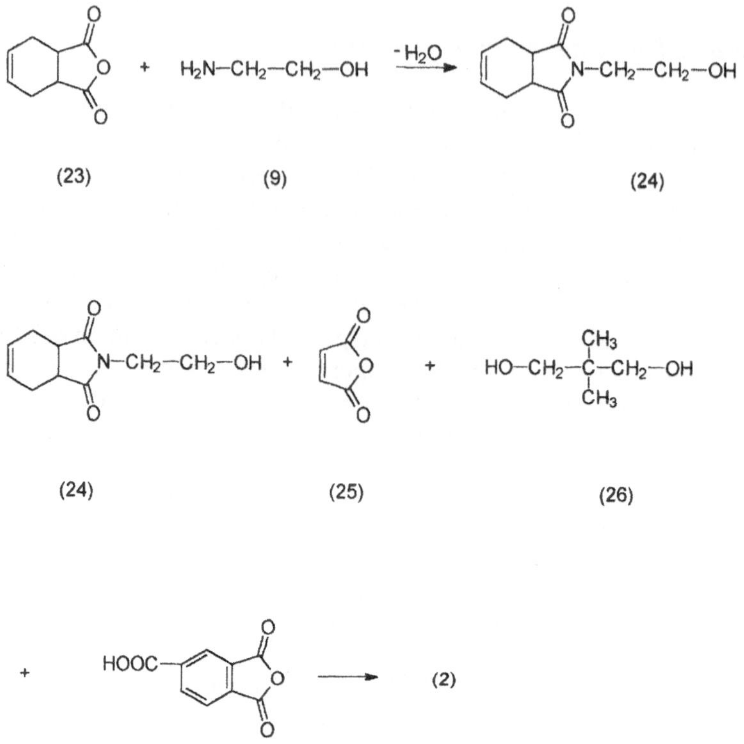

Fig. 8. Reaction scheme for unsaturated poly(ester-imide) preparation

Methyltetrahydrophthalic anhydride reacted with aminoethanol and maleic anhydride is used for addition to a low molecular epoxy resin [208]. The resulting material, diluted with styrene and polymerised, is claimed to have improved short term thermal stability.

Endomethylene tetrahydrophthalic anhydride and the methyl derivate are also claimed for imide formation with aminoethanol [209]. The reaction of the endomethylene tetrahydrophthalic anhydride with hexahydro-p-aminobenzoic acid followed by a esterification with ethylene glycol and terephthalic acid yields an unsaturated imidoester curable by Michael addition with polyamines [210].

The use of phthalic anhydride in imide forming reactions was also claimed [211], but this structure does not contribute to a crosslinking of the molecule during the cure.

The imide structure generated from trimellitic anhydride (7) and 4,4'-diaminodiphenylmethane (8) are used for the preparation of unsaturated poly(ester-imide)s having the imide in the polymer backbone. Thermal properties are superior, but the high intrinsic viscosity caused by the hard segment of the DID acid (11) is undesired. The resins known are suitable for thermal [212] and radiation curing [213].

When trimellitic anhydride (7) is reacted with aminoethanol (9) and this intermediate with propylene glycol and fumaric acid, a linear unsaturated poly(ester-imide) is obtained, soluble in 6,6-dimethylfulvene, claimed by the inventors to have good thermal stability after curing [214].

Another imide modified unsaturated polyester is obtained when trimellitic anhydride is reacted with epsilon-caprolactam, followed by esterification with THEIC, neopentyl glycol, and maleic anhydride. The product is claimed for coating motor windings [215].

The bicyclo(2,2,2)-oct-2-ene-2,3,5,6-tetracarboxylic dianhydride (15) was also used for making unsaturated poly(ester-imide)s [216].

Another category of unsaturated ester-imides has the crosslinkable double bonds at the ends of the molecules. So the diimide made from the bisphenol-A dianhydride (12) and allylamine is claimed for giving heat resistant polymers [217]. The same is valid for the diallylester made from the trimellitimide of p-phenylendiamine [218]. Another compound, also not a polymer, but polymerisable, giving good heat resistance and electrical properties is the allylester of trimellitallylimide [219]. Heat resistant polymers for electrical insulating purposes made from saturated poly(ester-imide) and endcapping the molecules with unsaturated structure are known. The transesterification product between a poly(ester-imide) and diallyl isophthalate, having allyl isophthalate endgroups, is claimed as an insulating coating [220]. A poly(ester-imide) made from benzophenone tetracarboxylic dianhydride (12), an aminoacid, and glycol is acrylated and used as insulating material [221]. The acrylated DID acid glycol based poly(ester-imide) is self curing in the presence of a photosensitizer [222].

Impregnation with suitable impregnating agents like unsaturated poly(ester-imide)s protects a magnet wire coil permanently against damage by outside influences. Such impregnation produces two effects of essential importance to the safe operation of electrical equipment. The winding is protected against mechanical stress, especially electrodynamic forces, as the individual layers of the winding are bonded together and the coil is bonded to the core and the core insulation. The wire is sealed and thus protected against dust, moisture and other damaging influences, mostly of a chemical nature. In such a combination of insulating materials the impregnating agent may be considered compatible with the enameled wire if the winding wire is not damaged by the impregnating agent during the impregnation process, the combination of materials has adequate mechanical strength, and the components harmonize with each other as they should in the long term. These criteria can be tested and evaluated by internationally standard methods, like measurement of the hardness of the enamel film after immersion in the impregnating agent (IEC 851-4), measurement of the bond strength of the impregnating agent/ enamelled wire combination (IEC 1033, Method A), and measurement of the temperature index, using specimens of impregnated winding wire.

The chemical resistance of an organic material depends to a very large extent on the molecular weight of the polymer concerned. When optimally cured, poly(ester-imide) wire enamels are resistant to the styrene used in impregnating and trickle resins. Optimum curing is dependent, when all the other parameters of the enameling machine are held constant, on the enameling speed of the wire.

The mechanical strength of the insulating material combination depends on the bond strength of the impregnating compound, the adhesion between enameled wire and impregnating material, the adhesion between enamel film and copper conductor, and the strength of the copper. The bond strength of the unsaturated poly(ester-imide) is dependent on the branching of the resin, the degree of crosslinking after curing, the mesh size in the cured polymer, and the glass transition temperature.

Long-term behavior of insulation systems can be estimated after the temperature index has been determined by the impregnated twisted pair test (IEC 172, criterion breakdown voltage) or helical coil test (IEC 1033, Method B, criterion bond strength). The two tests lead to differing results that do not correlate and the end user of the system enameled wire/impregnating resin has to decide which test is applicable for his electrical appliance. It is important to choose a suitable combination of materials and to ensure close cooperation between manufacturers of electrical insulating materials and those who later process them.

5
Imide Modified Alkyd Resins

The old technology for impregnating the magnet wire coils of electrical motors is based on using alkyd varnishes. These varnishes are still used today in many countries of the world and also in Europe in repair shops. The varnishes have as resins air drying alkyds and alkyds or polyesters which are crosslinked during curing by melamines, phenolics, or isocyanates. Solvents used are mostly hydrocarbons for conventional systems or water for newer products. To improve thermal properties of the impregnating varnishes the resins have been modified, similar to wire enamels and impregnating resins. A linseed oil alkyd, modified with THEIC, DID acid and solved in a xylene-cyclohexane mixture was patented [223]. Other THEIC and DID acid modified alkyds, containing tall oil fatty acid and being formulated with a phenolic resin are claimed as withstanding temperatures higher than 180 °C. They are dissolved in conventional solvents [224] or in water [225]. A polyurethane varnish, where the polyester is DID acid modified, is claimed to have good flexibility and thermal resistance [226].

The processing of impregnating varnishes requires curing cycles from 5 to 24 h, and sometimes two impregnating and curing steps. The solvents (accounting for about 50 wt% of the products) evaporate during the processing and must be burned. All these facts and the superior quality of a coil impregnated with an imide modified unsaturated polyester make the market share of the impregnating varnishes world-wide shrinking.

6
Imide Modified Coatings

The good properties of poly(ester-imide) based varnishes led to patents where this type of product is claimed for use as industrial coatings. Formulated with fluorinated resins, films with good adhesion and low friction coefficients were obtained [227–229]. Good adhesion [230] and good corrosion resistance combined with excellent long term temperature resistance were also claimed [231].

Formulated with pigments, like graphite and carbon black [232] or metal powders [233, 234], conductive films were obtained at low cost, used as electromagnetic shielding layers or for capacitive energy-storing inductive coils, when coated on an insulated magnet wire.

The use of a poly(ester-imide) as resin for a printing ink formulation was also claimed [235].

From hexahydrophthalic anhydride, aminoethanol, neopentylglycol, isophthalic acid, and trimellitic anhydride a carboxyl terminated poly(ester-imide) resin is made and formulated with triglycidyl isocyanurate, pigments, and additives to give a powder coating with excellent heat resistance and no chalking after 5 month at 120 °C [236].

7
Thermoplastic Poly(ester-imide)s

From the beginning, poly(ester-imide)s had been developed as thermo-setting polymers. In the 1970s a development was started to use imide modification for improving the properties of thermoplastic polyesters. Later on in the patent literature liquid crystalline polymers can be found as being claimed as molding resins.

Poly(ethylene terephthalate) was modified in 1971 with the DID acid ethylene glycol ester. The tensile strength of this poly(ester-imide) was after a storage for 20 days at 200 °C still 14.5 kg/mm, while a poly(ethylene terephthalate) film could not be measured due to the polymer degradation [237]. Using 1,6-hexamethylene diamine [238] or 1,3-diaminomethylene cyclohexane [239] to react with trimellitic anhydride, modified poly(ethylene terephthalate)s had been obtained and, being transparent and having superior gas barrier property, claimed for making bottles.

Another possibility for obtaining imide modified thermoplastic polyesters was to use as a monomer the hydroxy acid made from trimellitic anhydride and aminoethanol. Such a poly(ester-imide) was claimed for injection molding [240]. For the same use, poly(ester-imide)s containing aminophenol/trimellitic anhydride [241], imidised poly(butylene terephthalate) [242] and a wholly aromatic poly(ester-imide) made from trimellitic anhydride, p-aminobenzoic acid, p-acetoxybenzoic acid, diacetoxybiphenyl and terephthalic and isophthalic acids are known, which showing optical anisotropy [243].

Poly(ester-imide)s useful as molding plastics are made from terephthalic acid, trimellitic anhydride, 4-(aminomethyl)cyclohexanemethanol, and 1,4-cyclohexanedimethanol [244]. Blends of this poly(ester-imide) with polycarbonate are also patent protected [245]. An amorphous polymer containing units derived from N-3-hydroxyphenyl trimellitimide and units derived from p-hydroxybenzoic acid [246] and another having units derived from 9,9-bis(4-aminophenyl)fluorene [247] have similar claims.

Liquid crystalline poly(ester-imide)s having high glass transition temperatures and good dimensional stability are made from diphenols like t-butylhydroquinones [248], phenylhydroquinone [249] and bisphenol A [250].

Poly(ester-imide) elastomers have been prepared, and contain the reaction product of trimellitic anhydride with polyoxyalkylene diamines like polypropylene oxide diamine. Having excellent tensile strength they are used for making automobile parts [251], or when highly filled they are a suitable replacement for ceramics [252].

Useful in injection blow or extrusion molding of windscreens, structural parts, etc., are thermosettable modified phenolic resins containing ester-imide structures. They were made by reacting novolacs with trimellitic acid anhydride chloride. The trimellitic anhydride capped polyester was than reacted with 4,4-diaminodiphenyloxide to obtain the phenolic poly(ester-imide) [253].

Another area where poly(ester-imide)s have found an application are fibers, having excellent heat resistance. A linear poly(ester-imide) prepared from trimellitic anhydride, hexamethylene diamine, and the tetrahydrofurane ethylene oxide copolymer in the presence of tetra n-butyl titanate is giving a polymer with rubber elasticity. Fibers made by melt spinning have an elongation of 570% and good resistance to water [254]. The reaction product between pyromellitic anhydride and p-aminobenzoic acid was converted into the dimethylester and further reacted with a tetrahydrofurane ethylene oxide copolymer and ethylene oxide to give also an elastic polymer with a melting point of 190 °C. The fibers made by melt spinning have excellent heat resistance [255]. A fully aromatic poly(ester-imide) made from p-phenylene-bis-trimellitic dianhydride and 4,4'-diaminodiphenylether in dimethylformamide and spun into a coagulation bath of castor oil and acetone gives after thermal treatment at 370 °C for 10 min fibers with improved mechanical properties [256]. There are known thermotropic poly(ester-imide)s having high temperature stability and good workability, claimed for fibers made, e.g., from decamethylene-1,10-bis(trimellitimide) and 4,4'-diacetoxy-biphenyl [257]. Hexamethylene diamine is reacted with trimellitic anhydride in phenol and a poly(ester-imide) is made by adding and reacting a mixture of hydroquinone and bisphenol A and diphenyl carbonate, a material useful for fibre manufacturing, is obtained [258].

Poly(ester-imide)s are claimed as adhesives with high service temperatures. For general usage a poly(ester-imide) adhesive made from bisphenol A bistrimellitate [259] is claimed. Others are known for brake liner cementing [260], as adhesives for metal laminates [261], and as electroconductive adhesive films for die bonding [262].

8
Other Application Areas

Other uses of poly(ester-imide)s in high tech applications include wafer fabrications in microelectronics and printed circuit boards manufacturing in electronics. The adhesion of polyimides to the inorganic substrates of the wafers is improved by coating this substrates with a poly(ester-imide) made from ethylenebistrimellitate (16) and bis(3-aminopropyl)tetramethyldisiloxane in N-methylpyrrolidone, giving after curing at 150 °C for 1 h a 0.5 my thick film [263]. A poly(ester-imide) based varnish is used for coating a rectifier diode prior to be cast, giving improved temperature resistance and good adherence to the surface [264].

The use of poly(ester-imide)s as impregnating, laminating and adhesive agents giving excellent flexural strength and good temperature stability is known. So a glycerine branched poly(ester-imide), where the imide structures were generated by the reaction between melamine and trimellitic anhydride, is used for glass mat impregnation for prepreg manufacturing [265]. Another process comprises impregnating heat resistant fibres with a poly(ester-imide) resin made from 1,2,3,4-butanetetracarboxylic acid. The prepreg made has good adhesion to copper foil [266]. Poly(ester-imide) varnishes, claiming ethynylaniline as one of the starting compounds, are used for coating reinforcing materials like carbon or glass fibres or paper. The prepregs were clad on both sides with copper by heating under pressure to form a laminate used in electronic devices [267].

A poly(ester-imide) powder is coated on an electrolytic copper foil of 35 my thickness by an electrostatic spray coating machine and baked at 200 °C for 30 min. The substrata obtained is useful for printed circuit board manufacturing [268].

A film made from a poly(ester-imide) containing trimellitic anhydride, an aliphatic diamine and substituted hydroquinone is made by extrusion and is then laminated onto a metal layer. Flexible printed circuit boards were made from this [269].

For poly(ester-imide)s more uses were known, and patents claiming very different fields of applications have been filed.

Saturated poly(ester-imide)s made from polyethyleneglycol or polypropyleneglycol diamines were claimed in toner compositions. Some show good fusing and fixing characteristics [270], others high resolution and good deinkability [271]. Unsaturated products, containing fumaric acid and crosslinked with benzoyl peroxide were used in toner formulations giving no background deposits [272].

Besides other linear polycondensates, poly(ester-imide)s with benzophenone structures were claimed [273] as useful for producing photographic images by exposure through a photographic mask. The varnishes were dried at 80 °C, giving on a copper substrata a 1.2 my layer. Exposure is made with a 1000 W UV lamp. The exposed areas, crosslinked by benzophenone hydrogen abstraction, are insoluble and the unexposed areas of the film are removed with N-methylpyrrolidone.

Liquid crystalline polyesters with 3-methylphthalimide and chromophore side chains, having good orientation stability and mechanical properties, were claimed for making nonlinear optical devices like filters, polarisers, waveguides and others [274].

A poly(ester-imide) containing soft segments of poly(ethyleneadipate) and DID acid hard segments was used for making membranes claimed for separation of aromatics from saturates in a pervaporation process of a petroleum stream [275].

Coating the inner wall of a silicic acid capillary chromatography column with a poly(ester-imide) is known. After curing the stationary phase is applied. Claimed are improved mechanical strength, faster application of the stationary phase and better quality of the column [276].

9
Conclusions and Outlook

Thermosetting poly(ester-imide)s have been known for more than 35 years and are a class of mature polymers. Good thermal and mechanical properties, easy processing and acceptable prices are the reasons for the large tonnage of this polymers used as electrical insulating materials world-wide. A replacement by other polymers in this application field is unrealistic for now and the near future.

Poly(ester-imide)s having good heat resistance, high glass transition temperatures, good dimensional stability and excellent tensile strength have been developed and are used in liquid crystals, elastomers, fibres, etc. For continuous growth of this business it is important to have these products available at competitive prices around the world, whereupon continuous improvement of the processing will occur.

The economic importance of poly(ester-imide)s led and leads to the developments of new compositions having superior properties and better processing characteristics. Information on these developments can be found in patents and are presented at congresses. It is believed that similar big efforts were made by companies to reduce their costs and to improve the manufacturing processes of poly(ester-imide)s. The information on this progress is unfortunately not published.

10
References

1. Dr Beck & Co GmbH (1961) DE 1,445,263
2. Bornhaupt B von, Rating W (1971) Kunststoffe 61:46
3. Dünnwald W (1972) Kunststoffe 62:347
4. General Electric (1954) DE 1,033,291
5. Dr.Beck & Co GmbH (1956) DE 1,199,909
6. EJ DuPont de Nemours & Co FR 1,239,491; FR 1,256,203 (1962)
7. Schenectady Chem Inc (1961) US 3,342,780
8. Schenectady Chem Inc (1965) US 3,426,098
9. Dr Beck & Co GmbH (1958) DE 1,108,428
10. Dr Beck & Co AG (1965) DE 1,570,273

11. Dr Beck & Co AG (1965) FR 1,436,978
12. Dr Beck & Co AG (1965) FR 1,437,746
13. Mitsubishi Electric Corp (1969) JA 7,227,154
14. Tejin KK (1975) J5 2029–825
15. Toyo Spinning KK (1977) J5 2128–991
16. Toyobo KK (1977) J5 4008–697
17. Totoku Toryo KK (1980) J5 6100–830
18. Schenectady Chem Inc (1980) EP 55–085
19. Toshiba Chem KK (1986) J6 2288–672
20. Nitto Electric Ind KK (1986) J6 3159–481
21. Furukawa Electric Co (1989) J0 3007–775
22. Hitachi Chemical Co (1989) J0 3026–757
23. Showa Electr Wire KK (1989) J0 3115–477
24. Pirelli Cabos (1992) BR 9,203,611
25. Mitsubishi Petrochem Ind KK (1985) J6 2018–461
26. Dajichi Denko Ltd (1972) J4 9092–198
27. Mitsubishi Electric Corp (1974) J7 6036–866; J7 6036–867; J7 8007–240
28. Hitachi Chem Co Ltd (1995) JP 08,188,713
29. Dr Beck & Co GmbH (1964) BR 1,070,364
30. Dr K Herberts & Co (1964) FR 1,456,995
31. Hitachi Wire and Cable Ltd (1968) JA 7,240,716; JA 7,240,718
32. Hitachi Wire and Cable Ltd (1970) JA 7,320,439
33. Institut Francais Du Petrole (1966) FR 1,540,930
34. Idemitsu Kosan KK (1983) J6 0133–024
35. Nippon Teleg & Teleph (1985) EP 195–402
36. Inst Chem Przemyslo (1990) WO 9203–831
37. Asahi Electrochemical Co (1972) J4 8090–395
38. Mitsubishi Chem Ind KK (1983) J6 0004–531
39. Idemitsu Kosan KK (1984) J6 0190–422
40. Dr Beck & Co (1962) DE 1,495,113
41. Cable Ind Res Pln (1981) SU 943 859
42. P D George Co (1983) US 4446–300
43. Allied Chem Corp (1968) NE 6,909,459
44. VEB Leuna-Werk Ulbricht (1987) DD 258–986
45. Allied Corp (1986) US 4757–118
46. Relay Cons Des Techn Inst (1982) SU 1,086,764
47. Toshiba Chem Corp (1992) JP 06,150,721
48. Standard Oil Comp of Indiana (1967) US 3,377,321
49. P D George Co (1976) US 4069–209
50. Schramm Lack und Farbenfabrik AG (1967) NE 6,813,374
51. Dr.Beck & Co. AG (1968) DE 1,928,934
52. Alsthom-Atlantique (1977) DE 2,822,610
53. Hitachi Chem Co Ltd (1994) JP 07,292,244
54. Furukawa Electric Co Ltd (1969) JA 7,317,837
55. Nitto Electric Ind KK (1974) DE 2519–672
56. Furukawa Electric Co Ltd (1969) JA 7,317,838
57. Teijin Ltd (1971) JA 4,734,490
58. Dainichiseika Colour and Chem. Mfg Co Ltd (1968) JA 7,317,840
59. Nitto Denko Corp (1988) J01,207–362
60. Teijin Ltd (1973) J4 9132–116
61. Showa Elec Wire KK (1976) J5 2085–234
62. Toyobo KK (1984) J6 0166–454
63. Eastman Kodak Co (1992) US 5,266,429
64. Nippon Petrochemical KK (1973) J5 0025–693

65. General Electric Co (1985) US 4609–702
66. Furukawa Electric Co Ltd (1969) JA 7,317,839
67. Standard Oil (Ind) (1975) US 3976–665
68. Ciba Geigy AG (1991) EP 504,109
69. Tokyo Tokushu Densen (1971) J4 80011–396; JA 4,843,796; JA 4,840,895; JA 4,851,992; JA 4,851,993; (1980) J5 6122–834
70. Nitto Electric Ind KK (1972) J4 8072–296; (1974) DE 2519–671; (1979) J5 6062–823
71. Nitto Electric Ind Co Ltd (1971) JA 4,851,996
72. Showa Elect. Wire KK (1981) J5 80,007–464
73. Mitsubishi Electric Corp (1968) JA 7,240,720
74. Barvy Laky NP (1979) DL 142,056
75. Teijin Ltd (1971) J4 8025–794
76. General Electric (1981) US 4356–297
77. P D George (1976) US 4113–706
78. Idemitsu Kosan KK (1984) J6 1113–621
79. Inst Francais du Petrole (1974) FR 2264–013
80. Asahi Glass KJK (1976) J5 3024–393; J5 3024–395
81. Dr Kurt Herberts and Co (1969) DT 1,937,310
82. Inst Francais du Petrole (1975) DT 2610–307
83. Eastman Kodak Co (1964) Fr 1,450,704
84. Tokushu Densu Toryo KK (1969) JA 7,222,337
85. Furukawa Electr Co Ltd (1970) J7 4020–050; J7 4014–560
86. BASF AG (1974) DT 2412–471
87. General Electric Co (1986) US 4740–564
88. Dr Kurt Herberts & Co (1962) BR 1,067,541
89. Dynamit Nobel AG (1973) BE 811–685
90. Sumitomo Elec Ind KK (1973) US 3922–465
91. Cella-Lackfabrik Dr C Schleussner GmbH (1965) DT 1,790,228
92. Nitto Electric Ind Co Ltd (1968) JA 7,240,719
93. Tototuku Toryo KK (1979) J5 6081–337
94. Cella-Lackfabrik Dr C Schleussner GmbH (1970) DT 2,052,330
95. Furukawa Electric Co Ltd (1970) JA 7,341,879
96. Tokyo Tokushu Densen (1971) J4 8034–996
97. Chem Werke Albert (1964) DE 1,494,419
98. Phelps Dodge Magnet Wire Corp (1969) US 3,699,082
99. Mitsubishi Electric Corp (1970) J7 4011–277; J7 4014–528
100. Kirstein L, Richter D, Hartmann W (1970) DL 152,244
101. Nitto Elect Ind Co (1972) J4 8087–399
102. Dr Beck & Co GmbH (1963) Fr 1,451,228
103. General Electric Co (1978) US 4233–435
104. Schenectady Chem Inc (1977) ZA 7702–522
105. Hitachi Kasei Industries Ltd. (1966) JA 7,018,316
106. Nippon Teleg & Teleph (1987) J0 1123–819
107. Wells LM, Strunk MH (1993) Proceedings Electrical Electronics Insulation Conference, p 172
108. Anaconda Wire and Cable Co (1968) US 3,493,413
109. Fujikura Cable Works KK (1982) J5 8,204,064
110. General Electric Co (1968) US 1,903,739
111. Showa Elect Wire KK (1979) J5 6055–465
112. Toshiba Chem Corp (1993) JP 07,141,917
113. Hitachi Chemical KK (1971) J8 0004–155; (1984) J6 1136–550; (1987) J0 1098–661
114. Showa Electric Wire (1981) J5 8059–269
115. Hitachi Chemical KK (1989) J0 3033–121
116. Tokyo Shibaura Elect Ltd (1981) J 5 7195–167

117. Hitachi Chemical KK (1988) J0 2004–880; J0 2058–567
118. Nippon Shokubai Co Ltd (1994) JP 07,316,425
119. Merten C (1971) Angew Chemie 10:339
120. Furukawa Electric Co (1977) J5 3101–692
121. Toshiba Chem KK (1977) J5 3130–750
122. Hitachi Cable KK (1983) J5 9174–654
123. Hitachi Cable KK (1983) J5 9174–662
124. Dainichiseika Color (1980) GB 2068–395
125. Totoku Toryo KK (1986) J6 3168–907
126. Fujikura Kable Works (1980) J5 6106–976
127. Dr Kurt Herberts and Co (1976) DT 2712–495
128. General Electric Co (1979) US 4267–231
129. General Electric Co (1980) US 4480–007
130. Cables Ind Res Dev Inst (1990) RU 2,021,297
131. Totoku Toryo KK (1980) J5 7098–573
132. Showa Elec Wire KK (1980) J5 7061–068
133. Sumitomo Elect Ind KK (1980) J5 7070–165
134. Showa Elec Wire KK (1983) J5 9170–163
135. Chem Fab Wiedeking (1975) FR 2311–831
136. Nippon Shokubai Kagaku (1983) J5 9168–031
137. Chem Fab Wiedeking (1977) DT 2747–456
138. Hitachi Chemical KK (1977) DT 2849–120
139. Dr Beck & Co AG (1989) DE 3938–058
140. Chem Fab Wiedeking (1975) DT 2522–386
141. General Electric Co (1980) GB 2075–998
142. Wiedeking GmbH (1991) DE 4,133,161
143. Lienert K-W (1989) 8th Fachtagung Elektroisolier-Systeme B, Beck Elektroislier-Systeme, Hamburg
144. General Electric Co (1974) DT 2509–048
145. Cean SPA (1975) NL 7600–341
146. Schenectady Chemical Inc (1976) DT 2724–913
147. Standard Oil Co (Ind) (1976) US 4116–941
148. Furukawa Electric Co (1977) J5 4071–187
149. Mitsubishi Elec Mac (1974) J5 0141–697
150. General Electric Co (1975) US 4247–429
151. Dr Beck & Co AG (1975) DT 2605–790
152. Schenectady Chemical Inc (1976) US 4179–423
153. Essex Group Inc (1979) US 4290–929
154. Dr Beck & Co AG (1967) GB 1184 139; (1981) EP 66–194; EP 75–239
155. Nitto Electric Ind Co (1973) DT 2439–385
156. Stollack AG (1975) DT 2630–758
157. Nitto Electric Ind Co (1974) DT 2542–866
158. Inst Ingin Tehn Ind (1984) RO 88–680
159. Dr Beck & Co AG (1970) DT 2,135,157
160. Dr Beck & Co AG (1974) DT 2401–027
161. P D George Co (1974) US 4081–427
162. Maiofis Im (1975) SU 558–306; (1979) SU 974–417
163. Sumitomo Elect Ind KK (1975) J5 1107–330; J5 1107–331
164. Essex Int Inc (1975) US 4075–179
165. Showa Electric Wire KK (1976) J5 3059–791; (1977) J5 3112–930
166. Furukawa Electric Co (1977) J5 4071–121
167. Totoku Toryo KK (1980) J5 6110–729
168. General Electric Co (1980) US 4296–229; US 4389–457

169. BASF AG (1973) DT 2351-078; DT 2351-076; (1974) DT 2412-470; DT 2460-472; (1977) DT 2706-768
170. Mitsubishi Electric Corp (1970) J7 9029-535; J7 5033-695; J7 4014–529; (1974) DT 2456-587; (1975) J5 1142-038; J5 1112-850; (1979) J5 5135-180
171. Nitto Electric Ind KK (1985) J6 2121–774
172. Mitsubishi Electric Corp (1977) J5 4058–731
173. Herberts GmbH (1980) EP 54–925
174. Mitsubishi Electric Corp (1975) J5 1130–440; J5 5040–754
175. Mitsubishi Electric Corp (1983) J6 0008–375
176. Sumitomo Electr Ind KK (1972) J8 5027–127
177. Mitsubishi Electric Corp (1976) J5 3075–235
178. Herberts GmbH (1979) EP 17–798
179. Nitto Electric Ind KK (1980) J5 7042–708; J5 7042–709; J5 7042–711
180. Toshiba Chem KK (1983) J6 0108–423
181. Toyo Boseki KK (1980) DE 3007–445
182. General Electric Co (1971) US 3853–817
183. Westinghouse Electric (1976) J5 3,040,096
184. Minnesot Mining and Manufacturing Company (1985) EP 0,183,779
185. Sumitomo Elec Ind KK (1973) US 3,917,892
186. Riken Densen KK (1974) J8 2042–923
187. Dainichiseika Color Chem (1988) J0 1225–671
188. Dainichiseika Color Chem (1987) EP 289–955
189. Dainichiseika Color & Chem MFG (1992) JP 06,111,627
190. Dainichiseika Color Chem (1987) J0 1038–478; J01093–005; (1990) J0 3236–107
191. Dainichiseika Color Chem (1988) J0 1225–677
192. Hitachi Chem Co Ltd (1994) JP 07,278,302
193. Hitachi Chem Co Ltd (1994) JP 08,134,215
194. Rea Magnet Wire (1975) US 3988–283; US 4115–342
195. Furukawa Electric Co (1975) J5 1110–677
196. Totoku Toryo KK (1982) J5 8145–720
197. Schweizerische Isola-Werke (1969) BE 705,385
198. Schenectady Chem Inc (1986) AU 8777–706
199. Standard Oil Co (Ind) (1976) US 4012–555; US 4012–556
200. Showa Electric Wire (1981) J5 8076–467; J5 8076–468
201. Dr Beck & Co AG (1967) FR 1,572,570
202. Dr Beck & Co AG (1974) DE 2,460,768
203. Hitachi Chemical KK (1987) J0 1060–617
204. Nitto Denko Corp (1987) J0 1129–013
205. Nitto Denko Corp (1987) J0 1129–014
206. Szmercsanyi IV, Maros K, Makay E (1961) Journal of Polymer Science 53:211; Gulbins E, Funke W, Hamann K (1965) Kunststoffe 55:6
207. Hitachi Chemical KK (1988) J0 1158–020
208. Sidorenko KS (1976) SU 759-561; SU 777-049; (1979) DD 144-554
209. Mitsubishi Electric Corp (1977) J5 4070–396
210. Teijin KK (1976) J5 3040–760
211 Sidorenko KS (1976) SU 641–733
212. Dr Beck & Co (1967) US 4146–703
213. Nitto Electric Ind KK (1981) J5 8013–657
214. Asahi Chem Ind Co (1972) J4 8083–194
215. General Electric Co (1972) US 3,769,366
216. Dr Beck & Co (1968) FR 2,010,170
217. Teijin KK (1972) J4 9061–291
218. Teijin KK (1973) J4 9087–656
219. Teijin KK (1970) JA 7,250,790

220. Mitsubishi Electric Corp (1975) J5 2030-828
221. Mitsubishi Electric Corp (1975) J5 1125-477
222. Mitsubishi Electric Corp (1975) J5 1132-292
223. Ryuden Kasei KK (1972) J4 9096-024
224. P D George Co (1974) US 3945-959
225. P D George Co (1974) US 4375-528
226. Alsthom (1966) FR 1,477,698
227. Panshin Yu A (1975) SU 559-936
228. Glyco Metallw Daele GmbH (1979) WP 8101-375
229. Nippon Oils & Fat (1981) J5 153-057
230. AS Belo Metallopoly (1980) SU 978-945
231. BASF Lacke & Farben AG (1987) EP 317-795; WO 8809-359
232. BASF Lacke & Farben AG, Lackdraht Union (1989) EP 384-505
233. Matsushita Elect Ind KK (1983) J6 0115-103
234. Fujikura Ltd (1992) JP 06,020,518
235. Hercules Inc (1967) CA 846,908
236. Ciba Geigy AG (1986) EP 256-988
237. Mitsubishi Electric Corp (1971) JA 4,800,793
238. Teijin KK (1985) J6 2007-734
239. Goodyear Tire & Rubber (1990) EP 452-604
240. Standard Oil Co (Ind) (1974) US 3880-812
241. Bayer AG (1984) EP 171-707
242. General Electric (1986) US 4740-563
243. Idemitsu Petrochem KK (1987) EP 314-173
244. Eastman Kodak (1986) US 4769-441; US 4673-726
245. Eastman Kodak (1986) US 4645-802
246. Hoechst AG (1992) EP 582,220
247. Dow Chem Co (1993) US 5,386,002
248. DuPont de Nemours & Co (1994) WO 9,531,496
249. Neste Oy (1994) WO 9,601,284
250. DuPont de Nemours & Co (1994) WO 9,627,629
251. General Electric Co (1990) US 5,306,785; WO 9,207,026
252. General Electric Co (1990) WO 9,207,026
253. Allied Corp (1986) US 4771-113
254. Asahi Chemical Ind Co Ltd (1965) JA 20,477/69
255. Asahi Chemical Ind Co Ltd (1965) JA 7002 193
256. Inst Acad Scie USSR (1970) SU 332,147
257. Bayer AG (1985) DE 3516-427
258. Teijin KK (1986) J6 2270-626
259. Hitachi Chemical KK (1989) J0 3195-732
260. Th Goldschmidt AG (1979) DE 2925-261
261. Nippon Steel Corp (1982) J5 9073-943; (1987) J6 3170-030
262. Hitachi Chem Co Ltd (1992) JP 06,145,639
263. Hitachi Chemical KK (1986) J6 2227-473; J6 2227-953
264. Robert Bosch GmbH (1970) 768,009
265. BASF AG (1974) DT 2412-472
266. Nitto Electric Ind KK (1979) J5 6036-192
267. Kaneka Corp (1990) JP 04,198,208; JP 04,198,219
268. Nitto Electric Ind KK (1976) J5 2140-862
269. Teijin KK (1985) J6 2124-937
270. Xerox Corp (1993) US 5,348,831
271. Xerox Corp (1994) US 5,427,882
272. Xerox Corp (1994) US 5,427,881
273. Ciba Geigy (1983) EP 138-768

274. Flamel Technologies (1993) WO 9,511,476
275. Exxon Res & Eng Co (1992) US 5,241,039; US 5,288,712
266. Vysoka Skola Chem-T (1983) DE 3425–924

Received: March 1998

Liquid-Crystalline Polyimides

Hans R. Kricheldorf

Institut für Technische und Makromolekulare Chemie der Universität, Bundesstr. 45,
D-20146 Hamburg, Germany

The present review reports on structure-property relationships of LC-main chain polymers containing imide groups. The comments focus on the role of aromatic imide moieties as mesogenic building blocks. All the pertinent literature has been cited. This review is subdivided into six sections dealing: A) with fundamental stereochemical and conformational aspects of aromatic imide groups, B) with structure-property relationships of fully aromatic poly(ester-imide), C) with the properties of LC-poly(ester-imide)s containing regular sequences of aromatic and aliphatic building blocks, D) with synthesis and characterization of cholesteric poly(ester-imide)s, E) with structures and properties of LC-polyimides free of ester groups, and F) with the definition and characterization of layer structures in the solid state regardeless whether the pertinent polyimides form a true LC-phase or not.

Keywords: Liquid-crystalline polyimides, Poly(ester-imide)s, Thermotropic, Nematic, Smectic, Cholesteric, Mesogen, Layer structure, Aliphatic spacer, Trimellitimide

1
Introduction

Syntheses and properties of polyimides including poly(amide-imide)s, poly(ester-imide)s and poly(ether-imide)s have been explored over a period of roughly four decades. Polyimides (in their broadest sense) have been developed as duromeric or thermoplastic materials, as thermosetting resins, as insulating laquers for electirc wires, as photoresists, as fibres and for a variety of other applications [1–10]. Characteristic properties of polyimides which render them useful for various applications are their high thermostability, which in the absence of aliphatic comonomers is combined with a low sensibility to oxidation and a low sensitivity to the attack of light or organic solvents. Furthermore, high glass-transition temperatures and excellent mechanical properties can be achieved in fully aromatic polyimides.

Despite intensive research activities of numerous research groups and chemical companies on polyimides no systematic study on liquid-crystalline (LC) polyimides had been published before 1985. Only two patents of E.I.Du Pont filed in 1979 [11, 12] reported on LC-copolyesters containing N-(4-carboxyphenyl) trimellitimide as a comonomer. However, the LC-character of these copolyesters did not depend on the presence of the imide monomers, and thus these patents did not carry any information on the mesogenic character of the aforementioned imide moiety. In 1985 Evans et al. [13] published a systematic study of various polyimides derived from pyrromellitic dianhydride (PMDA). Although, the PMDI-unit was expected to be mesogenic because of its flat, rigid and linear structure, none of the polymers described in that publication was thermotropic. Therefore, those authors concluded: "We are unaware of any low molecular weight liquid crystalline compounds which contain the imide functional group. This combined with the results of the studies reported here suggests that it may be not possible to prepare thermotropic polyimides".

Systematic studies by the author started in 1985 [14] and later contributions of other research groups have meanwhile confirmed that aromatic imide moieties are relatively poor mesogens despite their planarity, rigidity and polarity. Surprisingly, the symmetrical imide units which were expected to be rather good mesogens proved to be almost non-mesogenic. Only the lengthy terphenyl-3,3',4,4'- tetracarboxylic imide was found to be a good mesogen (Sects. 4 and 6). Another tendency of general validity is the observation that ester groups are highly favorable for an LC-character of polyimides, because they favour both flexibility and linear conformations. Therefore, it is a logical consequence that most LC-polyimides discussed below are poly(ester imide)s and the first patents [11, 12] and papers dealing with LC-polyimides also concerned poly(ester imide)s.

2
Stereochemistry of Aromatic Imides

All LC-polyimides discussed in this review possess at least one phthalimide unit in the mesogenic unit and most of them contain an N-phenyl phthalimide moiety. For a better understanding of the potential mesogenic character of aromatic imide groups it is useful to discuss briefly their stereochemical and conformational properties.

When compared to a biphenyl, which is the smallest mesogenic group, the N-phenylphthalimide unit is slightly longer (8.5 vs 10.2 Å calculated from C to C plus one O-bond) but also slightly broader. Consequently the length/diameter ratio which is responsible for a mesogen according to the theory of Onsager and Flory [15–18] is nearly the same. A characteristic difference between biphenyl and N-phenylphthalimide is the low degree of symmetry of the latter moiety, which possesses only one mirror plane. At first glance, this low level of symmetry may look pretty unfavourable for a mesogenic character. However, most symmetrical imide moieties were found to be even less favourable. This unexpected failure of symmetrical imide units to yield polymers or low molar mass materials showing LC-phases may be attributed to their relatively high melting temperatures (T_ms) which are a logical consequence of a high symmetry combined with strong local dipoles. The lower T_ms of derivatives of non-symmetrical imide units may at least partially explain their higher mesogenic potential.

In addition to the symmetry both linearity and planarity of the N-phenylphthalimide group are important aspects. Tertiary amines are well known to possess a pyramidal structure due to the partial sp 3-hybridization of the nitrogen. Such a pyramidal structure has also been calculated for N-phenylphthalimide by the Tripos-Forcefield programm (Fig. 1, top) [19]. However, the more sophisticated PCFF-91 forcefield programm of Biosym yielded a linear structure and a torsion angle of 17° for the N-phenyl bond (Fig. 1, bottom) [20]. Only the assumption of a linear structure agrees with the experimental observations that 1,4 disubstituted N-phenylphthalimides may be liquid crystalline.

Of particular interest is the low torsion angle in combination with a low activation energy required for the formation of an exactly coplanar conformation (Fig. 2) [20]. Onsager [15] and Flory [16–18] have treated mesogenic moieties as rigid cylinders or bars which interact with their neighbours exclusively by their steric demands via repulsive forces. However, at least in the case of LC-polyimides there is an increasing number of experimental findings which suggest that attractive (enthalpic) electronic interactions between the mesogenic moieties play an important role for the stabilization of the nematic or smectic order. The optimization (energy minimization) of such interactions between the π-electrons of two aromatic rings or between π-electrons and dipoles (donor-accepter interactions) requires a stacking of nearly coplanar mesogens (quite analogous to the arrangement of nucleobases in the DNS double helix) (Fig. 3). In this connection it is an interesting result that O- substituted N-phenyl phthalimide derivatives of structure I are far less mesogenic [21–23],although their length/di-

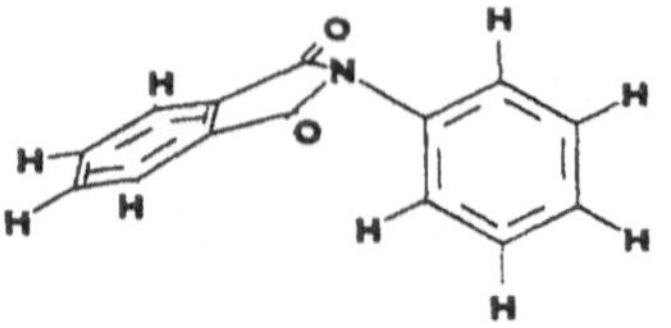

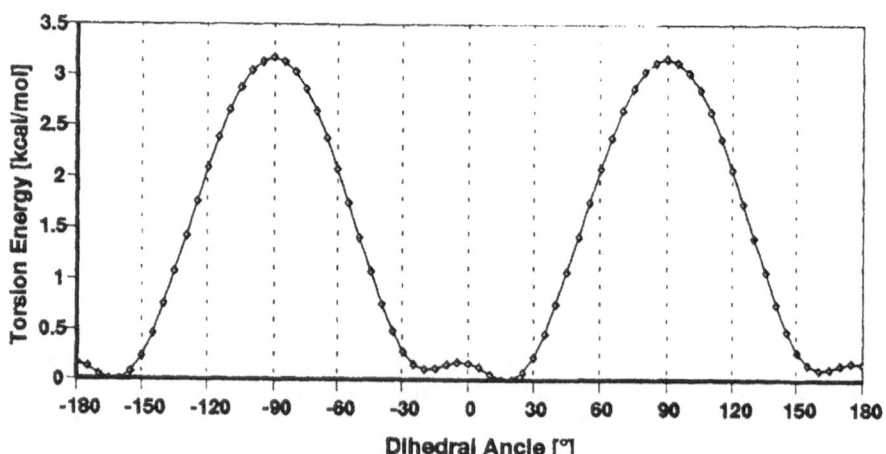

Minimized structure from Tripos-Forcefield
(torsion angle 90°, bond angle 109°)

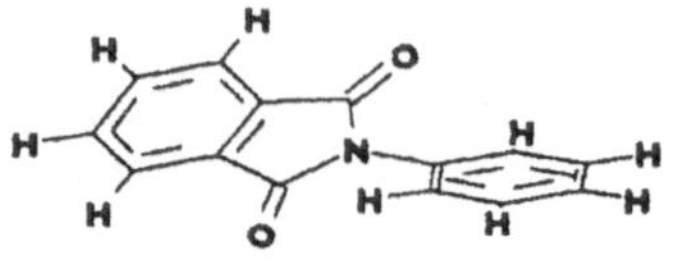

Minimized structure from CVFF-Forcefield
(torsion angle 23°, bond angle 180°)

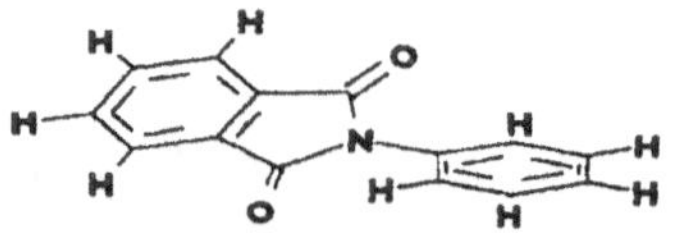

Minimized structure from PCFF91-Forcefield
(torsion angle 17°, bond angle 180°)

Fig. 1. Energy minimized conformations of N-phenyl phthalimide

Fig. 2. Energy profile of the rotation of the N-phenyl phthalimide group in the configuration of Fig. 1C as calculated by the PCFF 91 forcefield program of Biosym

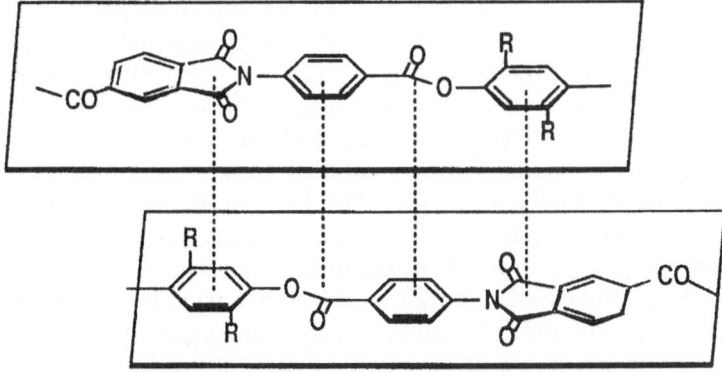

Fig. 3. Schematic illustration of the electronic interactions (dipole-dipole, donor-acceptor interactions and van der Vaals forces) between neighbouring chain segments having coplanar conformations

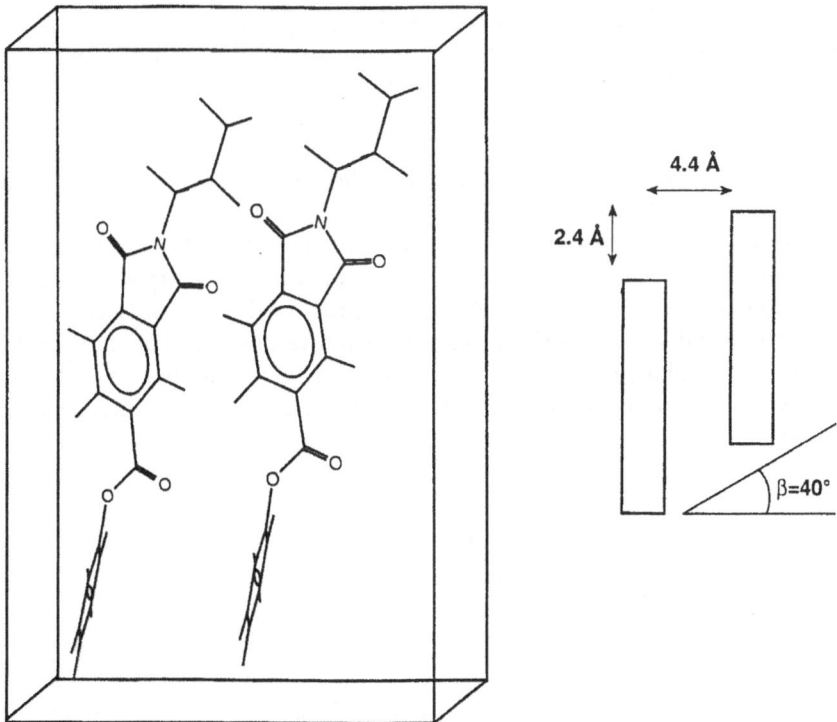

Fig. 4. Energy minimum calculation of a "docking experiment" for two *N*-propyl trimellitimide phenyl esters (from PCFF 91 forcefield of Biosym.)

ameter ratio is nearly identical with that of the unsubstituted unit. However, the methyl group ortho to the nitrogen raises the energy required for a coplanar position of phenyl- and phthalimide rings by a factor of 200–500 [22]. Further support for the electronic interaction between coplanar N-phenyl phthalimide groups comes from so-called "docking experiments". When two N-propyl trimellit-imide phenyl esters are brought together from a great distance so that a minimization of the total energy takes place, both imide units form a stack in coplanar position (Fig. 4) [24]. These preliminary results certainly improve a proper understanding of the mesogeneity of imide moieties but further experimental and theoretical studies are required for a full understanding. Nonetheless, the results gathered so far indicate that the theory of mesogeneity presented by Flory (although consistent in itself) is a poor adviser for the design of realistic LC-polyimides.

3
Aromatic Poly(ester-imide)s (PEIs)

Fully or mainly aromatic PEIs are of interest as melt processable engineering plastics. Compared to a neat polyimide such as Kapton, the incorporation of ester groups into the polyimide backbone has two main consequences. The negative consequence is a reduced thermostability, the positive consequence an improved flexibility which may result in a better processability either from the melt or from solution. However, an LC-character of the PEIs requires that most building blocks are linked in para position, and the combination of para-linked monomer units with the high polarity of the imide ring favours crystallization and high melting points of the crystallites. The following three strategies have been used to overcome this difficulty: (I) the use of substituted monomers, (II) the use of less symmetrical building blocks and (III) the use of flexible monomer, such as diphenyl ethers or aliphatic spacers. The incorporation of aliphatic spacers is in principle the most successful method to improve both solubility and meltability, but the penalty is a significant loss of oxidative and thermal stability. PEIs containing a regular sequence of shorter or longer aliphatic spacers will be discussed separately in the following section, whereas the first two strategies will be discussed below.

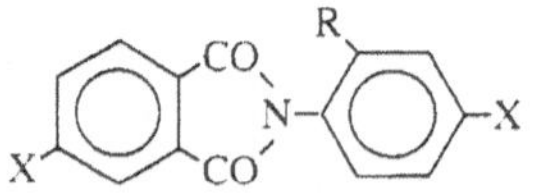

I

$X = O R'$, CO_2R'' $R = Cl$, CH_3
with $R' = $ Alkyl or Aryl

Structure 1

3.1
Symmetrical Imide Building Blocks

Symmetrical imide dicarboxylic acids such as **1a, 1b** or **2a–2f** can easily be prepared from the corresponding dianhydrides and meta or para aminobenzoic acid and a few (**1a, 1b, 2c**) of these dicarboxylic acids and their dichlorides have been described [25–29]. Nonetheless, LC-PEIs derived from these dicarboxylic acids have not been reported yet, presumably because they are not attractive starting materials. The solubility in any kind of inert solvent is extremely low and the melting temperatures (T_ms) are so high (400 °C) that the melting will automatically entail decarboxylation. Therefore it is difficult to bring these dicarboxylic acids into reaction. Furthermore, most of the hypothetical PEIs derived from the para-functionalized dicarboxylic acids should again possess extremely high melting temperatures or they should be isotropic. The best chance to obtain LC-phases below 400 °C might result from a polycondensation with substituted hydroquinones bearing bulky substituents or long side chains. The meltability and processability of polyesters derived from the meta dicarboxylic acids (e.g. **1b**) will be better, but the meta functionalization is unfavourable for the formation of LC-phases (see below).

The dicarboxylic acids of structure **3a–f** were prepared from trimellitic anhydride, the corresponding diamines and acetic anhydride. Numerous LC-polyesters containing such dicarboxylic acids were claimed in three patents of BASF AG [30–32] as fibre or film-forming materials. In principle these dicarboxylic acids do not possess mesogenic properties and are unfavourable as components of LC-polyester with exception of **3b** (see discussion of **4a–i** below). The dicarboxylic acids **3a–f** were used in smaller amounts (typically 5–25 mol%) as comonomers in LC-polyesters mainly composed of conventional monomers such as terephthalic acid., 4-hydroxybenzoic acid (4-HBA), hydroquinone or 4,4'-dihydroxybiphenyl. Short sequences of these "conventional monomers" are in fact the mesogenic building blocks, and the dicarboxylic acids **3a–f** have the purpose to reduce or to eliminate the crystallinity and to lower the melting temperatures. An additional effect may be an increase of the glass-transition temperature (T_g). The composition of most co PEIs was optimized to yield a nematic melt above 320 °C. Analogous LC-copolyesters were recently described by a Chinese research group [33] which was apparently not informed about the BASF patents. Their nematic copolyesters consisted of at least four comonomers, namely the dicarboxylic acid **3b**, terephthalic acid, hydroquinone and/or resorcinol and 4-hydroxybenzoic acid. A detailed study of the mesogeneity of the dicarboxylic acid **3b, 3d**, and **3f** has recently been published [34]. When homopolyesters derived from **3d** or **3f** and hydroquinone or substituted hydroquinones were characterized, it turned out that all these polyesters were isotropic. In other words the dicarboxylic acids **3d** and **3f** are, as expected, nonmesogenic.

A more complex picture was obtained from the homopolyesters **4a–i** derived from the dicarboxylic acid **3b**. Whereas **4a, 4b, 4c, 4f** and **4g** decomposed before melting, broad nematic phases were observed for **4c** or **4d** and a smectic LC-phase for **4h**. However, the polyester **4i** derived from 1,12-dodecanediol was iso-tropic, indicating that the dicarboxylic acid **3b** is a relatively poor mesogen. In this connection it should be mentioned that a Russian research group [35] had studied quite early (1977) the syntheses and some properties of the PEIs **5a–d**. All PEIs were crystalline and the crystal lattices were characterized, but phase transitions or LC-phases were not reported (PEI **4a** is identical with **5a** and the properties mentioned in [34, 35] are in good agreement).

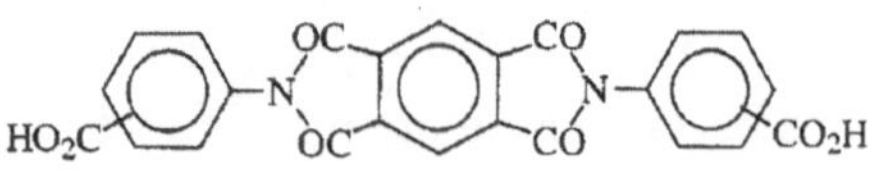

1a : meta,   **1b** : para

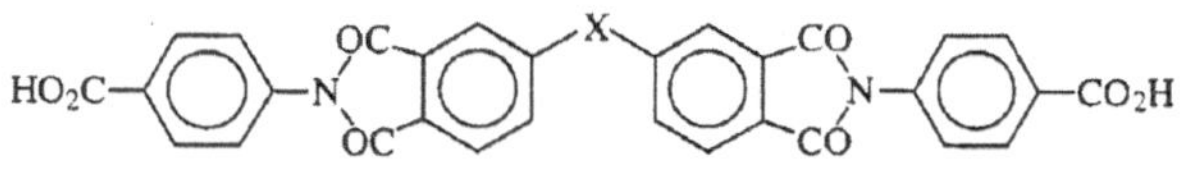

2a - f

a : X = σ-bond   c : X = CO

b : X = O   d : X = C(CF₃)₂

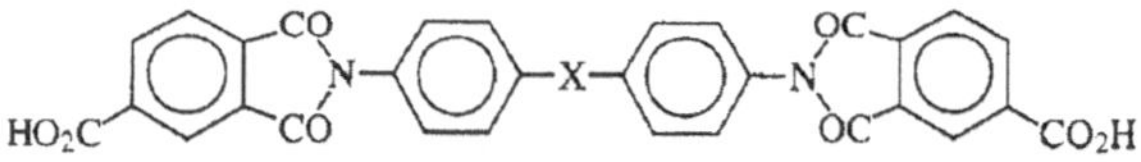

3a - f

a : X = σ-bond   c : X = CO   e : X = S

b : X = O   d : X = CH₂   f : X = SO₂

$$\longleftarrow \begin{cases} \textbf{4a - i} \\ \textbf{5a - d} \end{cases}$$

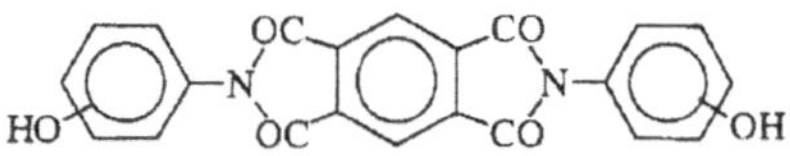

6a : meta,   **6b** : para

7a : meta   **7b** : para

Structure 2 (1)

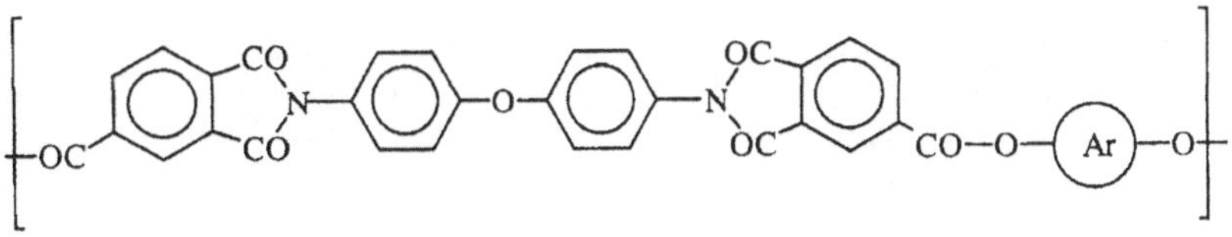

4a · i

a: $\left(Ar\right)$ =

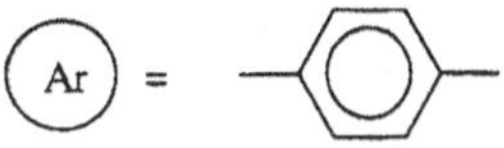

b: $\left(Ar\right)$ =

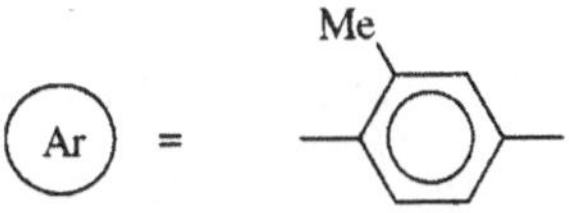

c: $\left(Ar\right)$ =

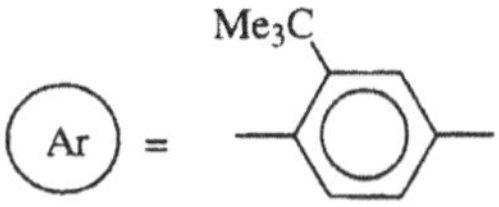

d: $\left(Ar\right)$ =

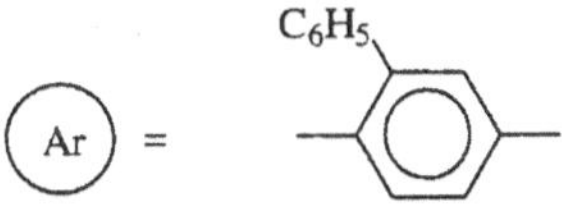

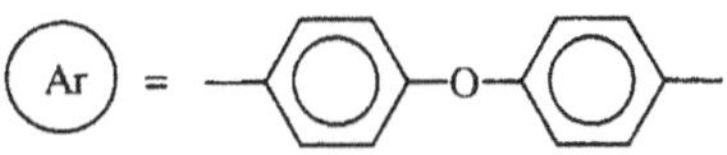

e $\left(Ar\right)$ =

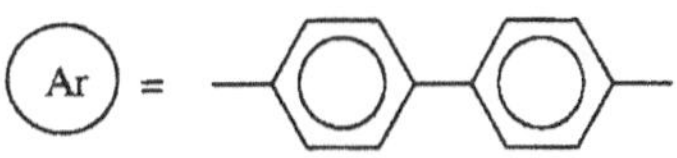

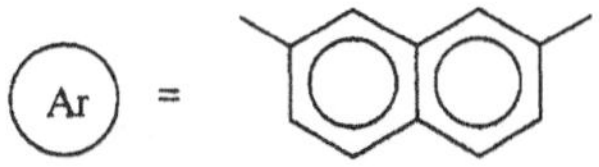

f $\left(Ar\right)$ =

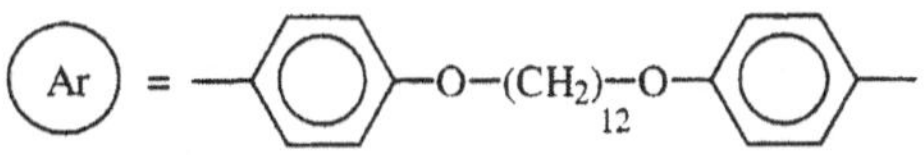

g: $\left(Ar\right)$ =

h: $\left(Ar\right)$ = $-O-(CH_2)_{12}-O-$

i: $\left(Ar\right)$ = $-(CH_2)_{12}-$

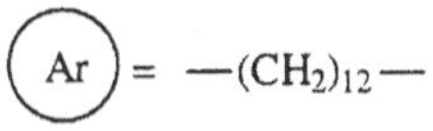

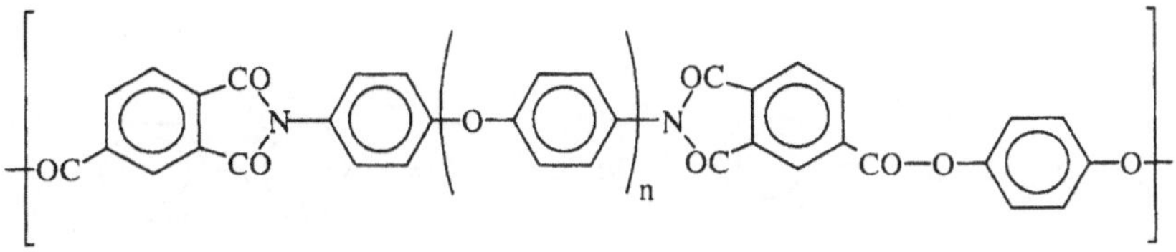

5a-d

a: $n = 1$ b: $n = 2$ c: $n = 3$ d: $n = 4$

Structure 2 (2)

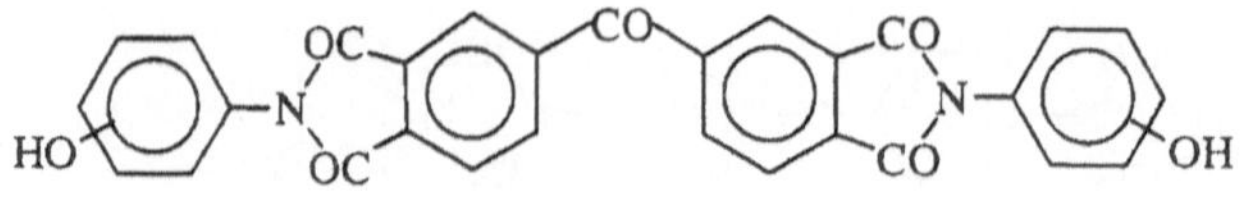

8a: meta   **8b:** para

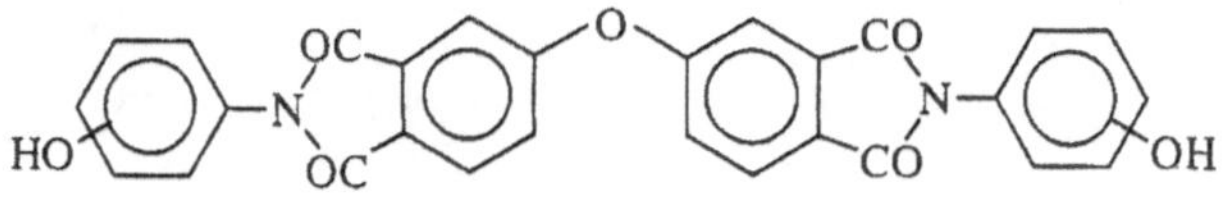

9a: meta   **9b:** para

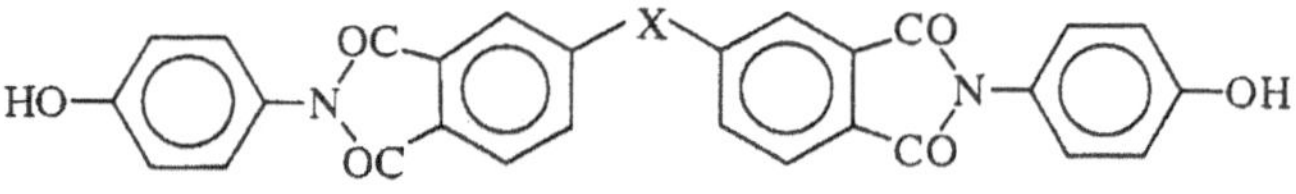

10a: X = SO$_2$ **10b:** X = C(CF$_3$)$_2$

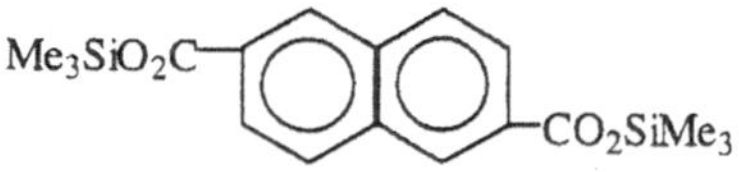

11

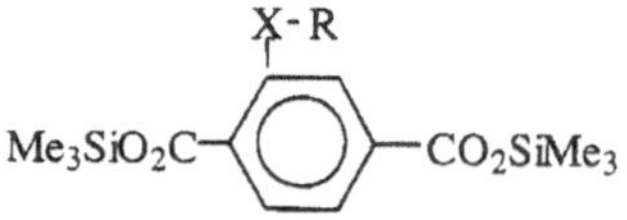

12a - d

a: X = O, R = 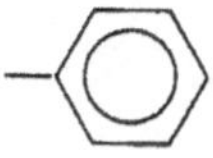   **b:** X = O, R =

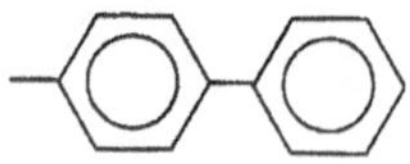

c: X = S, R = 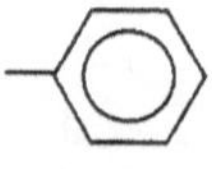   **d:** X = S, R =

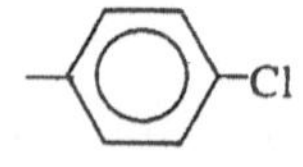

Structure 2 (3)

More useful as monomers than the dicarboxylic acids are the corresponding imide diphenols (**6a,b–10a,b**) and their bisacetates, because their T_ms are lower and their solubility in inert organic solvents is higher. Yet, even in the case of the bisacetates which are free of any H-bonds the T_ms may be as high as 500 °C (e.g. the bisacetate of **6b**, see Table 1), so that it cannot be used in melt polycondensations [36]. Even in the case of **7b** and **8b** the T_ms of the bisacetate are so high that they start to decompose in the melt mainly due to a Fries rearrangement [45]. However, the bisacetate of **5b** disolved readily in the melt of silylated dicarboxylic acids (**11** and **12a–d**) below 300 °C, so that successful polycondensations were feasible [36]. For all polycondensations of acetylated imide diphenols, silylated dicarboxylic acids were used as reaction partners to reduce the initial reaction temperature below 300 °C and to avoid proton catalysed side reactions.

The properties of the resulting PEIs can be summarized as follows. Regardless of the imide structure and regardless of the dicarboxylic acid all PEIs derived from "meta-diphenols" (**6a, 7a, 8a, 9a**) were isotropic [20, 36]. Therefore, the lower T_ms of the starting materials and of the PEIs themselves are useless, when an LC-character of the polyesters is the aim of the syntheses. In the case of para-diphenols unsubstituted terephthalic acid was avoided as reaction partners, because it was suspected that the resulting PEIs will be highly crystalline with a T_m above 400 °C and mostly even above 500 °C. This suspicion was confirmed by the properties of PEIs derived from silylated naphthalene-2,6-dicarboxylic acid (**11**). The PEI prepared from **7b** or **9b** and **10** had T_ms of 500 °C. In consequence, meltable PEIs forming an LC-phase were only obtained when substituted terephthalic acids (via the form of trimethylsilyl esters) were used as reaction partners. The yields and properties of the isolated LC-PEIs are compiled in Table 2. The above discussion concerns homopolyesters composed of one imide diphenol and one dicarboxylic acid. However, diphenols such as **9a** and **9b** may be useful components of thermotropic copoly(ester imide)s consisting of three or more comonomers. Such LC-copolyesters containing terephthalic acid and 4-HBA were described in a patent[37].

Nevertheless remarkable is the failure of the diphenols **10a** and **10b** to yield LC-homo PEIs. Computer modelling of **9b, 10a** and **10b** in their energy minimum conformations (Fig. 5) revealed the following differences. The diphenyl ether moiety has a relatively wide bond angle (124±1°) and requires little activation energy to adopt a nearly coplanar conformation of the phenylene rings.

Table 1. Melting temperatures of acetylated imide diphenols used as monomers for PEIs

Bisacetate of diphenol	T_m(°C)	Bisacetate of diphenol	T_m(°C)
4a	314	6b	374–376
4b	>500	7b	205–206
5a	295–297	7b	336–338
5b	341–342	8	258–260
6a	264–266	9	241–243

Table 2. Yields and properties of nematic LC-PEIs prepared from the acetylated diphenols 7b or 9b and substituted terephthalic acids (12a–d)

LC-PEIs prepared from	Yield (%)	$T_g{}^a$(°C)	$T_m{}^a$(°C)	$T_i{}^b$(°C)
7b and 12a	84	200	367 (1.H.)	390–400 (dec.)
7b and 12b	72	202	358 (2.H.)	395–405 (dec.)
7b and 12c	74	221	368	>470 (dec.)
7b and 12d	85	–	(332) 377	>470 (dec.)
			375	
9b and 12a	87	195	(354) 376	380–385
9b and 12b	83	185	342	385–400
9b and 12c	85	206	370/390	410–415
9b and 12d	84	200	360/378	405–415

a From DSC measurements with a heating rate of 20 °C/min
b Optical microscopy with a heating rate of 10 °C/min

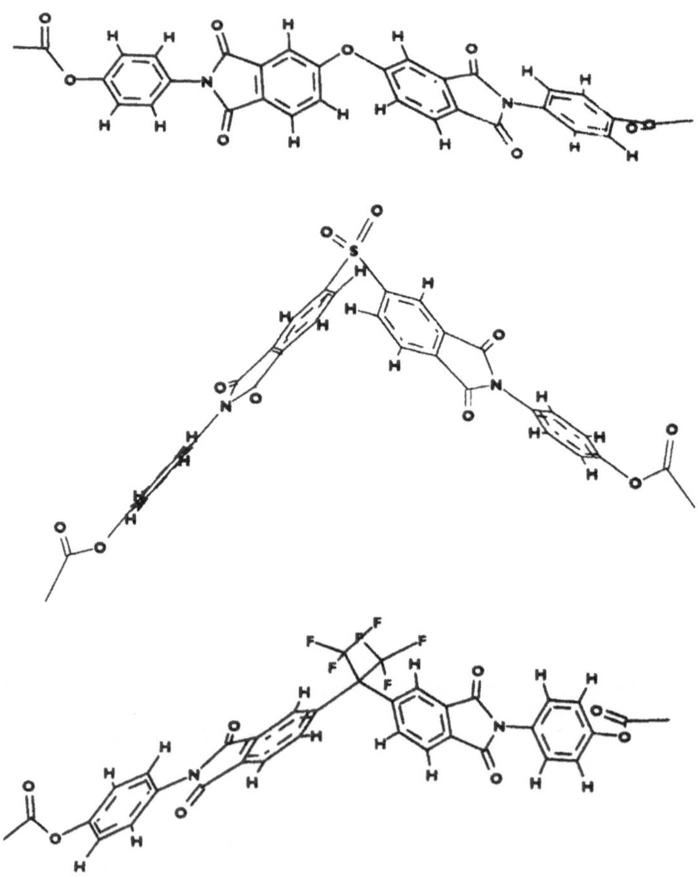

Fig. 5. Computer modelling of the energy minimum conformations of the imide diphenols 7b, 8 and 9 (in the form of bisactates)

A wider bond angle and a coplanar conformation favour an overall linearity of the PEI chain and favour a stacking of neighbouring chain segments due to donor-acceptor (DA) interactions. In the case of **10a** and **10b** smaller bond angles (109°) and high activation energies of the coplanar conformations were found [35]. Taken together, the successful syntheses of LC-PEIs from the symmetrical imide diphenols **8b** and **9b** in combination with **12a–d** is an example of strategy (I): the use of substituted comonomers.

A series of symmetrical diphenols were prepared from 4-hydroxy phthalic anhydride and various aromatic diamines (**13a–f**) [38]. Among these diphenols **13c** and **13d** seem particularly suited for the synthesis of LC-PEIs. However, to the best of our knowledge LC-PEIs derived from these diphenols have not been described yet. In this connection and with respect to the next section it should be mentioned that also PEIs derived from the diphenol **14** have not been synthesized yet.

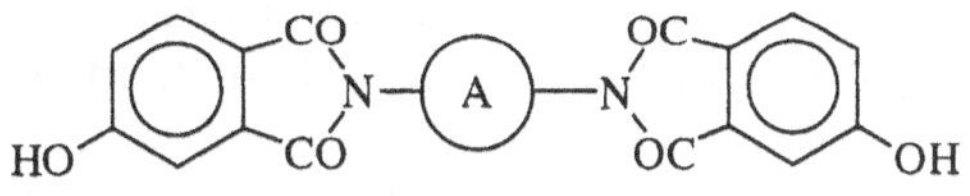

13 a - f

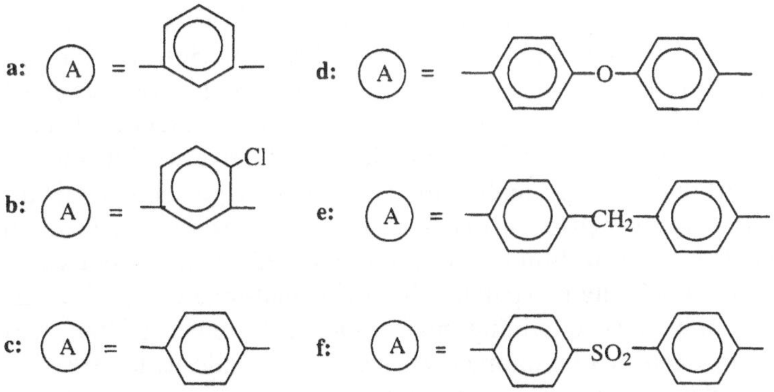

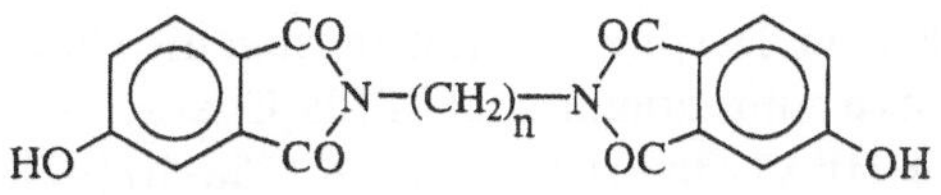

Scheme 3

3.2
Non-Symmetrical Imide Building Blocks

The strategy (II) is based on less symmetrical imide building blocks such as the hydroxy acids **15a, b** and **16a, b**. Since 4-hydroxyphthalic acid is not commercially available relatively little work has been reported on monomers and PEIs derived from this starting material. For instance no PEIs containing **15a** have been described so far.

The homo PEI of **15b** was prepared by polycondensation of its acetyl derivative in an inert reaction medium, and copolyesters containing 4-hydroxybenzoic acid in various molar ratios (**17**) were prepared analogously [39, 40]. Both the homo PEI and the co PEIs **17** proved to be highly crystalline with T_ms above 350 °C. Particularly interesting is the finding that these co PEIs can form single crystals with a whisker like morphology (Fig. 6) over a wide range of molar compositions. A partial explanation of this phenomenon consists of the fact that **15b** and the dimer of 4-hydroxybenzoic acid have nearly identical lenghts, and thus a good compatibility in the same crystal lattice[39, 40].

The homo PEI of **16b** (i.e. **18a**) was prepared by several research groups[14, 39–43]. It is again a highly crystalline material which melts and decomposes immediately above 550 °C. Furthermore, numerous co PEIs of **16b** were prepared but only few of them showed enantiotropic properties. For instance the co PEIs of **16b** and 4-hydroxybenzoic acid (**18a–f**) were again highly crystalline and infusible quite analogous to their isomers **17a–f**, but they did not form whisker-like crystals [14, 40]. In order to reduce both crystallinity and T_m, a ternary copolyester containing 3-chloro-4-hydroxybenzoic acid (**19**) was synthesized [14], and another ternary copolyester from cyclohexane-1,4-dicarboxylic acid and acetylated hydroquinone (**20**) [14]. Even these ternary copolyesters were more or less crystalline and did not melt below 360 °C [14]. However, the ternary copolyester **21** was found to soften above 315 °C and fibres were spun at 360 °C [2]. A more detailed characterization of this material was not reported. A lower melt viscosity and a greater flexibility (but a lower oxidative stability) was achieved by the incorporation of a small amount of ethylene glycol. For this purpose a mixture of 4-acetoxybenzoic acid and acetylated **16b** was reacted with PET [44]. No detailed characterization of the resulting co PEI *22* was reported in a Japanese patent [44]. Finally, two patents should be mentioned [45, 46] which reported on LC-copolyesters containing small amounts (10 mol%) of **16b**. These copolyesters mainly consist of monomers such as terephthalic acid, 4-HBA, hydroquinone 2-styryl hydroquinone or 4,4'-dihydroxybiphenyl and do not justify the label poly(ester-imide).

When compared to the homopolyester of **16b** the homo PEI of **16a** is a lower melting material (T_m=430–433 °C) [14, 47] but its melt is isotropic as expected for a meta functionalization. Despite the non-mesogenic character of **16a** this hydroxy acid can serve as a comonomer in LC-PEIs. Characteristic examples are the co PEIs containing 4-HBA (**23a–h**) or 2,6-HNA (**25a–h**) [47]. Two samples of LC-co PEIs of structure **23** were first described in a patent of Bayer AG [40].

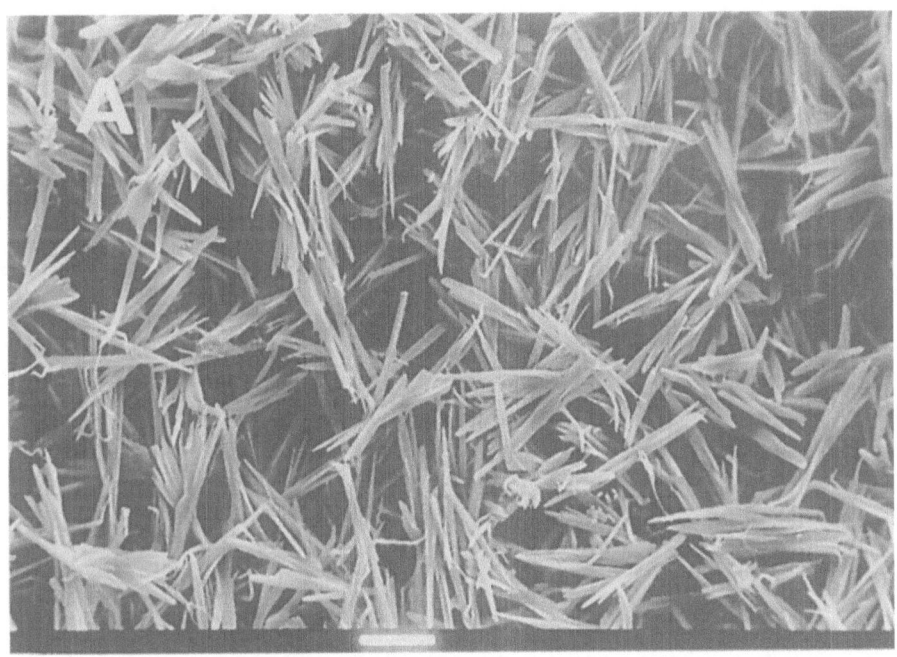

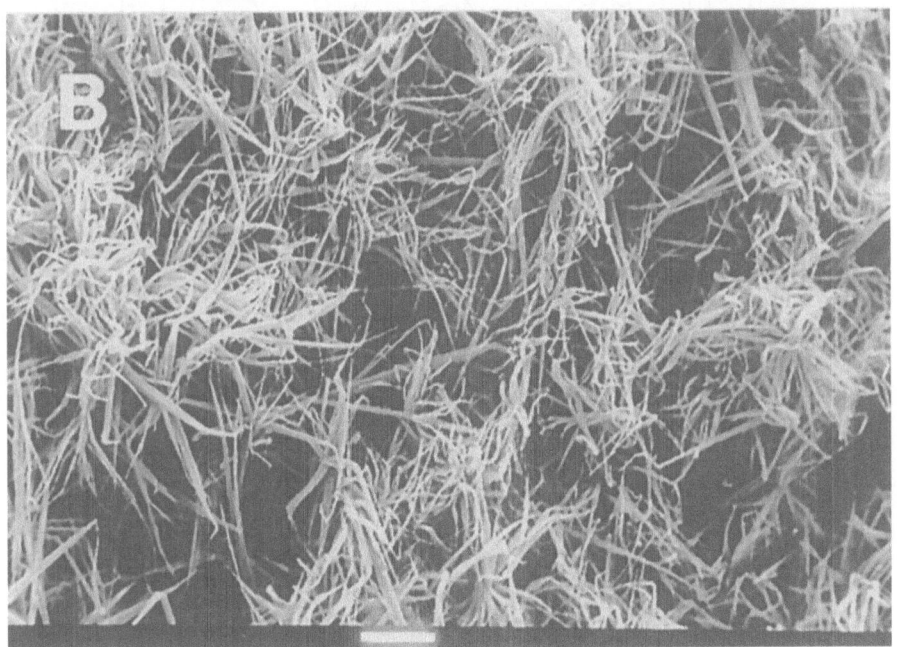

Fig. 6 A,B. SEM micrograph of the "whisker morphology" of: **A** the copoly(ester-imide) **17c**; **B** the homopoly(ester-imide) **15b**

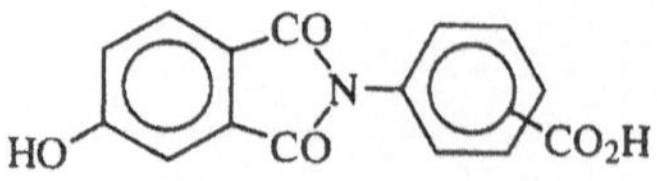

15a : meta

b : para

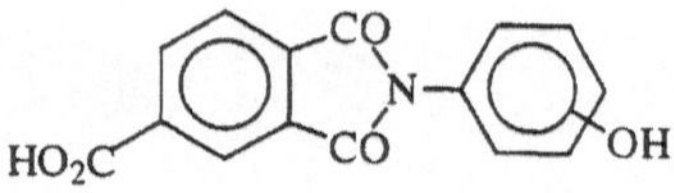

16a : meta

b : para

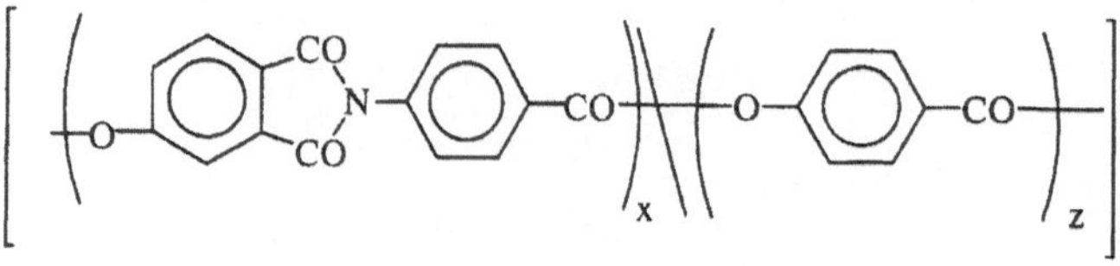

17a - f

a   : x/z = 10/0   c   : x/z = 4/6   e   : x/z = 2/8

b   : x/z = 5/5   d   : x/z = 3/7   f   : x/z = 1/9

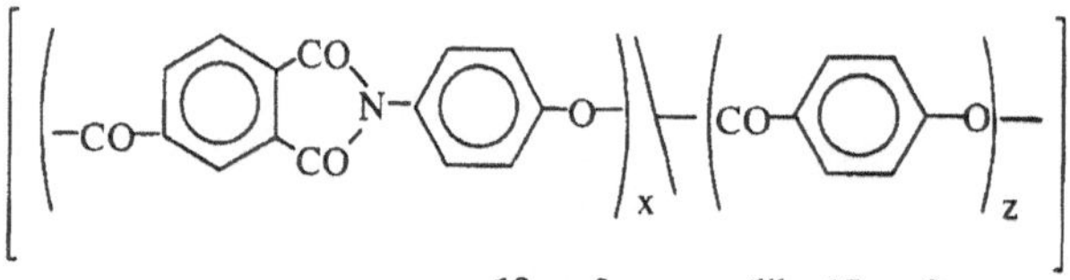

18a - f   (like 17 a - f)

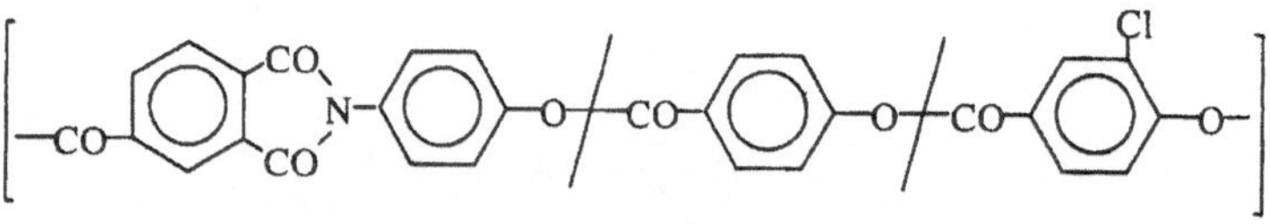

19

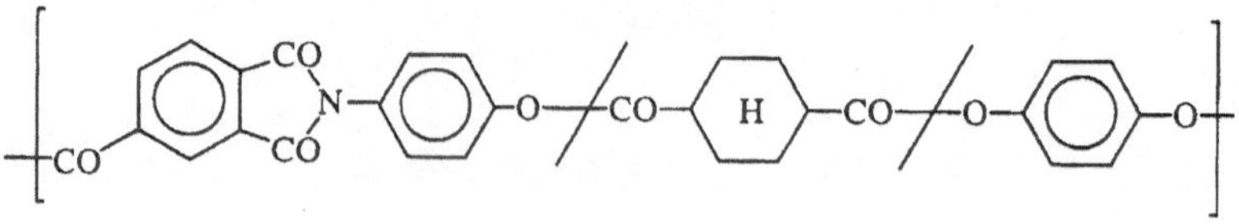

20

Structure 4 (1)

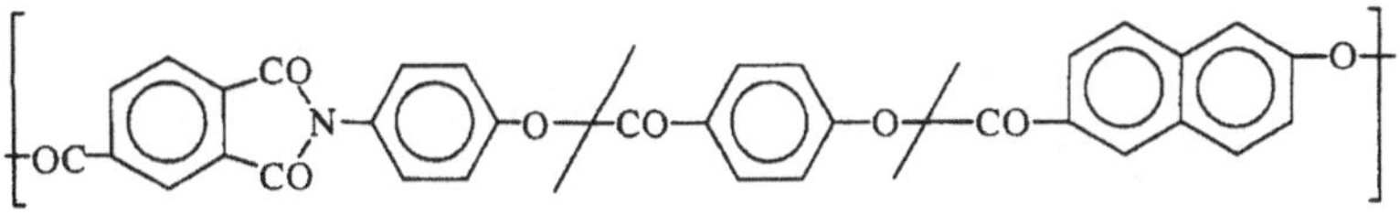

21

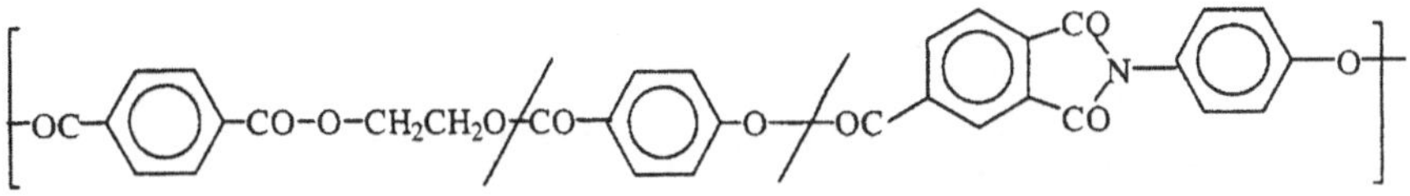

22

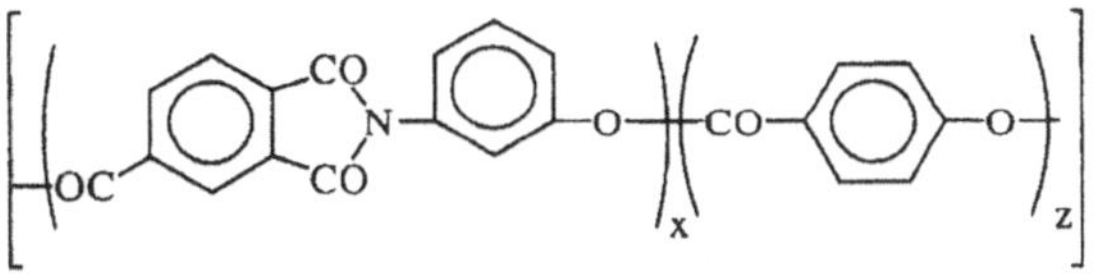

23 a - h

a	: x/z = 10/0	**d**	: x/z = 6/4	**g**	: x/z = 3/7
b	: x/z = 8/2	**e**	: x/z = 5/5	**h**	: x/z = 2/8
c	: x/z = 7/3	**f**	: x/z = 4/6		

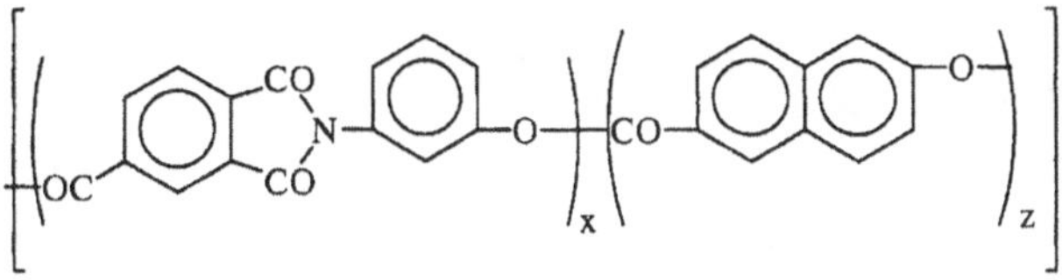

24a - h   (like 23a - h)

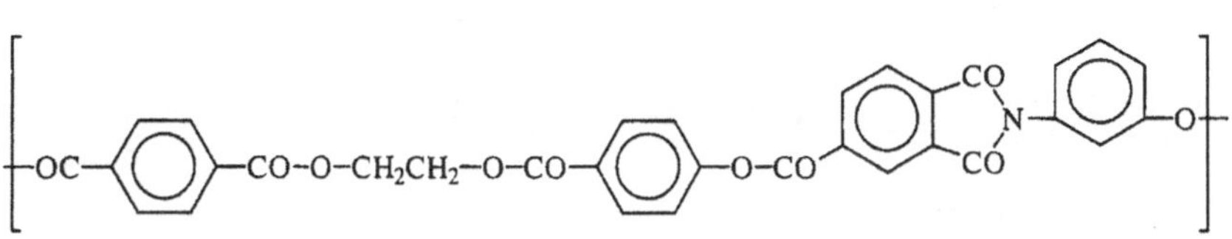

25

Structure 4 (2)

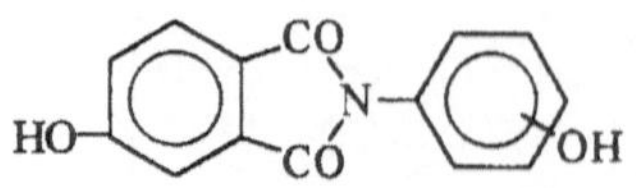

26   a : meta
         b : para

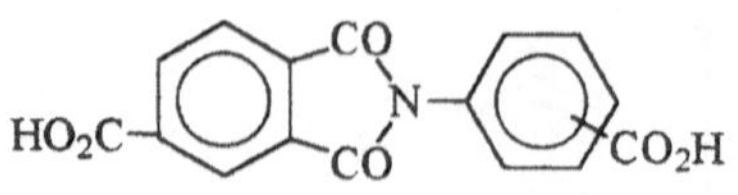

27   a : meta
         b : para

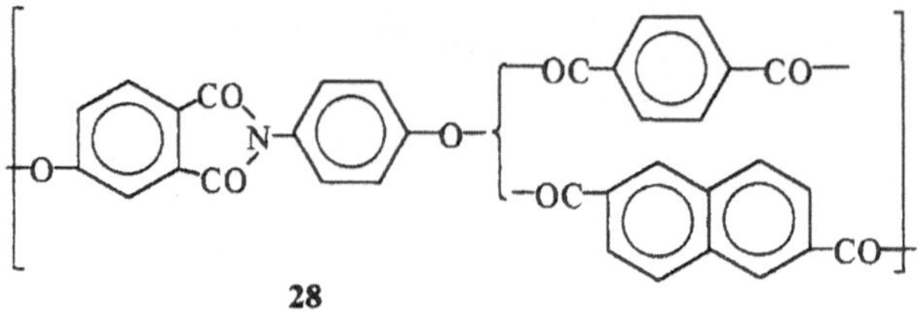

28

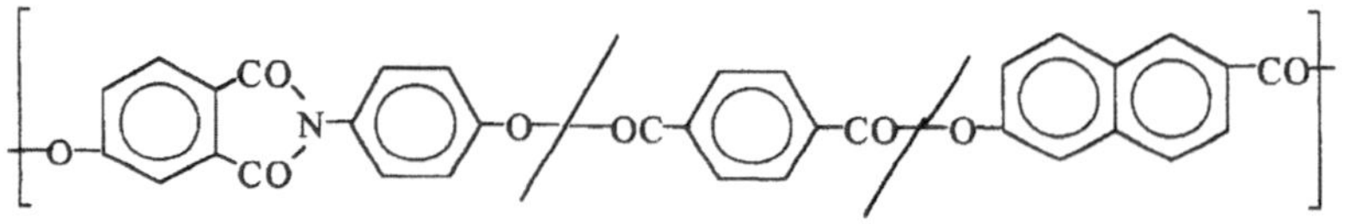

29

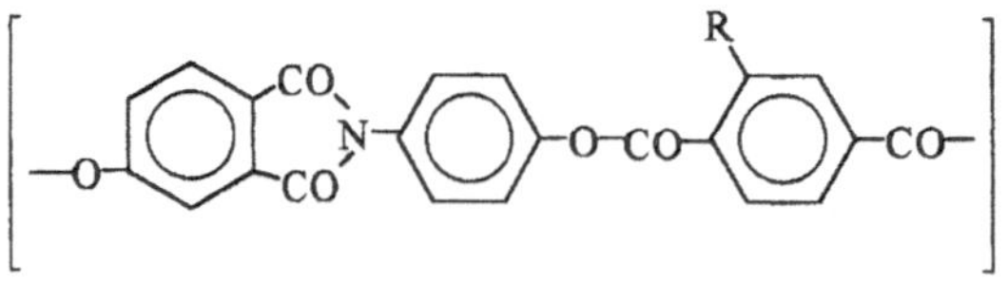

30 a - i

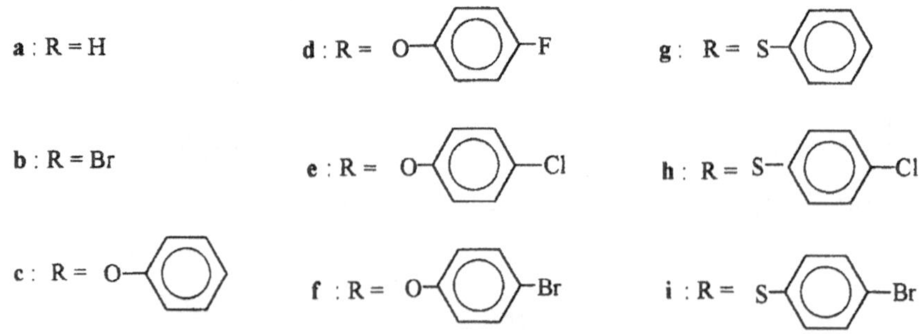

a : R = H

b : R = Br

c : R = O—⟨⟩

d : R = O—⟨⟩—F

e : R = O—⟨⟩—Cl

f : R = O—⟨⟩—Br

g : R = S—⟨⟩

h : R = S—⟨⟩—Cl

i : R = S—⟨⟩—Br

Structure 4 (3)

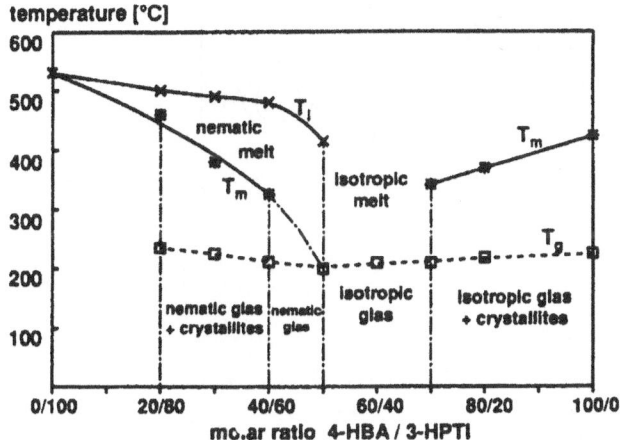

Fig. 7. Phase diagram of copoly(ester-imide)s prepared from the imide **16a** and 4-acetoxy-benzoic acid [47]

A detailed study yielding the phase diagram of Fig. 7 was elaborated by the author [47] and later confirmed by another research group [49]. The existence of a nematic phase depends very much on the molar ratio of the comonomers. A high molar fraction of **16a** is detrimental for the LC-phase. LC-PEIs containing ethylene glycol (**25**) were obtained by transesterification of PET with 4-acetoxy-benzoic acid and smaller amounts of acetylated **16a** [50].

Another group of interesting non-symmetrical imide monomers are the diphenols **26a, b** and the dicarboxylic acids **27a, b**. To the best of our knowledge, LC-PEIs derived from the "meta monomers" **26a** and **27a** have never been reported. However, several LC-PEIs were prepared from the diphenol **26b**. High melting co PEIs were obtained by polycondensations of acetylated **26b** with terephthalic acid, naphthalene-2,6-dicarboxylic acid (**28**) or 2,6 HNA (**29**) in bulk at 355–365 °C [11, 12]. Fibres were spun from the nematic melt at 382 °C but no detailed characterization of these co PEIs was given. Nonetheless, this work is particularly noteworthy because it reported for the first time on LC-polyimides. A series of homo PEIs prepared in bulk from acetylated **26b** and various terephthalic acids was studied by the author [51]. The properties of the resulting PEIs are listed in Table 3. ThePEI of unsubstituted terephthalic acid proved to be a semicrystalline material which did not form a mobile melt below 500 °C. However, all PEIS of substituted terephthalic acids were non-crystalline materials yielding a broad nematic melt above the glass-transition temperature (T_g). A characteristic structure-property relationship of the LC-PEIs **30c–i** are the higher T_gs and higher isotropization temperatures (T_is) resulting from arylthio substituents when compared to the corresponding aryloxy substituents(Table 3) (a similar trend was observed for the LC-PEIs **33h, i** and **34e, f** derived from **27b**).

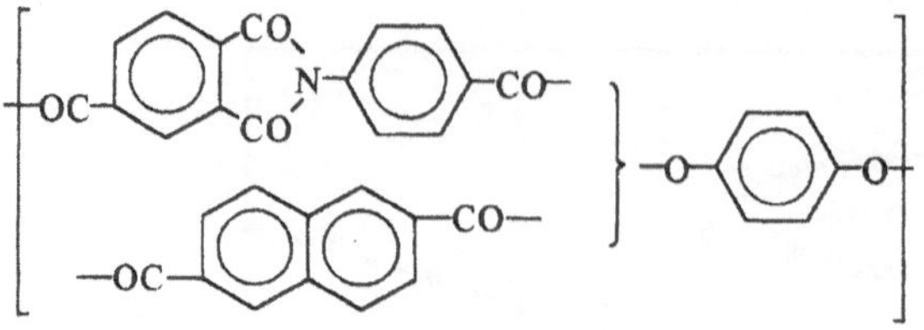

31

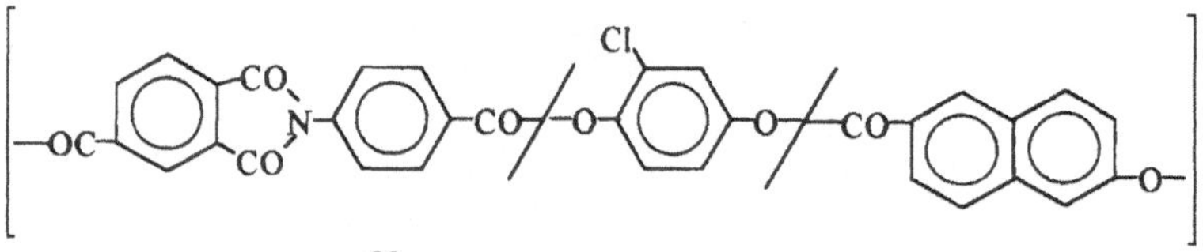

32

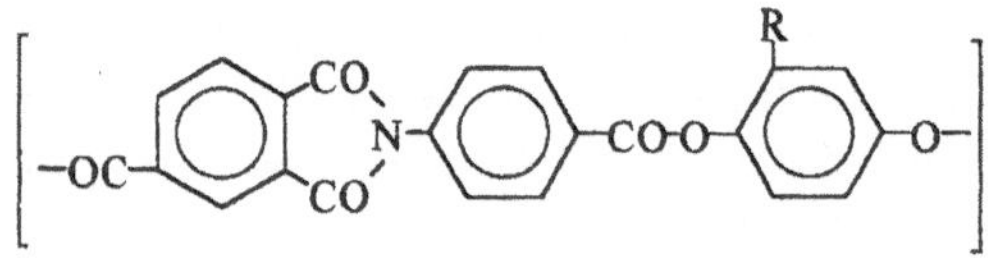

33 a - i   **a** : R = Cl   **d** : R = C(CH$_3$)$_3$ **g** : R = —C$_6$H$_5$

              b : R = Br   **e** : R = S—(CH$_2$)$_{11}$CH$_3$ **h** : R = —SC$_6$H$_5$

              c : R = CH$_3$ **f** : R = S—(CH$_2$)$_{15}$CH$_3$

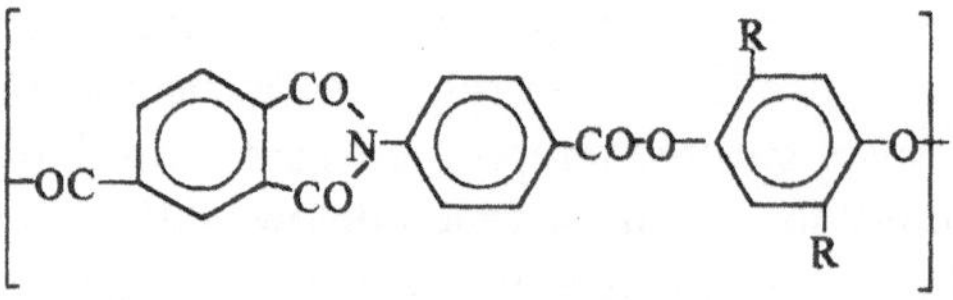

34 a - f   **a** : R = —S (CH$_2$)$_{11}$CH$_3$ **c** : R = —C (CH$_3$)$_3$ **e** : R = —O C$_6$H$_5$

              b : R = —S (CH$_2$)$_{15}$CH$_3$ **d** : R = —C$_6$H$_5$ **f** : R = —S C$_6$H$_5$

Structure 5 (1)

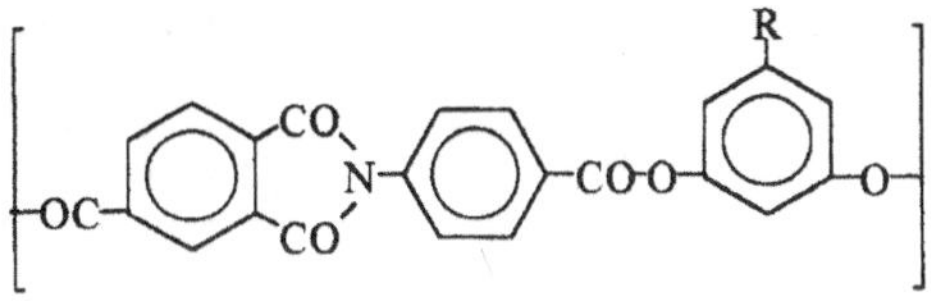

35a , b   a: R = H

b: R = CMe₃

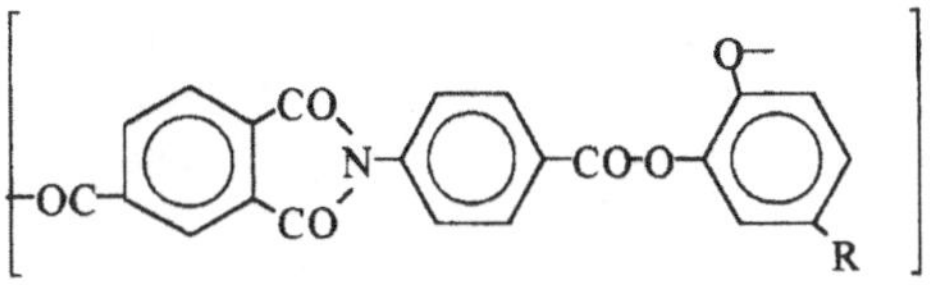

36a - c   a : R = H

b : R = CH₃

c : R = NO₂

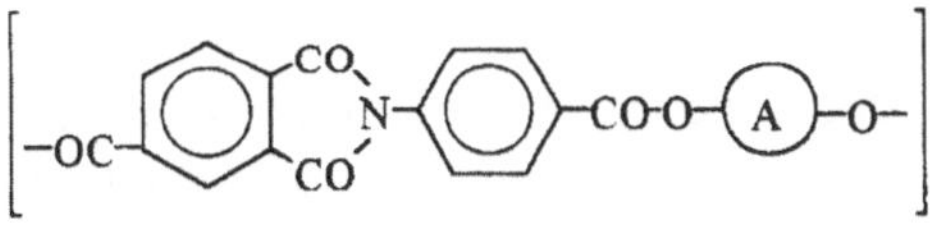

37a - e

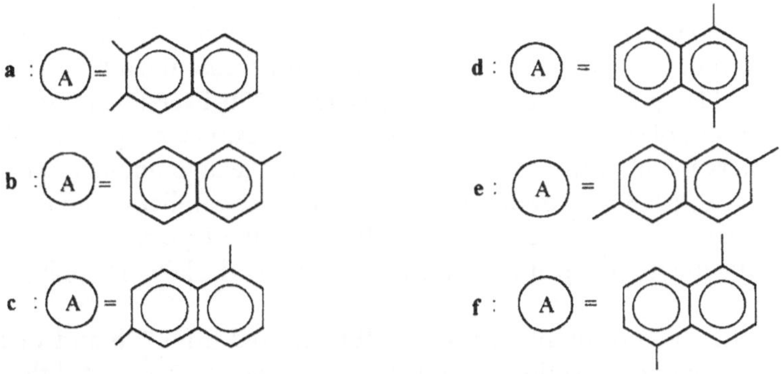

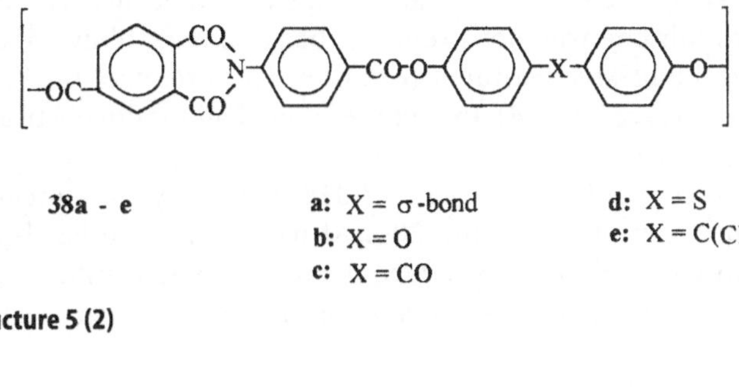

a : (A) =

b : (A) =

c : (A) =

d : (A) =

e : (A) =

f : (A) =

38a - e   a: X = σ-bond   d: X = S

b: X = O   e: X = C(CH₃)₃

c: X = CO

Structure 5 (2)

Table 3. Yields and properties of poly(ester-imide)s prepared from the imide diphenol **26b** and various terephthalic acids

I PEI No	Yield (%)	$\eta_{inh.}$[a](dl/g)	T_g[b](°C)	T_i[c](°C)
30a	68	insol.	163	500
30b	76	insol.	144	500
30c	90	0.26	140	370–375
30d	95	0.63	169	375–385
30e	92	0.92	140	375–380
30f	93	0.68	167	375–380
30g	98	0.98	180	460–465
30h	95	1.02	170	455–460
30i	96	0.97	176	435–445

a Measured at 20 °C with c=2 g/l in CH_2Cl_2/trifluoroacetic acid (4:1 by volume)
b From DSC measurements with a heating rate of 20 °C/min
c From optical microscopy

The dicarboxylic acid **27b** can easily be prepared from inexpensive commercial starting materials (in contrast to **26b**), and thus this monomer was used for the preparation of many PEIs in general and LC-PEIs in particular. Its mesogeneity (and that of the diphenol **26b**) is also discussed in the following section. The first aromatic LC-PEIs containing **27b** (**31** and **32**) were mentioned in two patents of Du Pont [11, 12] as early as 1979 and 1983. The properties of the PEIs **31** and **32** were not characterized, but fibres were spun from the nematic melt at temperatures around 370 °C. Unfortunately these few pieces of information do not say much about the mesogenic character of **27b**, because the other comonomers may yield LC-polyesters even in the absence of **27b**. Even less informative were the mentions of several copolyesters in two Japanese patents [45, 46]. Those LC-copolyesters mainly consist of conventional mesogenic monomers, such as terephthalic acid, 4-HBA, hydroquinone or 4,4'-dihydroxybiphenyl and the content of **27b** was in most cases 10 mol%. Obviously the incorporation of **27b** mainly had the function to justify a patent claim under the label LC-poly(ester-imide).

A systematic study of numerous homo PEIs derived from **27b** and various diphenols was conducted by the author [21, 52, 53]. This study revealed that the dicarboxylic acid has good mesogenic properties despite its lack of symmetry, and some unexpected structure property relationships were detected. The PEIs of unsubstituted hydroquinone was not prepared, because a T_m above 450 °C was expected. Even the smallest substituent (e.g. Cl or CH_3) reduces the T_m to values below 400 °C (**33a, c**, see Table 4). In fact the PEIs of all monosubstituted hydroquinones (**33a–i**) were thermotropic with broad nematic phases. Remarkable is the reluctance to crystallize in the case of **33g**. Three days of annealing were required to obtain a crystallinity below 20%. Also other more or less bulky substitutes, such as those of **33d, h** and **i** reduce the tendency to crystallize. As discussed below, (and in Sect. 4), this is an interesting aspect for the synthesis of

Table 4. Yields and properties of the poly(ester-imide)s **33a–h** prepared from *N*-(4-carbox-yphenyl) trimellitimide and various monosubstituted hydroquinones

Polym. No.	R	Yield (%)	$\eta_{inh.}$[a] (dl/g)	T_g[b] (°C)	T_m[b] (°C)	T_i[c] (°C)
33a	–Cl	>95	insol.	177	362	>500 (dec.)
33b	–Br.	>95	insol.	180	347	>500 (dec.)
33c	–CH$_3$	>95	insol.	175	351	>500 (dec.)
33d	–C(CH$_3$)$_3$	86	0.32	198	–	>400 (dec.)
33e	–S(CH$_2$)$_1$CH$_3$	91	0.72	183	–	430–440 (dec.)
33f	–S(CH$_2$)$_{15}$CH$_3$	96	0.80	160	199	400–410 (dec.)
33g	–C6H5	97	1.14	196	263	460–480 (dec.)
33h	–SC6H5	96	1.06	168	326	460 (dec.)

a Measured at 20 °C with c=2 g/l in CHCl$_2$/trifluoroacetic acid (volume ratio 4:1)
b From DSC measurements with a heating rate of 20 °C/min
c From optical microscopy with a heating rate of 10 °C/min

non-crystalline LC-polyesters. However, the most interesting and surprising finding is the fact that the PEIs derived from symmetrically disubstituted hydroquinones (**34c–f**) are amorphous and isotropic, when the substitutents are somewhat bulky. Orginally, the symmetrical substitution seems to be favourable for both crystalline and liquid-crystalline phases. A hypothetical explanation of this phenomenon assumes that the ordering in the crystalline and LC-phase is based on DA-interactions between neighbouring coplanar chain segments (Fig. 3).

If this assumption is correct, two bulky substituents of the hydroquinone hinder this interaction in two ways. Firstly, they raise the energy barrier for a co-planar conformation of hydroquinone and imide rings, as confirmed by computer modelling. Secondly, if rotation around the -bonds (Fig. 8) is taken into account at high temperatures (>200 °C in the melt) two substituents hinder a par-allelization and DA-interaction of neighbouring chain segments more efficiently for statistical reasons.

Further interesting structure-property relationships concern the PEIs derived from ortho and meta diphenols (**35a, b, 36a–c** and **37a, b**). The PEIs **35a, b** are amorphous and isotropic, as expected for the combination of a non-symmetrical monomer and a meta functionalized ones. However, the PEI **37b** derived from 2,7-dihydroxynaphthalene is thermotropic despite a kind of meta substitution. Since the electron density and the polarizibility of the naphthalene system is higher than that of a single benzene ring the LC-properties of **37b** can again be explained by DA-interactions between neighbouring almost coplaner chain segments involving the naphthalene group as π-donor (Fig. 3). The classical theory of "rigid-rod mesogens" proposed by Onsager [15] and Flory [16–18] is certainly not a more convincing alternative. This aspect is even more conspicious, when the PEIs **36a–c** and **37a** are taken into account. Despite the ortho-position of the involved diphenols all these PEIs form nematic melts over a

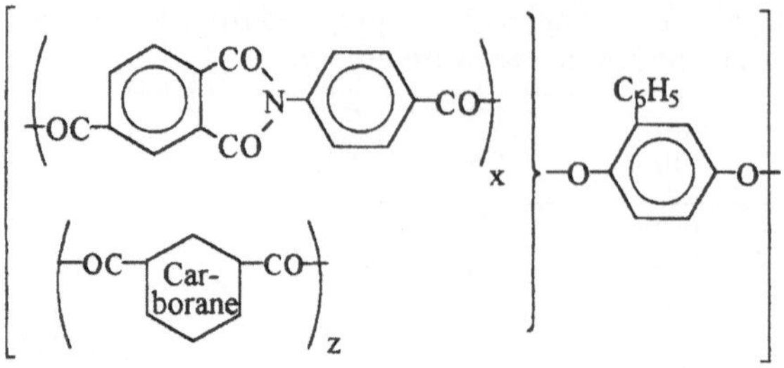

39 a - c   **a** : x/z = 9/1   **b** : x/z = 8/2   **c** : x/z = 7/3

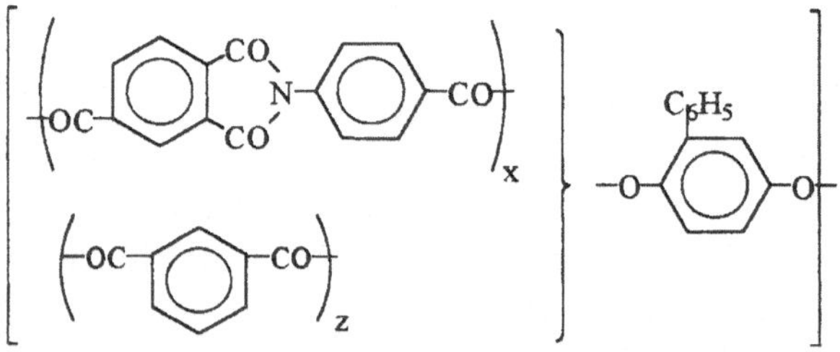

40 a - f   **a** : x/z = 9/1   **c** : x/z = 7/3   **e** : x/z = 5/5
                    b : x/z = 8/2   **d** : x/z = 6/4   **f** : x/z = 4/6

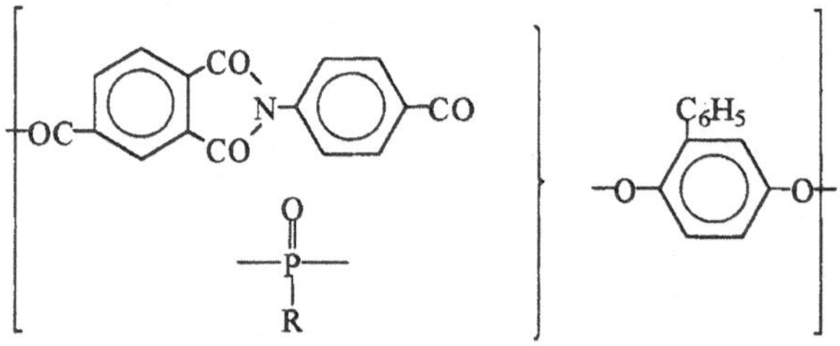

41 a - c   **a** : R = CH_3 **b** : R = C_6H_5 **c** : R = $O–C_6H_5$

Structure 6 (1)

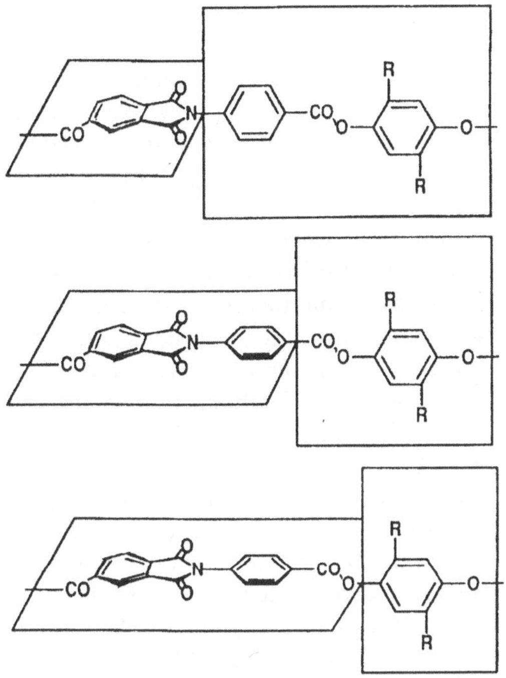

Fig. 8. Illustration of the ring planes which can rotate relative to each other in the case of PEIs **33a–h** and **34a–f**

broad temperature range (Table 4). The computer modelling of the energy minimum conformation of a single chain of **36a** indicates that a chain folding due to DA-interactions is the most favourable conformation (Fig. 9). Although the conformational behavior of a single chain in vacuo does not necessarily represent the reality of a nematic melt, it is obvious that attractive electronic interactions according to Figs. 3 and 9 play a key role for the formation of the LC-phase and not the repulsive forces between symmetrical, linear, rigid rods discussed by Flory. When the PEIs derived from dihydroxy naphthalenes (**37a–f**) are compared, crystalline, high melting ($T_m > 450$ °C) materials were found, whenever the OH-groups were in parallel position (**37d–f**), whereas broad LC-phases were only observed for the ortho and meta-functionalization in **37a,b**. The properties of the PEIs **38a–e** fit in with the structure property relationships discussed for the PEIs of the diphenols **7b, 8b, 9b, 10a** and **10b**. PEI **38a** has a $T_m > 500$ °C and decomposes immediately upon melting. Broad nematic phases were observed for **38b, c** whereas **38d, e** are isotropic. The benzophenone and diphenyl ether moieties favour coplanar conformations of their benzene rings, and this property seems to be decisive for the stabilization of the nematic phase. Taken together the structure-property relationships of the PEIs derived from **27b** make an important contribution to a better understanding of the mesogeneity of aromatic systems.

The finding that the PEIs of **27b** and monosubstituted hydroquinones form broad nematic phases, but show little propensity to crystallize, has prompted various modifications of their structures and properties. In this connection it should be stated that non-crystalline LC-polymers have found little interest in the past decades, but they may be attractive for various applications provided that the T_g can be varied between 90 and 250 °C. For instance, the absence of crystallinity has the advantage that the mechanical properties do not depend on the thermal history, and thus on the processing conditions. The temperature allowing a convenient processing may be reduced below 200 °C, which is of interest for the processing of LC-polymer reinforced blends and composites. Furthermore, non-crystalline nematic LC-polyesters are a useful basis for the synthesis of cholesteric lacquers, films or pigments (Sect. 5).

A modification of PEI **33g** with carborane-1,7-dicarboxylic acid (**39a–c**) or isophthalic acid (**40a–f**) gave the following results [54]. The carborane moiety enables thermal cross linking at temperatures >300 °C so that in principle thermosetting LC-resins may be prepared. Unfortunately, the carborane moiety proved to be extremely unfavourable for the existence of the nematic phase, and incorporation of 10 mol% sufficed to eliminate the nematic phase completely. In contrast, the LC-phase of **33g** was compatible with up to 40 mol% of isophthalic acid despite a similar meta funcionality. This result fits in with the LC-hypothesis of Fig. 3 because the isophthalic acid is a flat and the carborane a bulky building block. The incorporation of phosphate or phosphonate groups into polymers may have the purpose of reducing their flammability. The sp3-hybridization of the pentacovalent phosphorus is, of course, unfavourable for the stabilization of an LC-phase. Nonetheless, it was found [55] that up to 40 mol% of phosphate or phosphonate groups may be incorporated into PEI **33g** (or **41a–c**) before the LC-phase completely vanishes.

Whereas, the co PEIs **39a–c**, **40a–f** and **4a–c** are characterized by a partial replacement of the dicarboxylic acid **27b**, a series of co PEIs (**42**) was described in a patent [56] where the composition of the diphenols was varied. Higher contents of hydroquinone raise the melting point and the crystallinity; higher contents of *tert*-butylhydroquinone raise the T_g. These series of co PEIs which allow a broad variation of the T_g were prepared by transesterification of PET with combinations of the dicarboxylic acid **27b** and acetylated methyl hydroquinone, *tert*-butylhydroquinone or phenylhydroquinone (**43a–c**)[57]. Figure 10 displays the influence of the molar composition on the T_g and on the E-modulus of the co PEIs **43b**. All these co PEIs are noncrystalline and nematic up to a molar fraction of 75–80% PET. At higher PET contents crystallization of long PET blocks takes place and the LC-phase has completely vanished. A further modification of these LC-PEIs can be achieved by cocondensation of 4-acetoxybenzoic acid or 6-acetoxy-2-naphthoic acid [58].

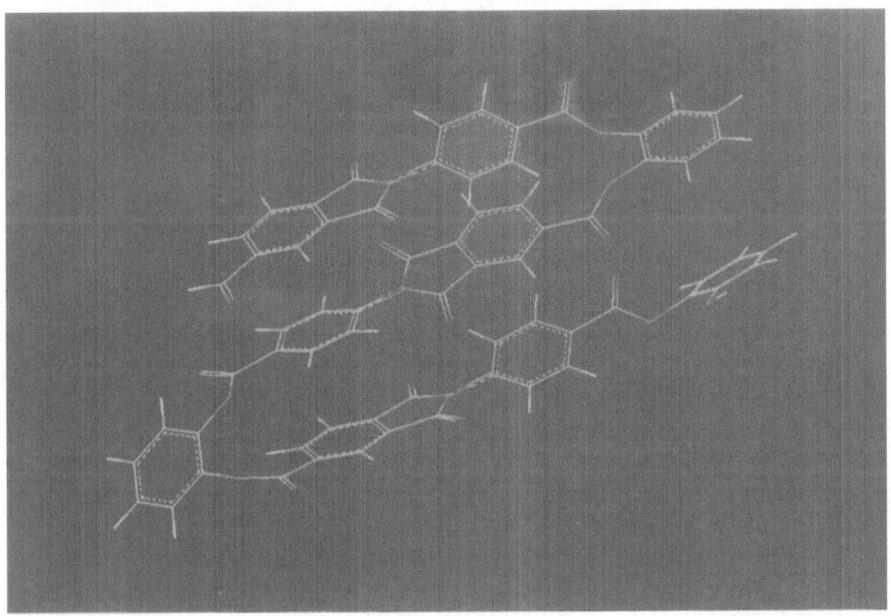

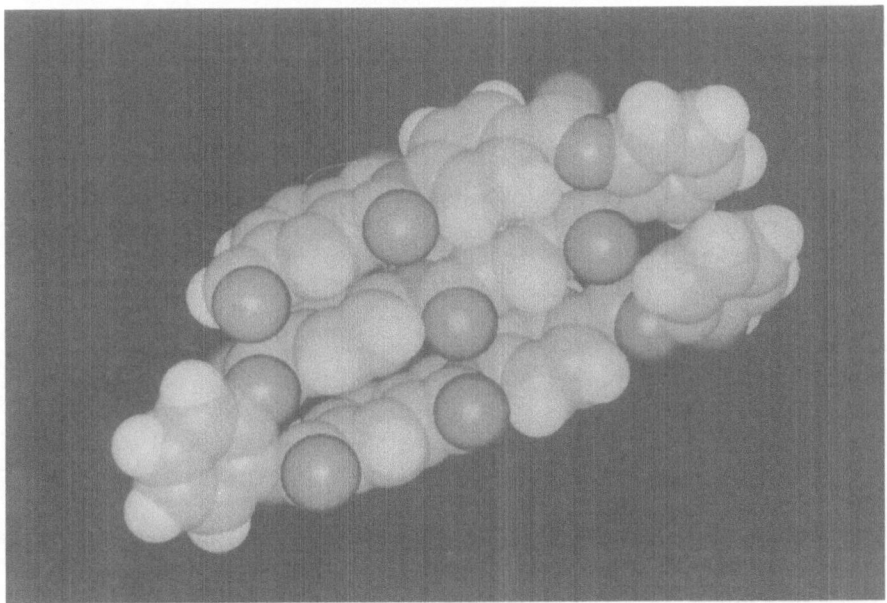

Fig. 9. Computer model of the energy minimum conformation of a single PEI **36a** chain

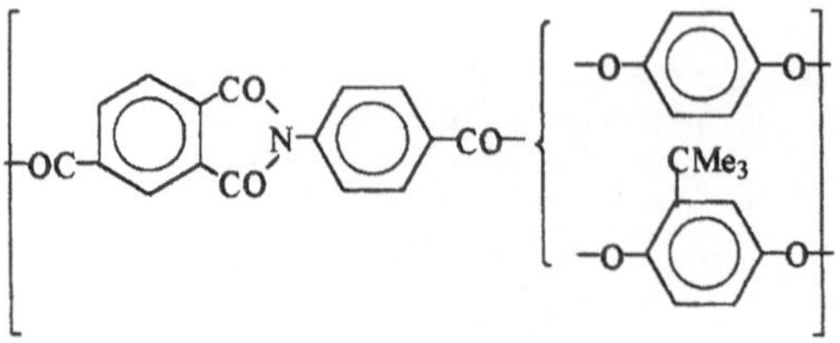

42

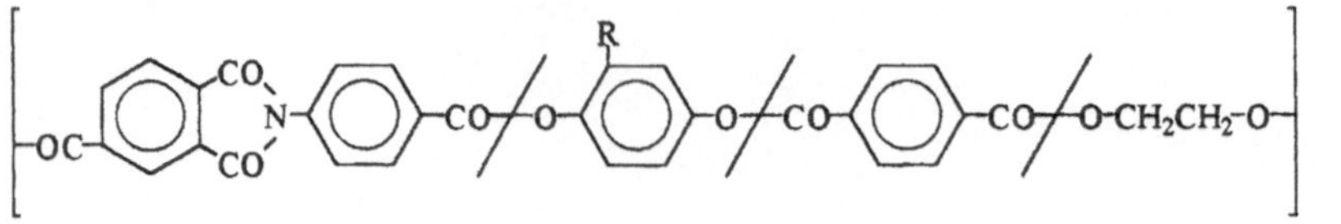

43 a - c   **a** : R = Me   **b** : R = CMe₃   **c** : R = —⬡

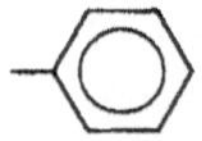

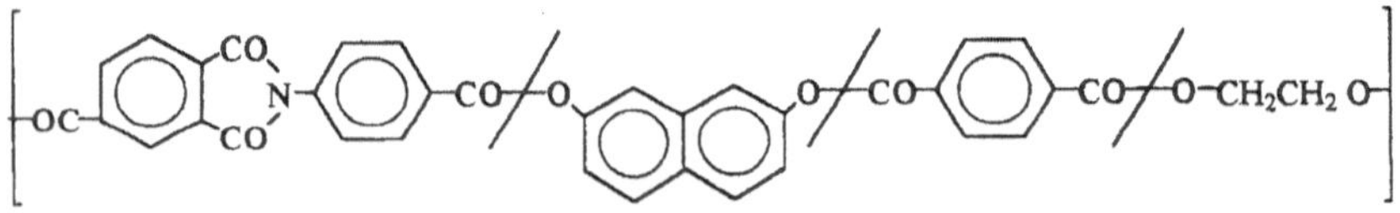

44

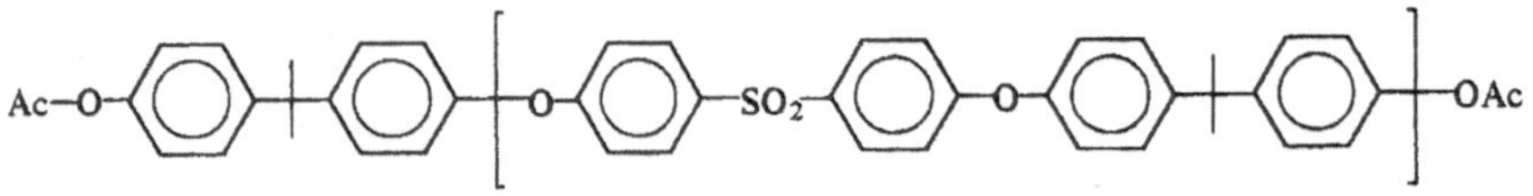

45

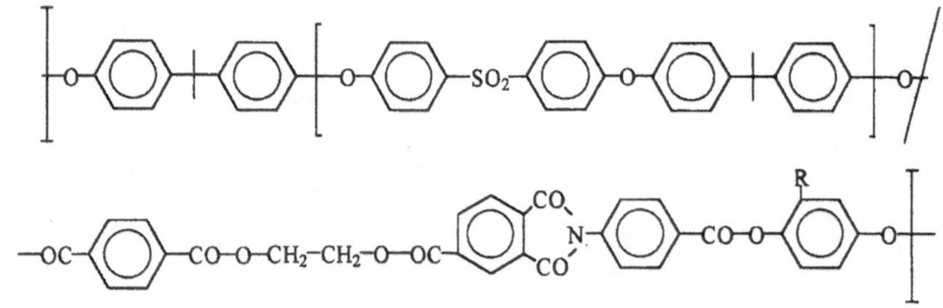

46

Structure 6 (2)

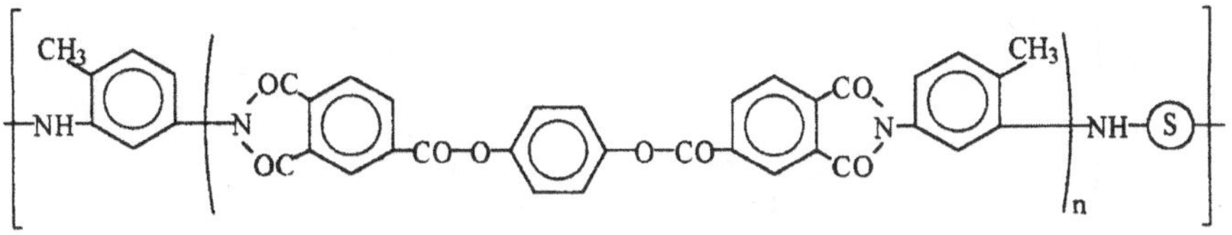

47

a: (S) = $\begin{array}{c} CH_3 \\ | \\ -Si- \\ | \\ CH_3 \end{array}$

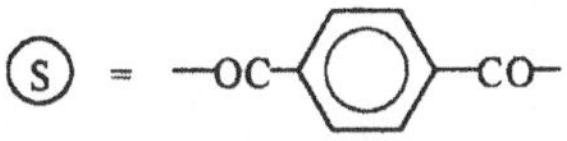

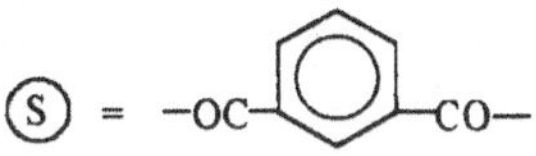

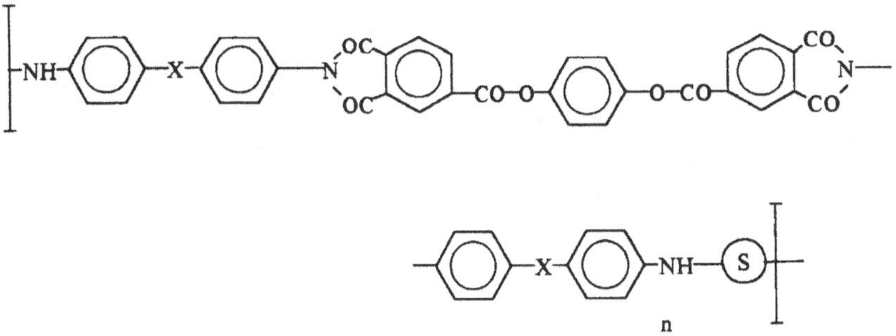

c: (S) = $-OC-\langle\bigcirc\rangle-CO-$

b: (S) = $-OC-\langle\bigcirc\rangle-CO-$

48 a - e   (X = CH₂ or O)

a: (S) = $\begin{array}{c} CH_3 \\ | \\ -Si- \\ | \\ CH_3 \end{array}$

b: (S) = $\begin{array}{c} CH_3 \quad CH_3 \\ | \qquad | \\ -Si-O-Si- \\ | \qquad | \\ CH_3 \quad CH_3 \end{array}$

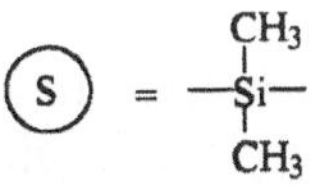

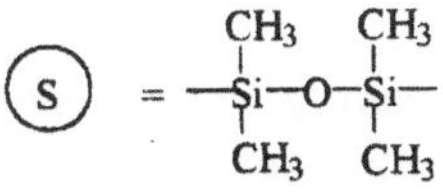

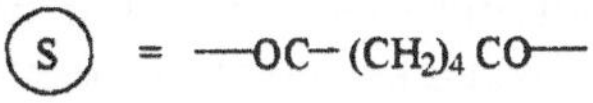

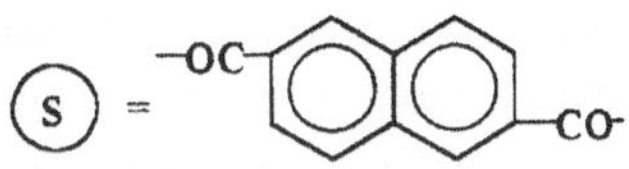

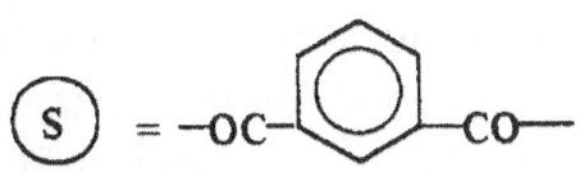

c: (S) = $-OC-(CH_2)_4\,CO-$

d: (S) = $-OC-\langle\bigcirc\bigcirc\rangle-CO-$

e: (S) = $-OC-\langle\bigcirc\rangle-CO-$

f: (S) = $-OC-\langle\bigcirc\rangle-CO-$

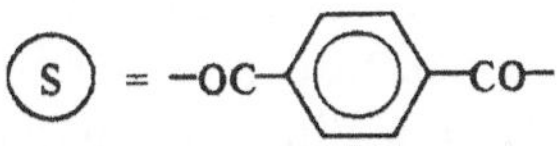

Structure 6 (3)

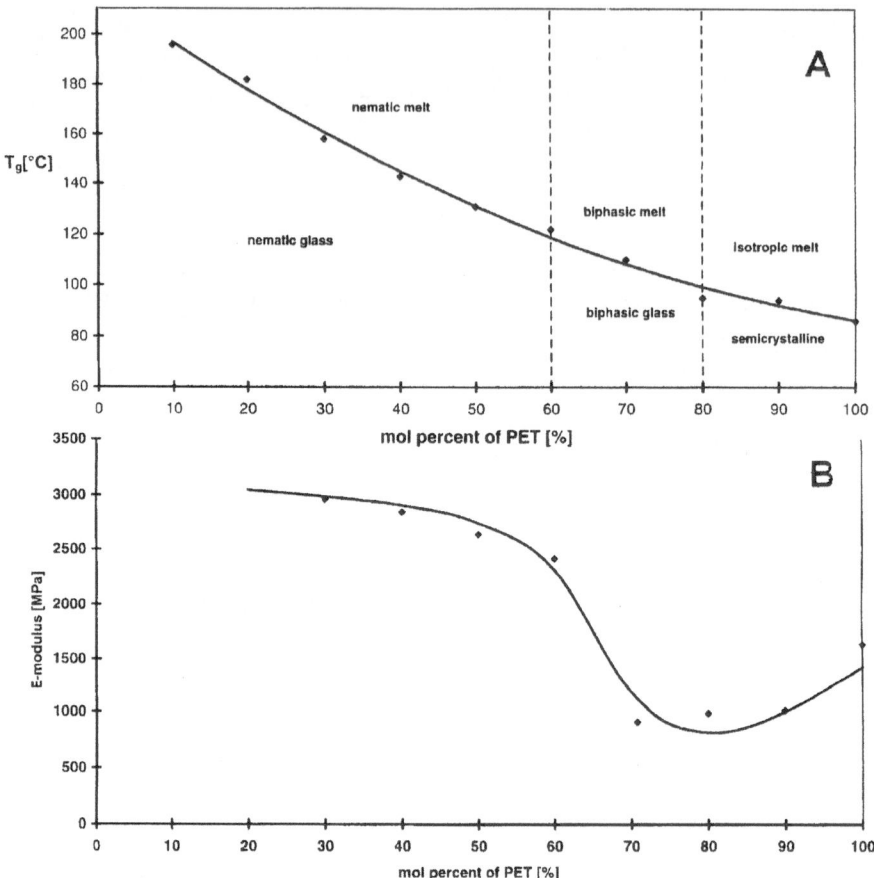

Fig. 10 A,B. Influence of the molar composition on: **A** the glass-transition temperature; **B** the E-modulus of the PET based co PEIs **43b**

Quite recently CO_2H-terminated telechelic oligo- and polyesters were prepared by transesterification of PET with acetylated *tert*-butylhydroquinone and an excess of *N*-(4-carboxyphenyl) trimellitimide. These blocks were polycondensed in bulk with oligo(estersulfone)s having acetate endgroups (**45**) [59,61]. Most of the resulting multiblock copolymers **46** were liquid crystalline. They possess high T_gs (around 200 °C) and showed excellent mechanical properties. Furthermore, a wide variety of aromatic poly(amide-ester-imide)s most of which have the structures **47a–c** or **48a–f** has recently been reported [62]. A great deal of these copolymers were not thermotropic. However, their high melting temperatures and high melt viscosities made a proper characterization of the molten state difficult. Therefore, the nature of the LC-phases and their temperature ranges were poorly defined.

4
Poly(ester-imide)s Containing Aliphatic Spacers

The combination of a more or less rigid aromatic mesogen with a flexible aliphatic spacer is an established strategy to improve the solubility and to lower the melting temperature (when compared to fully aromatic polymers) and thus, to improve the characterization and the processability. In the case of polyimides in general and poly(ester-imide)s in particular, regular sequences of stiff mesogens and aliphatic spacers have proven to possess an exceptionally high tendency to form layered supermolecular structures in the solid state. The present section provides a general description of the PEIs containing aliphatic spacers, whereas the problems related to the characterization of solid layer structures will be discussed in the last section of this review.

In this connection it may be useful for a better understanding of the following text to discuss shortly what the term "smectic" means in connection with LC-main chain polymers. The terminology of smectic phases and the physical properties related to individual smectic phases has been developed for low molar mass LC-materials. When the terminology of smectic phases is exclusively applied to a labeling of certain supermolecular structures, it can be taken over to LC-main chain polymers without any restriction. However, their meaning in terms of physical properties may change. As illustrated in Fig. 11A, layers of small molecules may be mobile relative to each other by gliding along their surfaces defined by the methyl endgroups.

Such a mobility does not exist in layered LC-main chain polymers (Fig. 11B). The mobility required for the definition and existence of a molten state results in LC-main-chain polymers exclusively from a gliding of chains along each other. Such a motion is only possible when the intermolecular forces between the mesogens are relatively weak. Therefore only smectic -A and smectic -C phases are true LC-phases. In contrast to small molecules smectic -B (and higher ordered smectic phases) are solid mesophases. The difference between a solid smectic mesophase and a smectic crystalline phase lies in the extent of the three dimensional order and is usually difficult to determine experimentally (see Sect. 7).

4.1
Short Symmetrical Imide Mesogens

The smallest symmetrical imide moiety which might play the role of a mesogen is the pyromellit imide group (PMDI, see structures **49–54** and **56–60**). Its mesogenic potential has been a matter of controversial reports. A first short comment on PEIs of structure **49a–g** was published by the author in the introduction of a paper dealing with benzophenone-tetracarboxylic imide (BTCI) [63]. The formation of a LC-phase was assumed only for the PEIs of hydroquinone (**49a**) and 4,4-'dihydroxydiphenyl ether (**49g**). Furthermore, Aducci et al. mentioned a thermotropic character of the PEIs **52a–d** and **53a–e** in two reviews [8, 64]. Yet

A   B

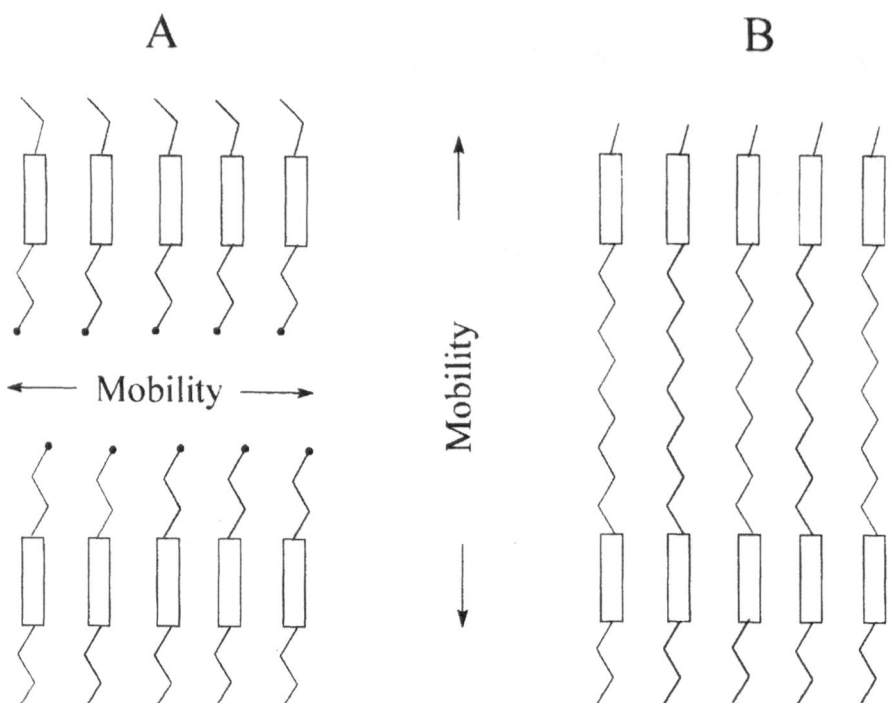

Fig. 11 A,B. Scheme of mobilities in smectic layer structures of: **A** a low molar mass material having alkyl substituents; **B** a LC-main chain polymer

in both papers any detailed description of the syntheses and the characterization is lacking . In a more recent systematic study of the PEIs **49a–g, 50a–e, 51a, b** and **52a–d** the author reached the conclusion that none of these PEIs forms a true LC-phase. All these PEIs are highly crystalline with a kind of smectic layer-structure in the solid state. With regard to the syntheses it should be mentioned that the PEIs derived from diphenols (**49a–g, 50a–c**) were prepared by poly-condensations of the PMDI dicarboxylic acids with the acetylated diphenols in bulk [65]. The PEIs **51a,b, 52a–d** and **53a–e** were either obtained by transesteri-fication of the free diols with the dimethyl esters of the PMDI dicarboxylic acids [65] or by polycondensations of the diols with the dichlorides in solution [8,64].

At the same time the polycarbonates **54a–f** (all of them with m=3, 4, 5, 6) were prepared and characterized by Sato and coworkers [66, 67]. All syntheses were conducted by transesterification of the corresponding diols with the biscar-bonates **55** in bulk. None of the four homo PEIs (**54a**, m=3–6) proved to be ther-motropic. In the case of the copolycarbonates only those containing more than 50 mol% of the biphenylester unit (e.g. **54d–f**, m=3) were apparently thermo-tropic. This thermotropic property was attributed to the mesogenic character of

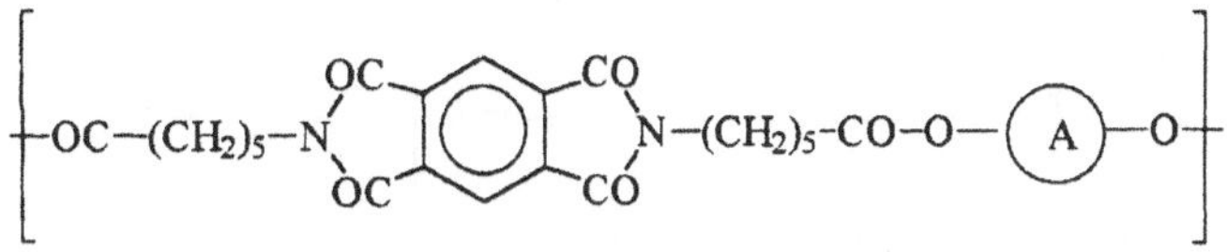

49 a - g

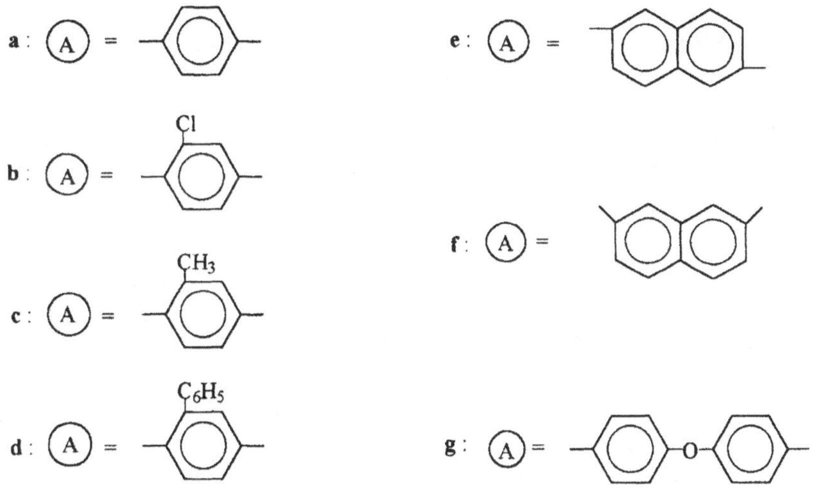

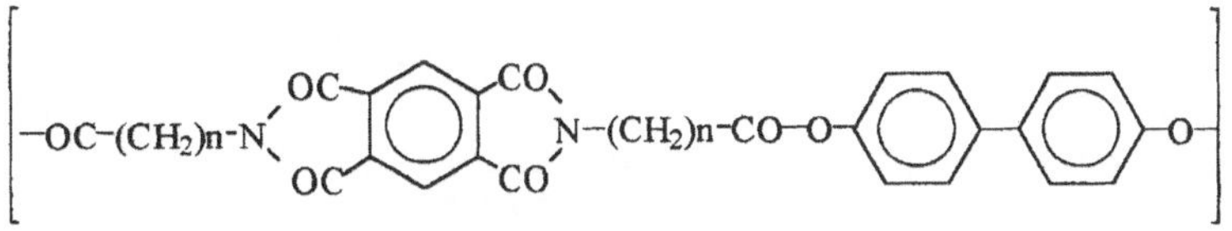

a : (A) = —⟨◯⟩—   e : (A) = ⟨◯◯⟩

b : (A) = —⟨◯⟩— (Cl)

c : (A) = —⟨◯⟩— (CH₃)   f : (A) =

d : (A) = —⟨◯⟩— (C₆H₅)   g : (A) = —⟨◯⟩—O—⟨◯⟩—

50 a - e	**a** : n = 3	**c** : n = 5	**e** : n = 11
	b : n = 4	**d** : n = 10	

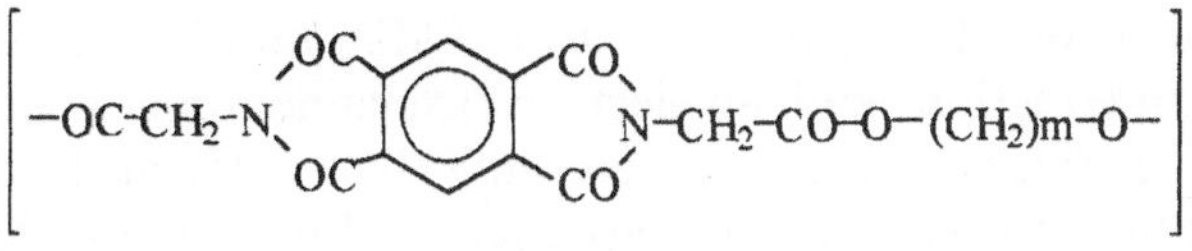

51 a - b   a : m = 6   b : m = 12

Structure 7 (1)

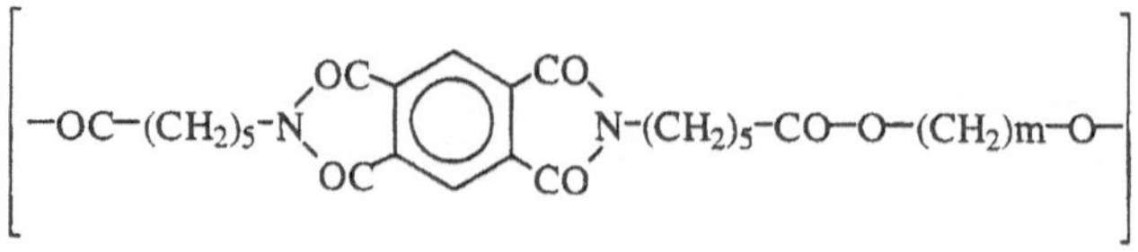

52 a - d   **a** : m = 6   **c** : m = 10
                      b : m = 8   **d** : m = 12

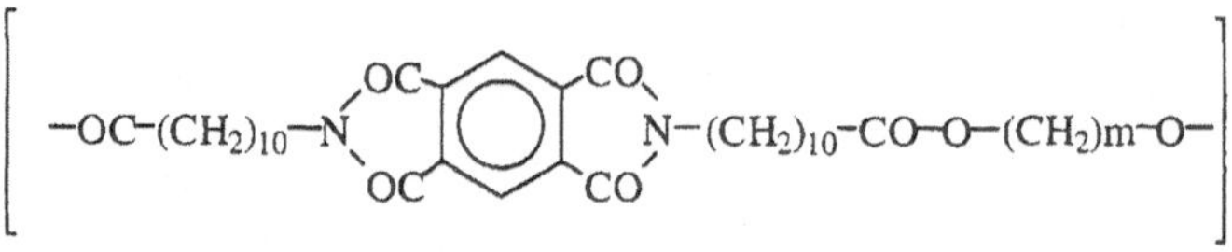

53 a - e   **a** : m = 6   **c** : m = 9   **e** : m = 12
                      b : m = 8   **d** : m = 10

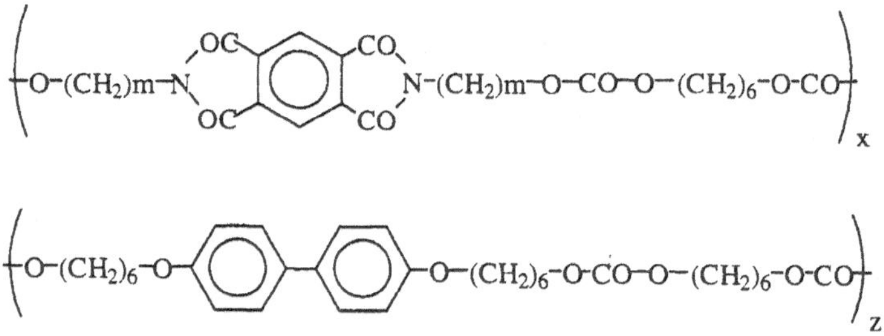

54 a - f   m = 3, 4, 5, 6

a : x / z = 10 / 0   **d** : x / z = 5 / 5
b : x / z = 8 / 2   **e** : x / z = 4 / 6
c : x / z = 4 / 6   **f** : x / z = 2 / 8

Structure 7 (2)

the biphenyl moiety. The absence of an LC-phase in the case of **54a–c** supports the aforementioned conclusion of the author, that the PMDI unit is not mesogenic unless strong DA-interactions with an electron rich comonomer are operating. Further support for this conclusion was presented by a study of the poly(carbonate-urethane)s **56** [68]. The ratio of carbonate and urethane groups was varied over a wide range, but an LC-phase was never observed. More recently Sun and Chang [69] reinvestigated poly(carbonate imide)s of the structure **57a, b,** part of which is identical with **54a** (m=3–6).

Again no LC-phase was found in agreement with the results of Sato and cow-orkers. However, for the copolycarbonates derived from various diphenols 58a–e Sun and Chang reported the existence of nematic LC-phases [69]. At least after annealing two endotherms were found in the DSC heating traces and, on the basis of microscopic observation, interpreted as T_m and T_i. A further interesting finding is the relatively broad range of the nematic phase in the case of resorcinol (58a, m=2, 3). The hypothetical explanation assumes that the meta functions of resorcinol and the bond angles of the carbonate groups compensate each other, so that an overall linear conformation of the main chain is favoured.

Several PEIs containing oligoethyleneglycol spacers were also reported. One sample (59a) was built up from diethyleneglycol, whereas all other members of this series (59b) were derived from commercial mixtures of oligo(ethyleneglycol)s with a moderate polydispersity [70]. Regardless of the lengths of the spacers, none of these PEIs was thermotropic. A quite different situation represent the PEIs 60a–d prepared by polycondensation of N,N'-dihydroxy-PMDI and various aliphatic dicarboxylic acid dichlorides (in organic solution in the presence of NEt_3) [71]. The DSC traces of these PEIs display one melting and one isotropization endotherm. Optical microscopy with crossed polarizers confirmed the existence of a mobile LC-phase between both endothermes. However, the absence of X-ray measurements and the published Schlieren texture do not allow a clearcut decision, if the LC-phase has a nematic or smetic-C order [71]. Taken together, the results discussed above demonstrate that the PMDI moiety is a poor mesogen which requires a special neighbourhood to form an LC-phase.

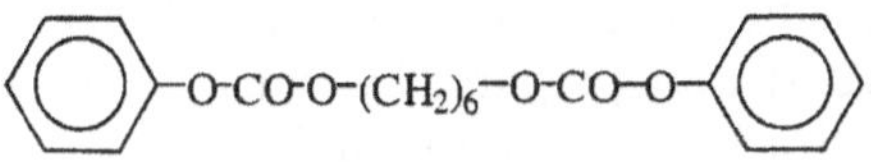

55

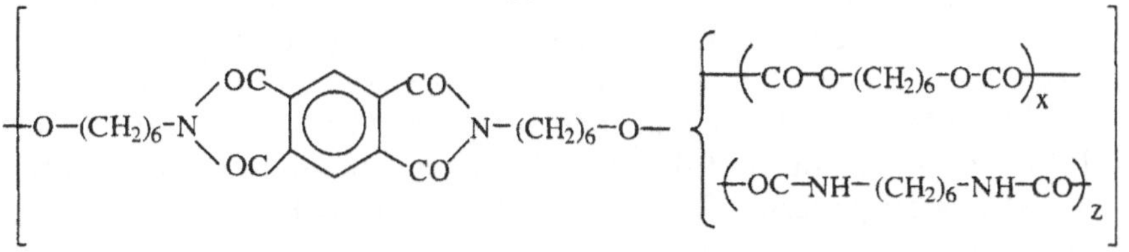

56

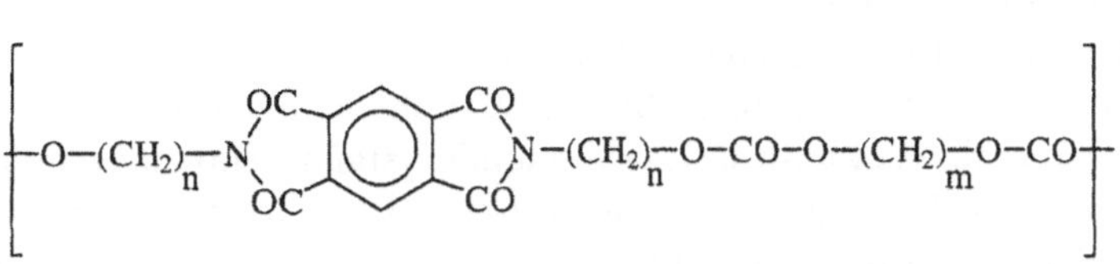

57 a - b

a : n = 2,   m = 2 - 4
b : n = 3,   m = 3 - 10

Structure 8 (1)

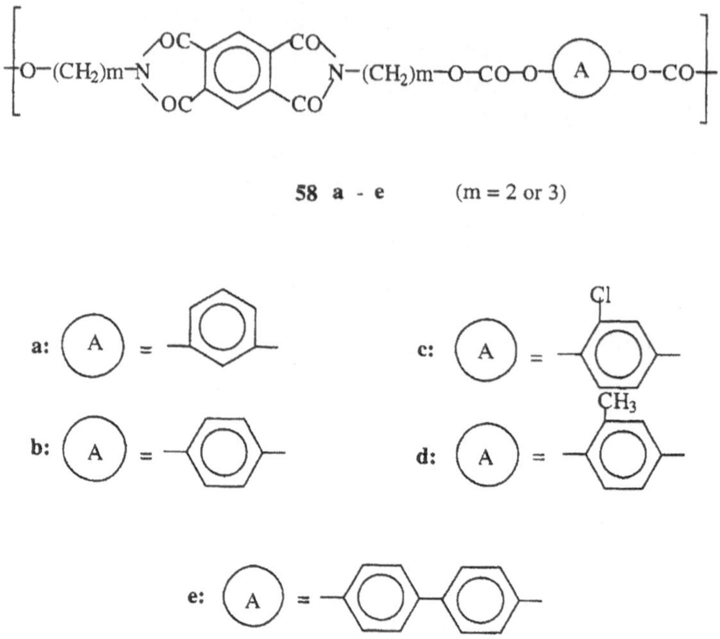

58 a - e   (m = 2 or 3)

Structure 8 (2)

Somewhat heterogeneous results were also reported for PEIs derived from bi-phenyl-3,3',4,4'-tetracarboxylic imide (BiTCI). Three different classes of PEIs based on BiTCI were studied by Sato and coworkers [68, 72, 73]. The homo- and co PEIs of structure **61a–f** were prepared by transesterification of the dimethyl esters of the aliphatic dicarboxylic acids with the corresponding diols catalyzed by ZnAc₂ [72]. Interestingly, all three homopolyesters (**61a–c**) and one co PEI (**61d**) proved to be crystalline, forming an isotropic melt. This means that the mesogeneity of the BiTCI moiety is almost as low as that of the PMDI unit. For the co PEIs **61e** and **61f** a narrow nematic phase (T=20–25 °C) was reported [72].However, the textures of the alleged nematic melt were not reported, and the WAXS powder patterns of the melt still show weak reflections, indicating the presence of an ordered solid state. Therefore, the identification of a nematic phase for **61e–f** is quite doubtful.

The polycarbonates **62a–f** and **63a–f** were prepared by transesterification of the corresponding diols with the biscarbonate **55** in the molten state. At least af-ter annealing all members of both series were semicrystalline showing a melting endotherm in the DSC heating traces. Unfortunately, DSC curves of **62a–f** were not published. An unidentified birefringent mesophase was mentioned for **62a–e**, whereas a nematic melt was reported for **62f**. However, the published texture of this melt is not a typical nematic Schlieren texture, but looks like a suspension of solid particles in an isotropic melt. For all members of series **63a–f** a nematic melt was postulated over a temperature range of 10–30 °C. No textures were published, but several endotherms in the DCS heating traces and the WAXS powder patterns recorded at variable temperatures support the identification of the LC-phase. The postulated LC-character of 62f or 63a–f, and the absence of the LC-phase in the case of 62a–e is somewhat inconsistent and unsatisfactory. In the case of the poly(carbonate-urethane)s 64a–f a nematic melt was described for 64b, c, along with an unidentified birefringent mesophase for 16a. For all other members of this series no LC-phase was detected. However, the published texture of 64b looks like a suspension of solid particles in an isotropic melt, and the WAXS powder patterns exhibit reflection indicating the presence of more or less crystalline particles in the melt. Therefore, the identification of a nematic melt in the case of 64b, c is again quite doubtful.

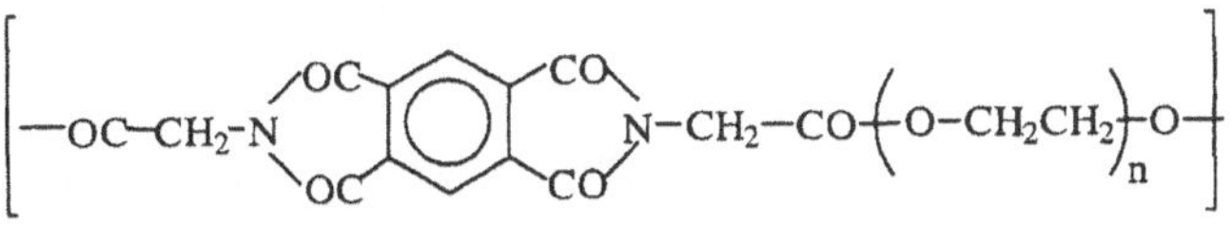

59 a - b   **a:** $n = 2$ **b:** $n = 7.$,   various samples with moderate mol. weight distribution

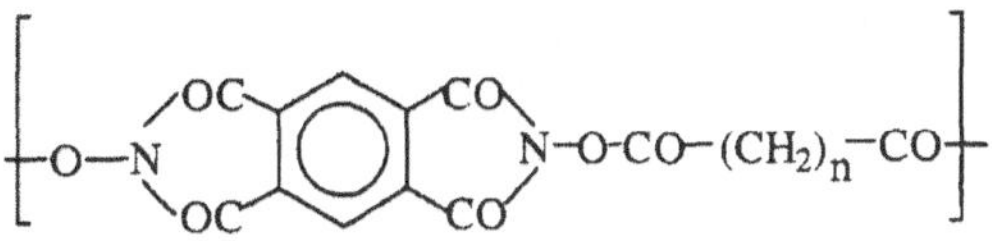

60 a - d   **a :** $n = 3$ **c :** $n = 7$
                b : $n = 4$ **d :** $n = 8$

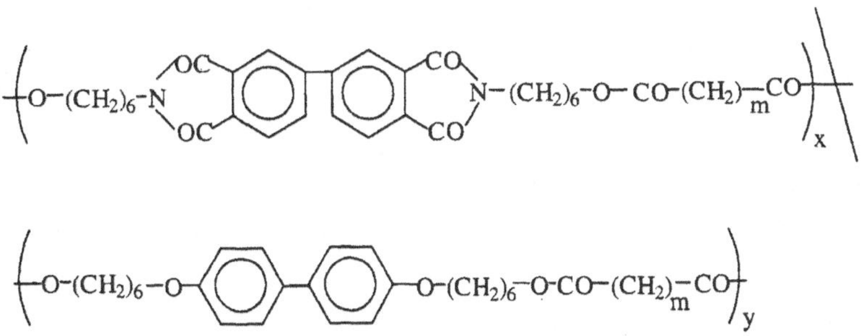

61 a - f   **a :** $x/y = 10/0$, $m = 6$ **d :** $x/y = 5/5$, $m = 6$
               b : $x/y = 10/0$, $m = 8$ **e :** $x/y = 5/5$, $m = 8$
               c : $x/y = 10/0$, $m = 18$ **f :** $x/y = 5/5$, $m = 18$

Structure 9 (1)

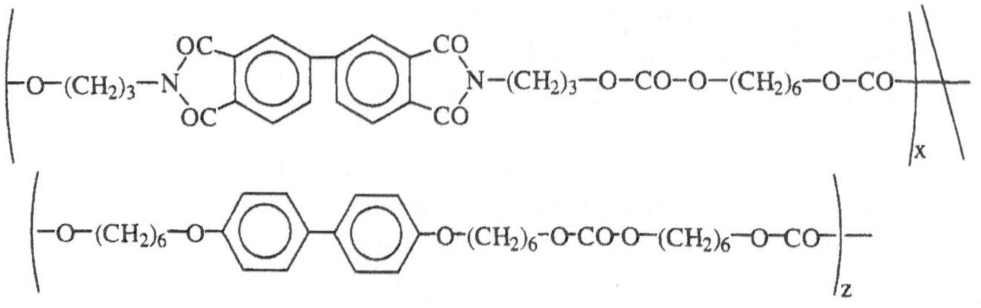

62a - f

a: x/z = 10/0   **c:** x/z = 6/4   **e:** x/z = 4/6
b: x/z = 8/2   **d:** x/z = 5/5   **f:** x/z = 2/8

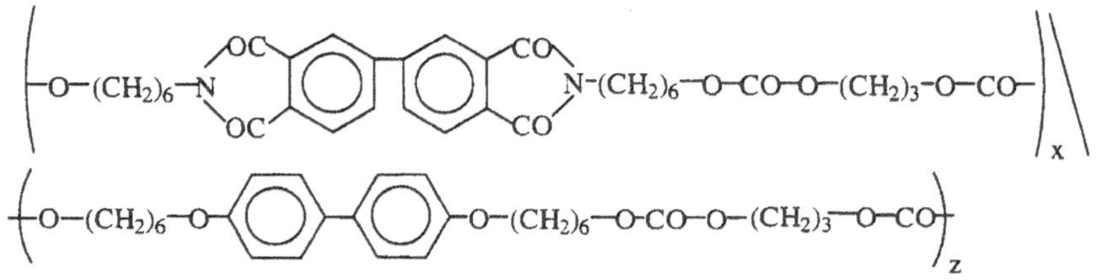

63a - f   (like <u>14</u>)

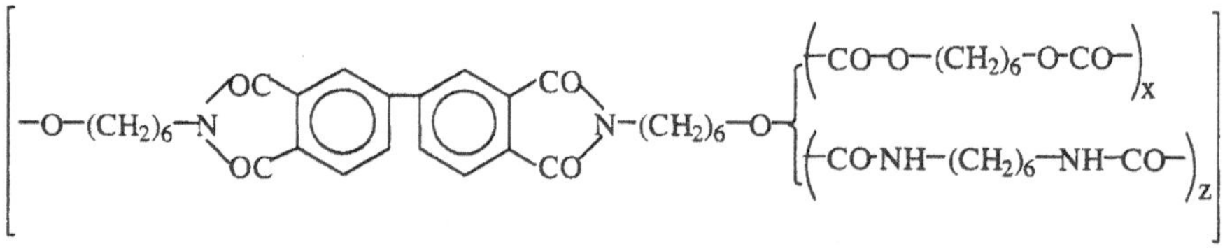

64a - f

a: x/z = 10/0   **c:** x/z = 6/4   **e:** x/z = 2/8
b: x/z = 8/2   **d:** x/z = 4/6   **f:** x/z = 0/10

Structure 9 (1)

When the length of the mesogen is further extended, as in the case of terphenyl 3,3', 4,4'-tetracarboxylic imides (TTCI), the mesogeneity does significantly
improve. Therefore, it is obvious that the PEIs of structure **65** form LC-phases,
which were unambiguously identified as nematic[74]. Furthermore, imide moie-

ties based on condensed aromatic systems, such as naphthalene 1, 4, 5, 8,-tet-racarboxylic acid or perylene-3, 4, 9, 10-tetracarboxylic acid, were used for the preparation of semirigid PEIs (**66a–i** [75] and **67a–h** [76]). Both classes of PEIs have in common that no LC-phase was detectable. Furthermore, both series of PEIs adopt a crystalline layer structure in the solid state. Penetration measure-ments demonstrated that this smectic crystalline state is as hard as any normal crystalline phase, and a penetration only occured at the T_m. These mechanical results together with the WAXS powder patterns indicate that these solid smectic phases are a kind of crystal modifications and not a highly viscous mesophase. Similar results were reported from a study of PEIs derived from benzophenone 3,3',4,4'-tetracarboxylic imide (BzTCI) **68a–f** and **69a–f** [63]. All members of these three series were smectic crystalline in the solid state and did not form an LC-phase. These findings indicate that the BiTCI moiety, like the napthalene and perylene units of the PEIs **66a–i** and **67a–h**, are very poor mesogens. This con-clusion was confirmed by the properties of the crystalline model compounds **70a–g** which did not form LC-phases. However, when 4,4'-dihydroxybiphenyl was used as comonomer, PEIs (**71a–f**) were obtained which obviously yielded a smectic LC-phase in a narrow temperature range [63]. Two endotherms were present in the first and second heating trace, and the WAXS powder patterns re-corded with synchrotron radiation proved the existence of a layer structure up to the T_i. However, the texture was not clearly identified, and it is difficult to dis-tinguish in a narrow temperature range a suspension of solid particles in a vis-cous isotropic melt from a poorly developed schlieren texture of a smectic-C phase.

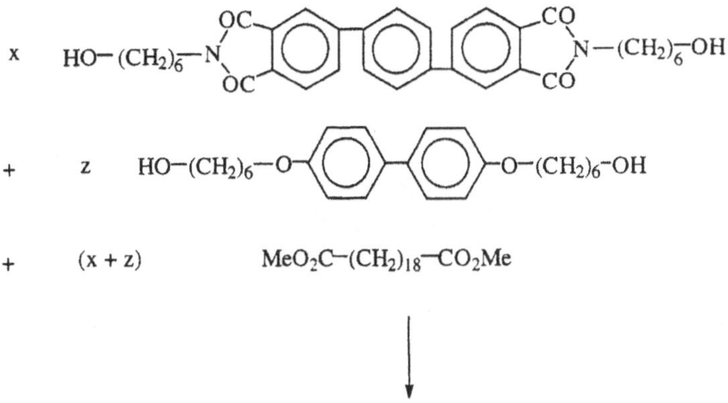

65a - g

a x/z = 10/0
b x/z = 8/2   d x/z = 5/5   f x/z = 2/8
c x/z = 6/4   e x/z = 4/6   g x/z = 0/10

Structure 10 (1)

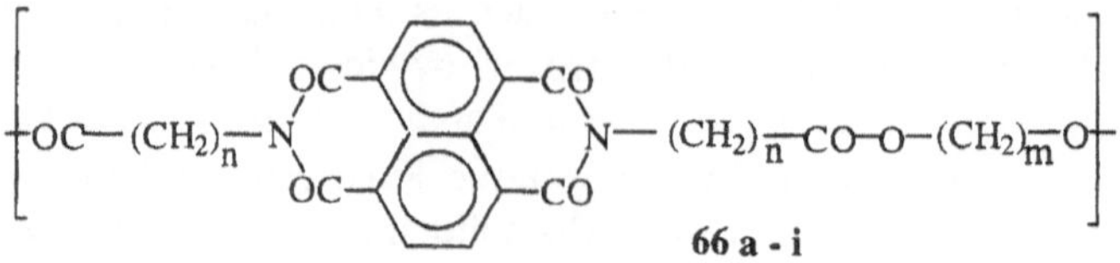

66 a - i

a : m = 3	d : m = 6	g : m = 9
b : m = 4	e : m = 7	h : m = 10
c : m = 5	f : m = 8	i : m = 12

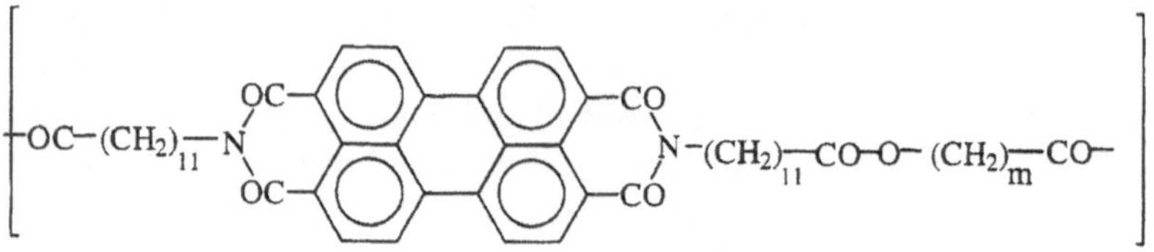

67 a - h

a : m = 4	d : m = 7	g : m = 10
b : m = 5	e : m = 8	h : m = 12
c : m = 6	f : m = 9	

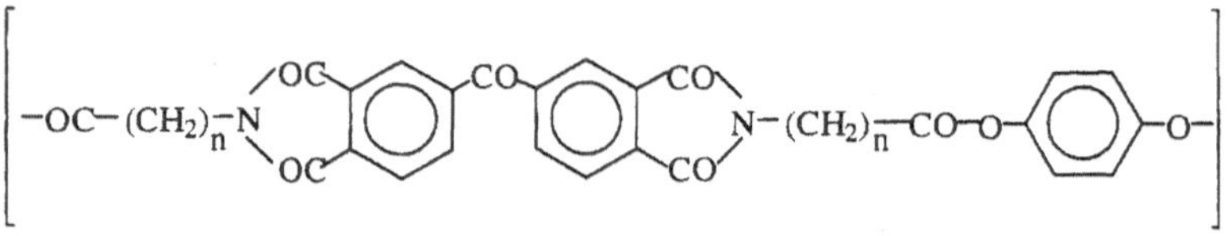

68a - f

a: n = 3	c: n = 5	e: n = 10
b: n = 4	d: n = 6	f: n = 11

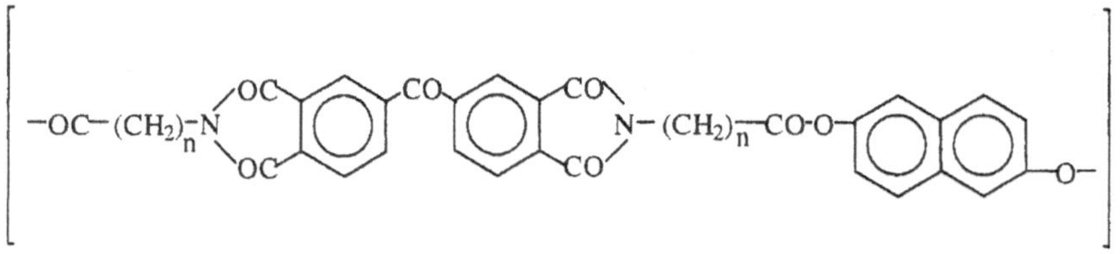

69a - f   (like 68)

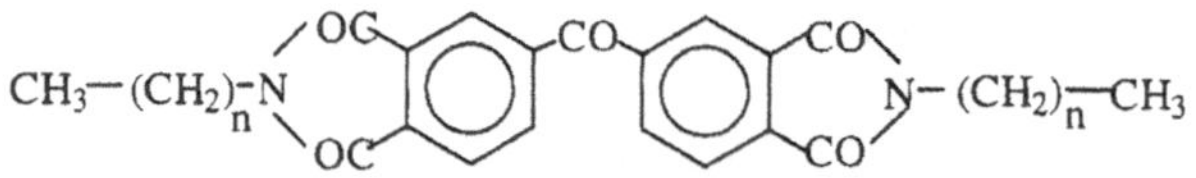

70a - g

a: n = 3	d: n = 9	f: n = 13
b: n = 5	e: n = 11	g: n = 15
c: n = 7		

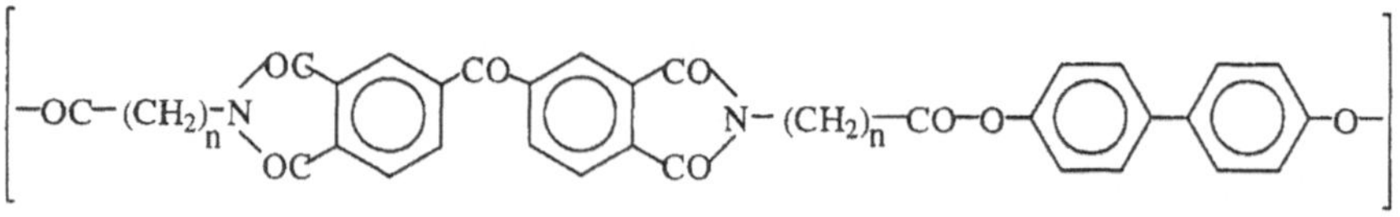

71a - f   (like 68)

Structure 10 (2)

A class of semirigid copolycarbonates containing the BiTCI unit (**72a–f**) was more recently studied by Sato and coworkers [77]. For all but one copolymer (**72e**) the existence of a birefringent mesophase was mentioned. Since birefringence may also result from a suspension of solid particles in an isotropic melt or from mechanical stress, this characterization does not provide any useful information on the formation of a true LC-phase. In the case of **72e** a nematic phase was postulated. However, the texture was not published and the WAXS powder patterns measured at elevated temperatures exhibit several reflections in contradiction to the existence of a nematic melt. In other words this study of Sato and coworkers does not give any clearcut information about the mesogeneity of the BiTCI moiety. Taken together, it may be concluded that the short symmetrical imide moieties are almost useless as mesogens.

4.2
Long Symmetrical Imide Mesogens

It is obvious that imide mesogens with high length/diameter ratio (e.g. **2a, b** or **7b**) may be good mesogens, but a practical problem is the high melting temperatures of the monomers themselves and of the PEIs derived from them. Nonetheless, when the bispropionate of the diphenol **7b** was polycondensed with

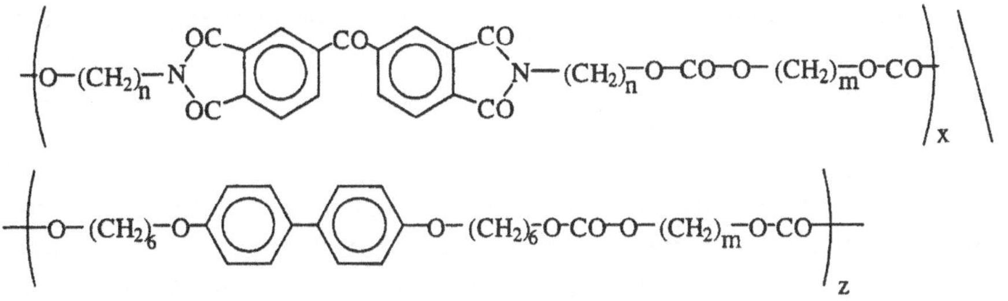

72 a - f

a:	$n = 3, m = 3$	**d:**	$n = 3, m = 3$
	$x = 1, z = 0$		$x = 0,5, z = 0,5$
b:	$n = 6, m = 3$	**e:**	$n = 6, m = 3$
	$x = 1, z = 0$		$x = 0,5, z = 0,5$
c:	$n = 6, m = 6$	**f:**	$n = 6, m = 6$
	$x = 1, z = 0$		$x = 0,5, z = 0,5$

Structure 11

aliphatic dicarboxylic acids, homogeneous melts were obtained below 300 °C and the PEIs **73a–c** were prepared with satisfactory yields and viscosity values [78]. Likewise successful proved polycondensations of the diethylester of **2a** with various aliphatic diols in the presence of transesterification catalysts. In this way the PEIs **74a–f** were synthesized [8]. Furthermore, the diphenol **7b** was alkylate with 11-bromoundecanol and the resulting diol **75** was polycondensed with various aliphatic dicarboxylic acid dichlorides in 1-chloronaphthalene. This procedure yielded the PEIs **76a–d** with extremely long spacers. Analogously, the dimethyl esters **77** was synthesized by alkylation of the diphenol **7b** with methyl bromoundecanoate. Its polycondensation with various α,ω-diols yielded the PEIs **78a–d** [80]. An interesting aspect for comparisons is the fact that the repeat units of **73a–c** and **74a–f** are isomeric like the repeat units of **76a–d** and **78a–d**.

All these PEIs derived from BiTCA proved to form a broad smectic-A type LC-phase and a smectic crystalline solid state (for definition see Sect. 7). The details of their syntheses and properties will be published in the near future [78–80].

Another interesting and versatile group of PEIs containing aliphatic spacers is based on polycondensations of the imide dicarboxylic acids **79a–i** with various acetylated diphenols. These polycondensations were conducted in bulk at temperatures up to 320 °C. The thermostability of the resulting PEIs **80a–i**, **81a–i** or **82a–i** allows these high reaction temperatures provided that oxygen is rigorously excluded. When diphenols such as bisphenol-A, 2,7-dihydroxy naphthalene, resorcinol or substituted hydroquinones were used as comonomers of the dicarboxylic acids **79a–i**, the resulting PEIs were isotropic [81]. When the unsubstituted hydroquinone was incorporated, the resulting PEIs **80a–i** showed the existence of an enantiotropic smectic-A phase over a narrow temperature range (10–15 °C). A broader temperature range of the smectic LC-phase was found with 2,6-dihydroxy naphthalene as building block (**81a–i**) [81]. Since neither hydroquinone nor 2,6-dihydroxy naphthalene are mesogenic moieties per se, the LC-phases of **80a–i** and **81a–i** prove that the mesogenic unit in the PEIs is a combination of the diphenol and both neighbouring trimellitimide units. Depending on the conformation of the ester groups, these mesogenic units contain at least two mirror planes as elements of symmetry.

Even broader LC-phases were observed for the homo PEIs containing 4,4'-dihydroxybiphenyl as component of the mesogenic unit (**82a–i**) [82]. The phase diagram of these PEIs is given in Fig. 12. The conspicious odd-even effect of the melting temperatures (T_ms) is a consequence of a different chain packing in the solid state. All PEIs of structure **80–84** and **86–88** form a smectic crystalline state (see also Sect. 7). In the case of the PEIs **80, 81** and **82** those PEIs containing spaces with an even number of CH_2 groups the mesogens are in upright position (smectic-E like), whereas the mesogens are tilted relative to the layer planes when the spacers are odd-numbered (smectic-H like) [24]. This latter mode of chain packing is energetically less favourable and results in lower T_ms. However, all PEIs of structure **80, 81** and **82** form a smectic-A phase immediately before their isotropization, and thus the T_is, do not show an odd-even effect [82].

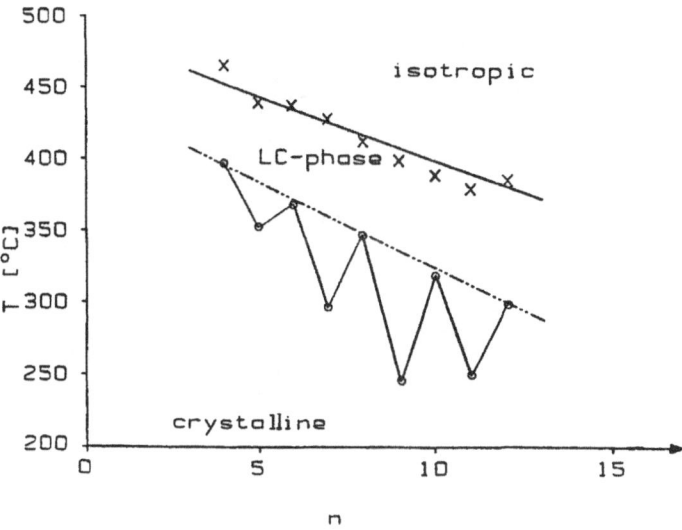

Fig. 12. Phase diagram of the PEIs **82a–i**: temperatures of phase transitions vs the number (n) of CH$_2$ groups in the spacer

Detailed X-ray studies of the PEIs **82a–i** and **83a–f** with synchrotron radiation up to temperatures around 400 °C revealed that the phase behaviour of these polyesters is more complex. Upon melting of the PEIs with even numbered spacers, a shrinkage of the layer distances takes place according to a smectic-E smectic-C transition [24]. In the case of odd-numbered spacers such a shrinkage does not occur, and a smectic-C phase is formed from the smectic-H structure without a significant change of the tilt angle. Upon heating of the smectic melt the layer distances increase and finally all homo PEIs **80a–i**, **81a–i** and **82a–i** form a smectic-A phase prior to their isotropization. This interpretation of the X-ray data (i.e. calculation of the layer distances from the middle angle reflections via the Bragg equation) was confirmed by optical microscopy. Upon slow cooling from the isotropic melt the typical bâtonet texture of a smectic-A phase becomes detectable (Fig. 13A.) [24] which gradually changes to a fan-shaped texture upon cooling (Fig. 13B). In agreement with this interpretation three endotherms are detectable in the DSC-heating trace of **82i** (or **82g** and **83b**); namely the smectic-E→C transition in the smectic-C→A transition and the isotropization. The corresponding three exotherms appear in the cooling traces [29].

In addition to the homopolyesters **80a–i**, **81a–i** and **82a–i**, three classes of co PEIs were studied: **83a–g** [24], **84a–d** and **85a–f** [84]. The combination of two different alkane spacers did not significantly change the properties of the co PEIs **83a–d**, when compared to **82i**. However, when the difference on their length increased, a destabilization of the smectic layer-structures became evident, with the consequence that a nematic phase was formed on top of the smectic-C phase

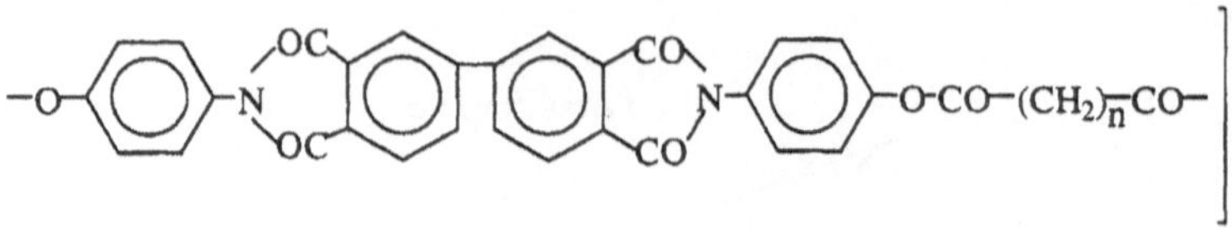

73 a - c

a : n = 10   b : n = 14   c : n = 20

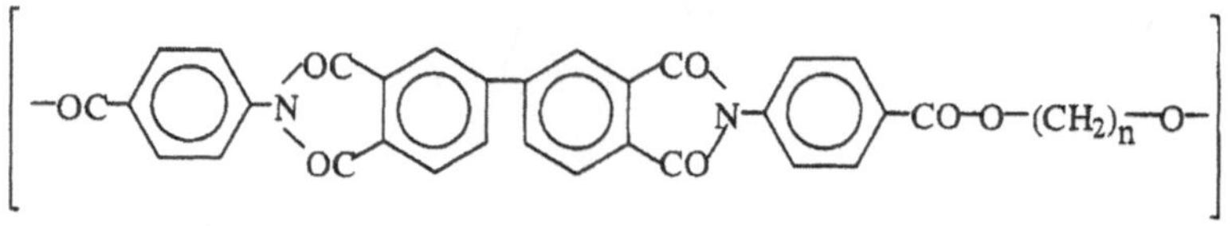

74 a - f

a : n = 7   c : n = 9   e : n = 12
b : n = 8   d : n = 10   f : n = 16

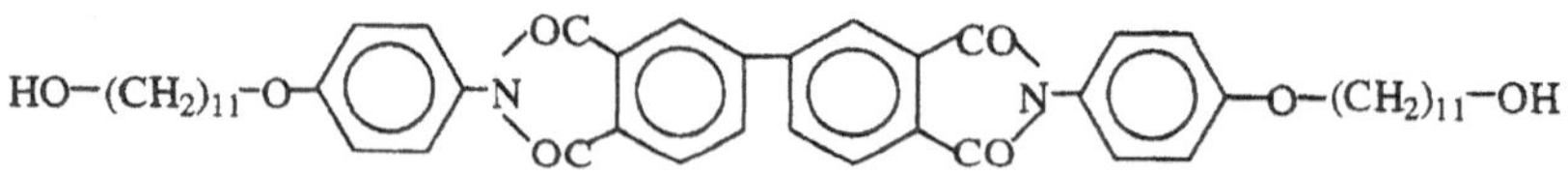

75

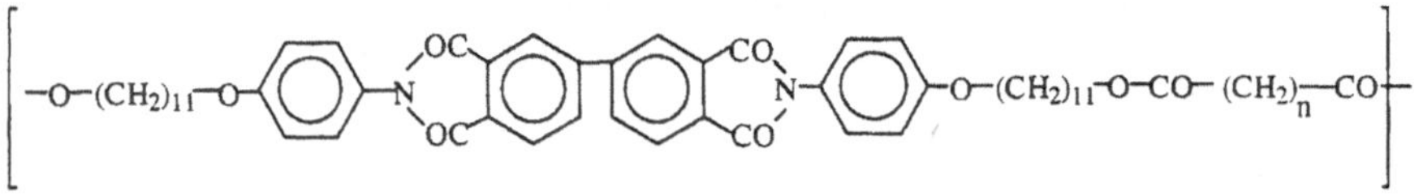

76 a - d   a : n = 4   d : n = 12
                       b : n = 8   e : n = 20

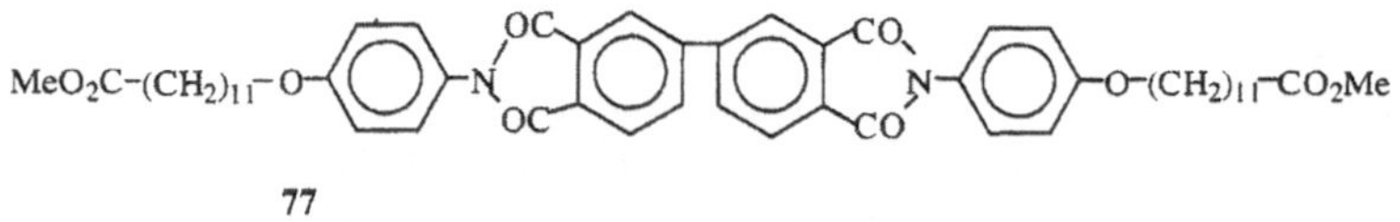

77

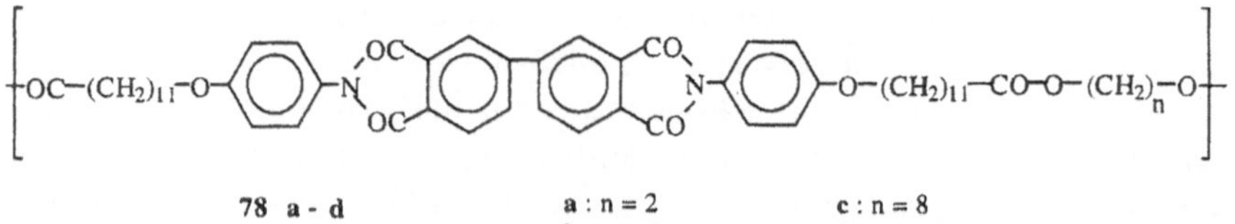

78 a - d   a : n = 2   c : n = 8
                       b : n = 4   d : n = 12

Structure 12

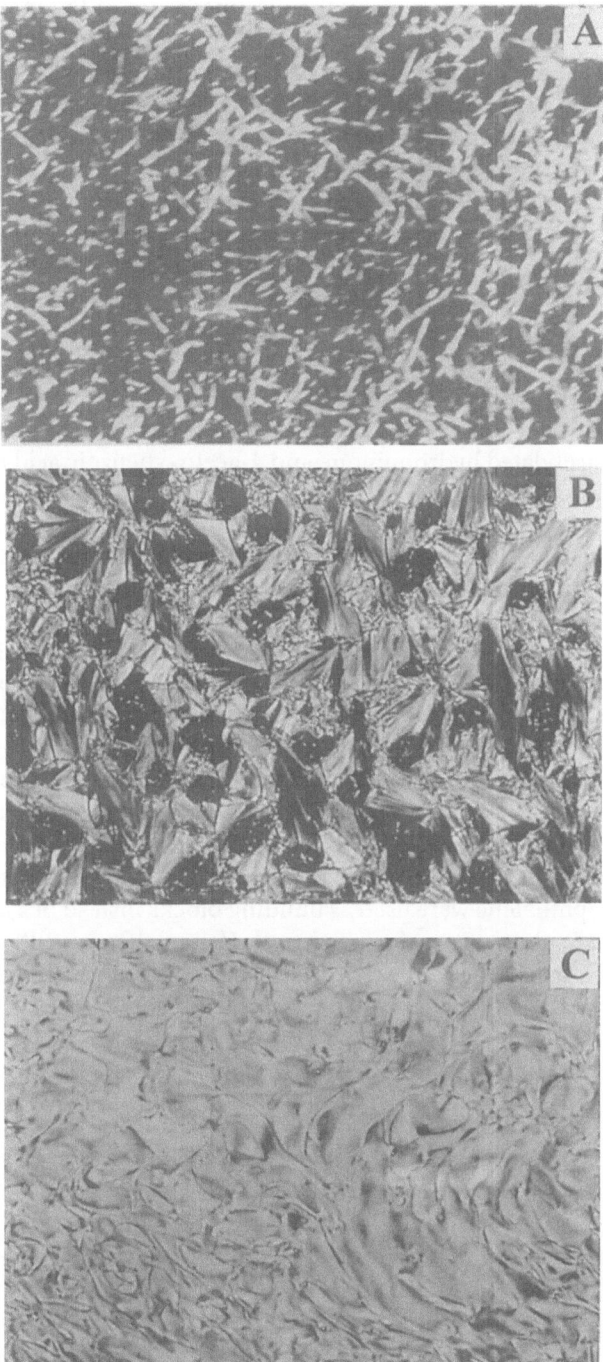

Fig. 13. A Bâtonet texture of the smectic-A phase of PEI **82i** upon cooling from the isotropic melt. **B** Smectic texture of **83a**. **C** Nematic texture of co PEI **83g**

(83e–g). This change was proven by the disappearance of the middle angle reflections and by the typical nematic schlieren texture (Fig. 13C). However, even the co PEIs 83e–g maintained a layer structure in the solid state [24]. Taken together, the homo PEIs 82a–i and the co PEIs 83a–g can form two different LC-phases just by variation of the temperature, a property which is rather rare among LC-main chain polymers [24].

The properties of the co PEIs 84a–d agree largely with those of their parent homo PEIs. They form a smectic solid phase and a smectic LC-phase. A differentiation between smectic-A and smectic-C has not been reported yet [83]. The co PEIs 83a–f and 84a have in common that when fibres were drawn from the melt the chain axes may be arranged parallel to the fibre axis as usual but also in a perpendicular array (Fig. 14). A detailed discussion of the reasons underlying this unusual behavior is presented in [83].

The co PEIs 85a–f were prepared by copolycondensation of the dicarboxylic acid 79i with acetylated hydroquinone and 4-acetoxybenzoic acid in bulk in the presence of transesterification catalyst, so that random sequences should have been formed [84]. The increasing molar fraction of 4-oxybenzoic acid units destabilizes the layer structure of the homo PEI 80i in the solid state and in the melt. The WAXD powder patterns (Fig. 15) demonstrate that the middle angle reflection MARs (1st through 5th order in the range of $\vartheta=0$–$6°$) gradually fade away, whereas a hexagonal lateral packing of the chains is maintained. A schematic illustration of this change from a solid smectic-E phase to a columnar mesophase is given in Fig. 16. For the molten state a change from the narrow smectic LC-phase of 80i to a broad nematic phase was observed [84].

A further modification of the PEI 82a–i consisted of the incorporation of ether groups into the aliphatic spacers [85]. Again, the PEIs 86a, b form a smectic crystalline state and a smectic A phase, but the temperature range of the smectic LC-phase is smaller than in the case of 80f–i. When hydroquinone or 2,6-hydroxynaphthalene were used as building blocks instead of 4,4'-dihydroxybiphenyl, no LC-phases were formed at all. Hence, these results indicate that ether groups in the aliphatic spacers of PEIs destabilize the LC-phase. A further confirmation of this hypothesis will be presented below.

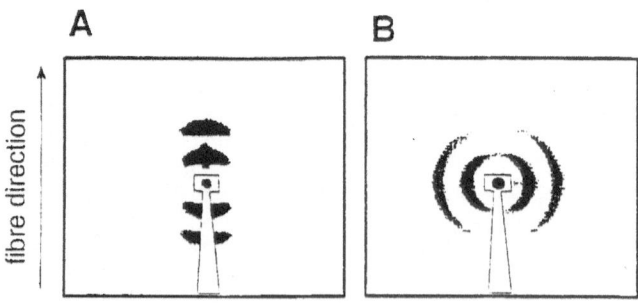

Fig. 14. X-ray fibre patterns of: **A** PEI 83f; **B** PEI 83g

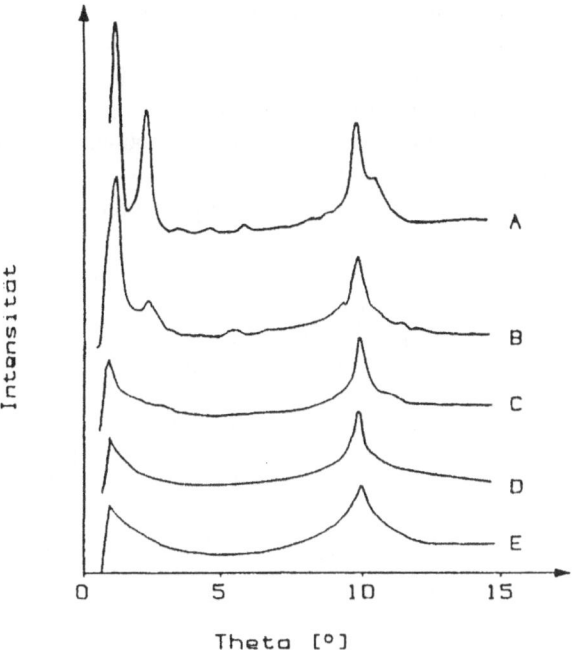

Fig. 15. WAXD powder patterns of: **A** PEI 80i; **B** PET 85a; **C** PEI 85b; **D** PEI 85c; **E** PEI 85d

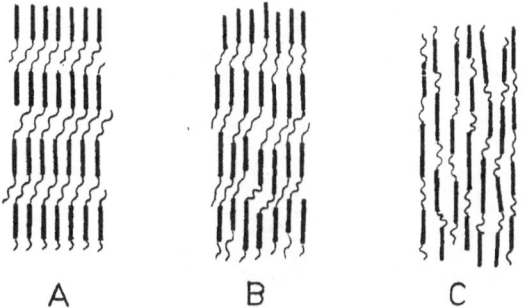

Fig. 16. Schematic illustration of the gradual change from a smectic layer structure to a co-lumnar mesophase (in the solid state) or nematic phase (in the melt) as discussed for the PEIs **80i** and **85a–d**

Another interesting structural variation of the PEIs **81a–i** and **82a–i** was obtained by using methyl substituted diamines as spacers (PEIs **87a, b** and **88a, b**) [86]. Again, a destabilization of the layer structures was found. The PEIs **86a** and **86b** still formed as smectic crystalline state, but the melt of **86a** was nematic and that of **86b** a combination of smectic and nematic. The solid state of **88a, b** turned out to be a smectic glass or in other words a frozen smectic-A phase. No

crystallization occurred upon annealing. Obviously the three methyl groups of
the spacers provided sufficient steric hindrance for the formation of a higher ordered solid state. Upon heating above the T_g the PEIs **88a, b** yielded a mobile
smectic-A phase which gradually changed to a nematic phase with increasing
temperature. Taken together, the PEIs of structures **80–88** proved to be an interesting family of LC-main chain polymers which allowed a broad study of the relationships between chemical structure and supermolecular order.

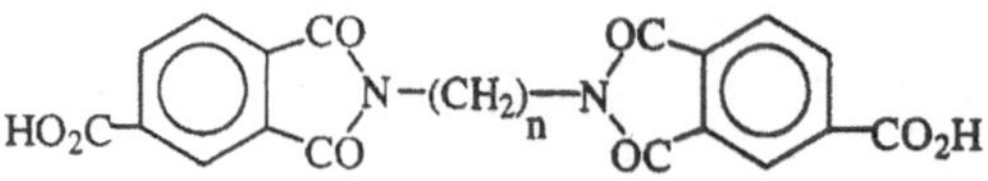

79 a - i

a : n = 4	d : n = 7	g : n = 10
b : n = 5	e : n = 8	h : n = 11
c : n = 6	f : n = 9	i : n = 12

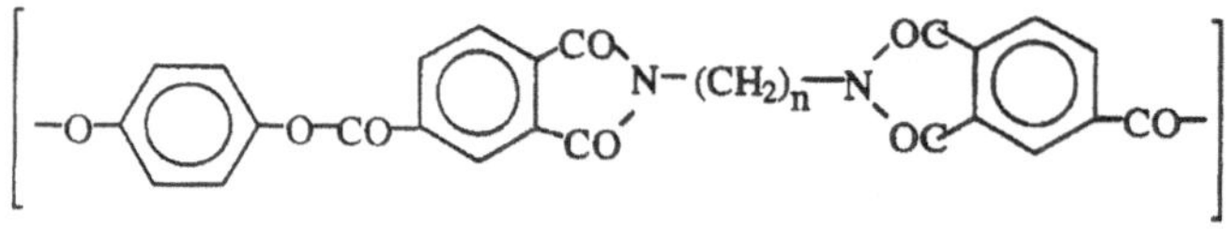

80a - i

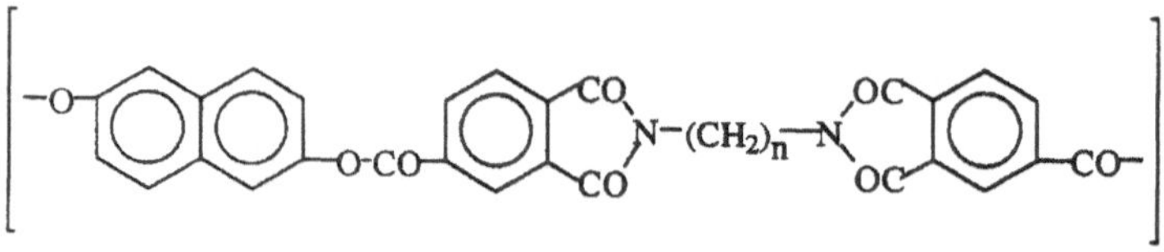

81a - i

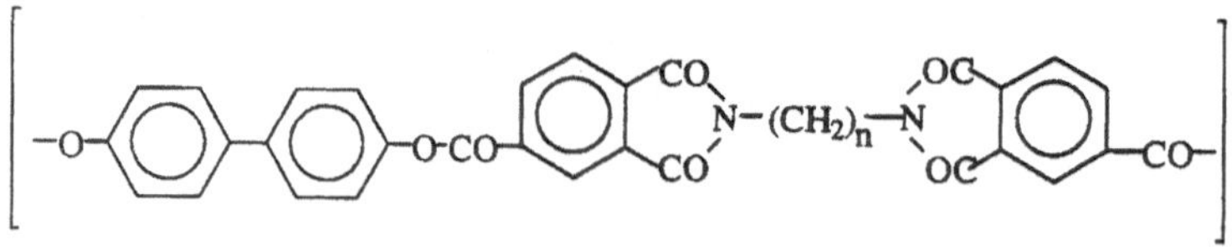

82a - i

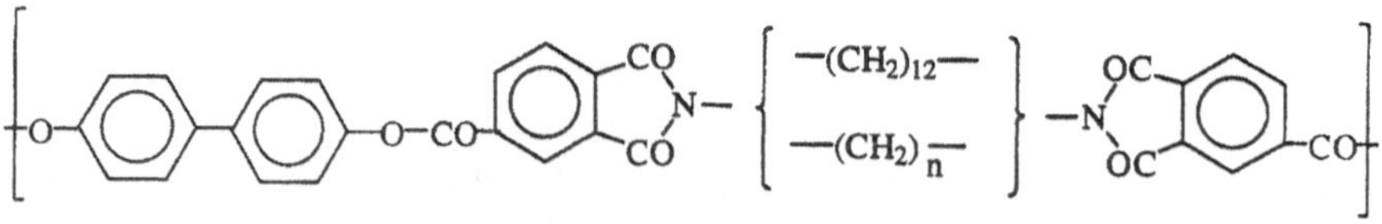

83a - g

a : n = 11	c : n = 9	e : n = 7
b : n = 10	d : n = 8	f : n = 6
		g : n = 5

Structure 13 (1)

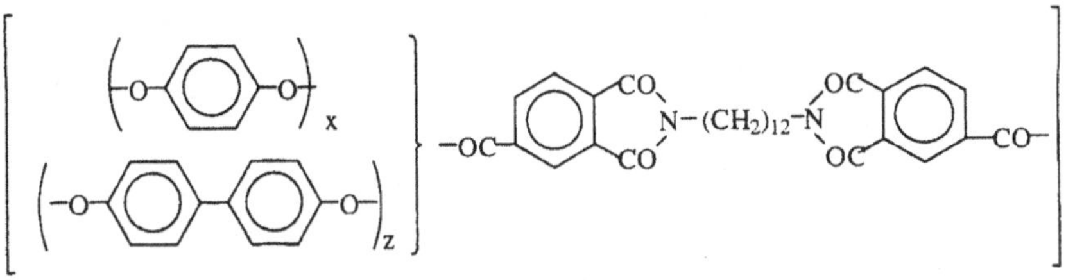

84a - d

a : x/z = 80/20   c : x/z = 40/60
b : x/z = 60/40   d : x/z = 20/80

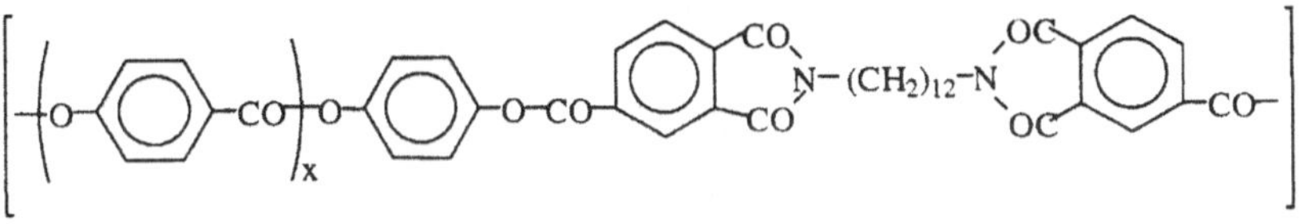

85a - f

a: x = 1   c: x = 3   e: x = 5
b: x = 2   d: x = 4   f: x = 6

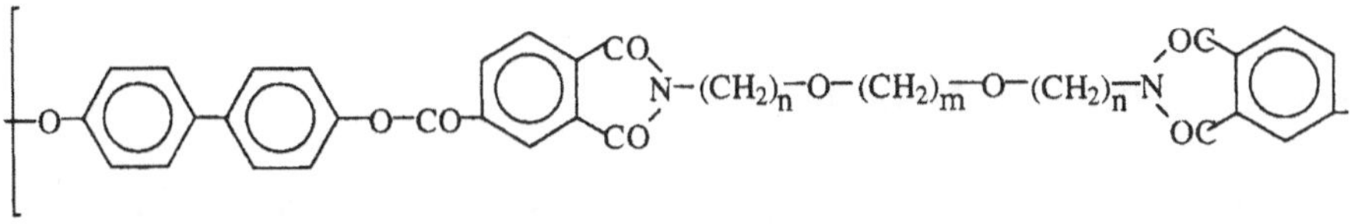

86a , b   a: n = 3, m = 3   b: n = 4, m = 3

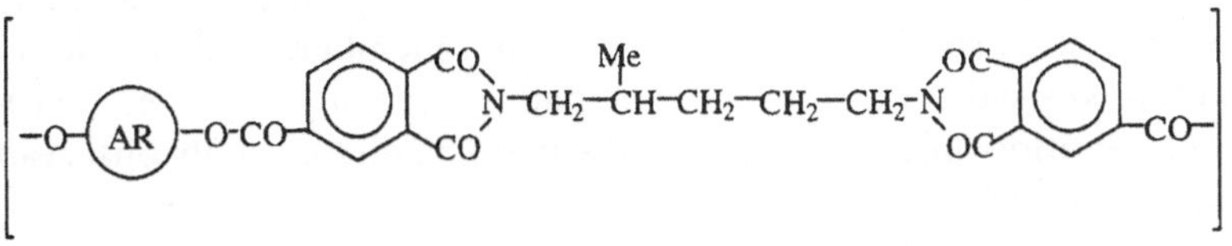

87a, b

Structure 13 (2)

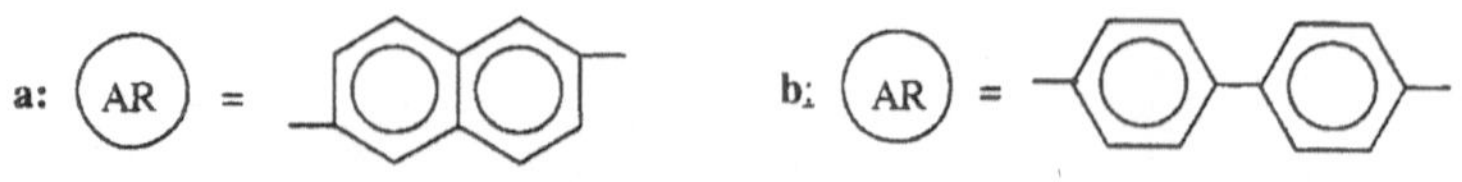

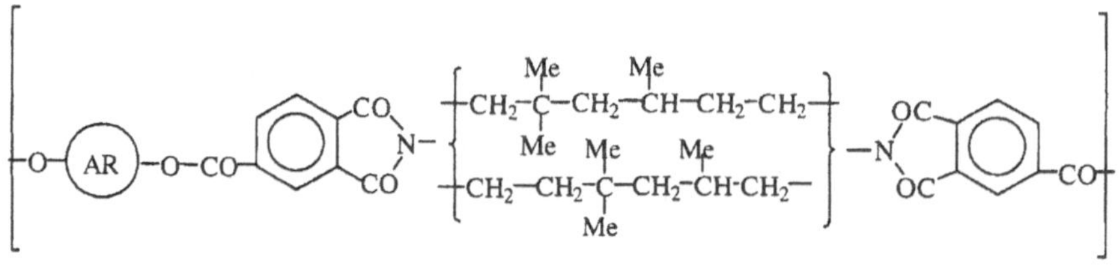

a: $\left(AR\right)$ = [naphthalene structure]   b: $\left(AR\right)$ = [biphenyl structure]

88a, b

a: $\left(AR\right)$ = [naphthalene structure]   b: $\left(AR\right)$ = [biphenyl structure]

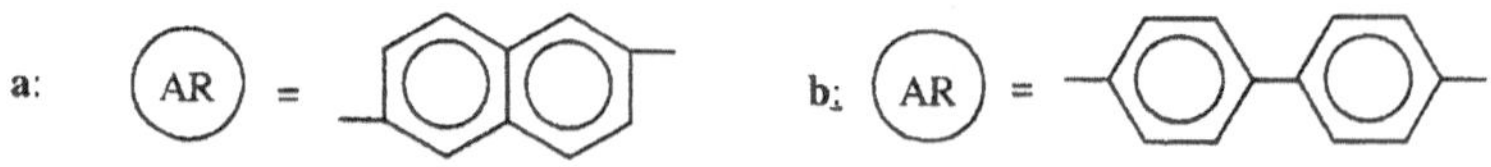

Structure 13 (3)

4.3
Non-Symmetrical Imide Mesogens

All non-symmetrical imide mesogens discussed in this section may be summarized under the label 4-substituted N-phenyl phthalimides. The first class of
such PEIs reported in the literature [87] were the polyesters **89a–m**. These PEIs
were either prepared by polycondensation of the N-(4-acetoxyphenyl) 4-acetox-
yphthalimide with aliphatic dicarboxylic acids or by polycondensation of the silylated diphenol with dicarboxylic acid dichlorides. Both methods gave nearly
identical results. Although, N-(4-hydroxyphenyl) 4-hydroxyphthalimide is a relatively short mesogen of low symmetry it proved to be an unexpectedly good
mesogen. All the PEIs **89a–m** have in common that they form a smectic crystalline phase in the solid state. Above their T_m the first members of this series (up
to 12 CH_2 groups) yield an enantiotropic nematic melt. Longer spacers (**89l, m**)
eliminate the enantiotropic character and the nematic phase is only observable
upon rapid cooling [88]. The phase diagram of Fig. 17 illustrates this trend. It is
quite normal for LC main chain polymers that longer spacers reduce the temperature range of the LC-phase, but this trend was not detectable for the PEIs **81a–
i** or **82a–i**. The properties of the copoly(ester-imide)s **90a, b** are discussed in
Sect. 7.

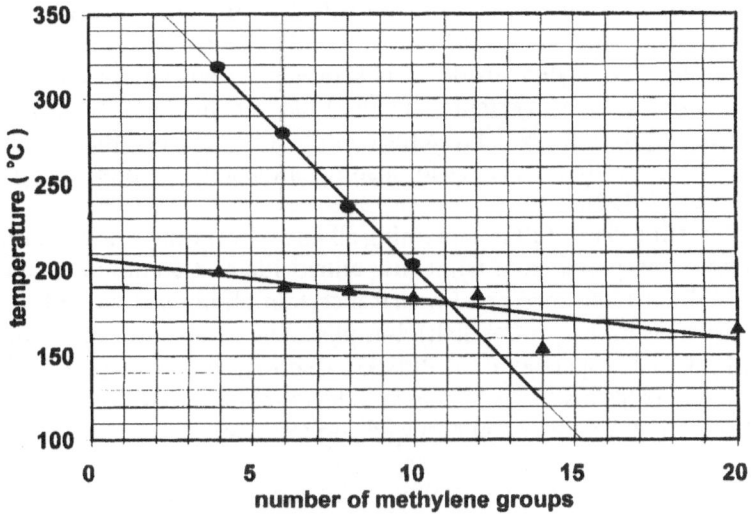

Fig. 17. Plot of the melting temperature (T_m) and of the isotropization temperatures (T_i) vs the number of CH_2 groups for the PEIs **89a–m** (even numbers only)

The homo PEIs **91a–k** and the co PEIs **92a–c** were at first prepared by Krichel-dorf et al. [89] via the transesterification of the corresponding diols (or equimolar mixtures of two diols) with the dimethylester of N-(4-carboxyphenyl) trimellit imide. Later Aducci et al. [90, 91] prepared the PEIs **91a–h** from the dichloride of N-(4-carboxyphenyl) trimellitimide by polycondensation with the free diols on refluxing 1, 2, 4-trichlorobenzene. Regardless of the spacer length neither the homo PEIs **91a–k** nor the co PEIs **92a–c** form an enantiotropic LC-phase. However, detailed studies of several research groups [89–91] agree in that the PEIs **91a–k** and **92a–c** show a monotropic smectic A-phase upon cooling from the isotropic melt. The lifetime of this smectic-A phase is only of the order of seconds or parts of seconds. The solidification of the short lived smectic-A phase may yield a smectic-A glass, a smectic B or a smectic-E-like crystalline state depending on the cooling rate and on the further thermal treatment [89, 91, 92]. A detailed discussion of layer structures of the chain packing of the crystallization kinetics and of morphological aspects is given in Sect. 7.

The formation of a monotropic smectic-A phase proves that the N-(4-carboxy phenyl) trimellitimide is a good mesogen despite its rather low length/diameter ratio and despite its low degree of symmetry. When the 4-aminobenzoyl unit in formula **91a–k** is replaced by a 3-aminobenzoyl unit, (i.e. when N-(3-carboxy-phenyl) trimellitimide is used as monomer) the resulting PEIs are amorphous and isotropic [90]. The structure of the PEIs **91a–k** was also modified by incorporation of heteroatoms such as sulfur or oxygen into the spacers [92]. The incorporation of sulfide groups (**93a, b** and **94a–c**) did not significantly change the properties. The temperatures of all phase transitions were lowered, but all sulfide group containing PEIs formed a smectic solid state and a monotropic smectic-A phase quite analogous to the parent PEIs **91a–k**. However, the ether groups in the oligoethylene oxide spacers of **95a–c** reduced the stability of the layer structures to such an extent that these PEIs were neither smectic in the solid nor in the liquid state [92]. From the co PEIs it was learned that roughly 50% of the spacers need to be alkanes for a sufficient stabilization of smectic layers (see also Sect. 7) [92].

An interesting class of PEIs containing alkane spacers and a non-symmetrical mesogen was prepared from trimellitic anhydride and 4-aminocinnamic acid (**97a–i**) [93]. This new mesogen imparts a certain photoreactivity into the PEIs (e.g. photocrosslinking in the melt), but photochemical studies have not been reported yet. The properties of the PEIs **97a–i** show a significant odd-even effect. Those members of this series having an odd-numbered spacer form an enantiotropic smectic-C phase and upon rapid cooling a smectic-C glass. In contrast, the PEIs built up by even numbered spacers only showed a monotropic smectic phase, but formed a smectic crystalline solid state. When compared to the PEIs **91a–k** the polyesters of the trimellitimidocinnamic acid differ by the tilted array of the mesogenic units and by a greater stability of the smectic LC-phase.

An even longer mesogenic building block was prepared from 4-nitrophthalic anhydride and ethyl 4 aminobenzoate followed by a nucleophilic substitution of the 4-nitrogroup with ethyl 4-hydroxybenzoate [94]. The resulting diethylester **98** was polycondensed with alkane diols or tetraethylene glycol in the presence of a transesterification catalyst. The PEIs **99a–c** adopted a smectic crystalline solid state, but the melt was isotropic and even a monotropic LC-phase was not detectable. The PEI **100** turned out to be completely amorphous and isotropic, in good agreement with the properties of the PEIs **95a–c** [94]. This result and the properties of the PEIs **86a–b** clearly confirm that ether groups in general and oligo(ethylene-oxide)s in particular are highly unfavourable for the existence of LC-phases in the case of polyimides (see Sect. 7). In this regard polyimides are quite different from other aromatic LC-polyesters.

Starting out from trimellitic anhydride chloride Reineke et al. [95] synthesized a series of PEIs containing a regular sequence of imide, amide and ester bonds (**102a–h**). When compared to the PEIs **91a–h** and **99a–c** the incorporation of 4-aminobenzoic acid significantly improved the mesogenic properties. All members of the series **102a–h** showed an enantiotropic smectic-A phase with bâtonet texture. Yet an odd-even effect was observed for the solid state. A smectic crystalline character was found in the case of even-numbered spacers and a smectic-A glass for the PEIs derived from odd numbered spacers. Unfortunately these poly(ester-amide-imide)s are quite unstable upon heating to temperatures above 250 °C. Due to ester-amide interchange reactions, a scrambling of the regular sequence takes place, and the first heating-cooling cycle is not reproducible.

Non-symmetrical and non-mesogenic dicarboxylic acids can easily be prepared from trimellitic anhydride and ω-amino-acids or their lactams [96]. The polycondensation of these dicarboxylic acids with acetylated 4,4-dihydroxybiphenyl yielded the PEIs **104a–f**. Quite analogous to the PEI **82a–i** the mesogenic units were a result of the polycondensation reaction and not a preformed build-ing block. Due to a random cocondensation at high temperatures in bulk these PEIs contain three different mesogenic units (including **104A, B**). Despite a ran-dom sequence of three different repeating units the PEIs **104a–f** were capable of forming a smectic crystalline solid state with upright mesogens (smectic-E type) and an enantiotropic smectic-A phase. Yet, when compared to the more regulary structured PEIs **82a–i** the T_ms and T_is of **104a–f** were 30–50 °C lower.

An interesting family of PEIs was prepared by Chang and coworkers from *N,N'*-bis (-hydroxyalkyl) PMDI and either *N*-(3-carboxyphenyl) trimellitimide (**105a–e**) or *N*-(4-carboxyphenyl) trimellitimide (**106a–e**) [97, 98]. These polycondensa-tions were performed with diphenylchlorophosphate as condensing agent in pyri-dine. Yet despite a variation of the reaction conditions only low molecular weights were obtained in most cases. The PEIs derived from 3-aminobenzoic acid (**105a–e**) were reported to be amorphous and isotropic [97]. In contrast, the PEIs of 4-aminobenzoic acid (**106a–e**) formed a smectic solid state and in the case of **106c, e** an enantiotropic smectic LC-phase was observed which according to its texture was most likely a smectic-A phase. These observations clearly demonstrate that the *N*-(4-carboxyphenyl) trimellitimide and not the PMDI unit carrys the mes-ogenic properties. The same research group also synthesized three classes of co PEIs with similar structure (**107a–e, 108a–f** and **109a–h**) [98]. The PMDI was par-tially replaced by phenylhydroquinone or other diphenols (**109a–h**). Co PEIs showing broad enantiotropic nematic LC-phases were obtained, when substituted hydroquinones were used as comonomers. With 4,-4'dihydoxydiphenylether or 4,4'-dihydroxydiphenylsulfide isotropic melts were observed. Obviously, the prop-erties of these PEIs were dominated by the fully aromatic ester-imide blocks ac-cording to the properties of the homo PEIs **33a–i**. In other words the co PEIs rep-resent a direct connection between the fully aromatic PEIs described in Sect. 3 and the spacer containing PEIs of this section.

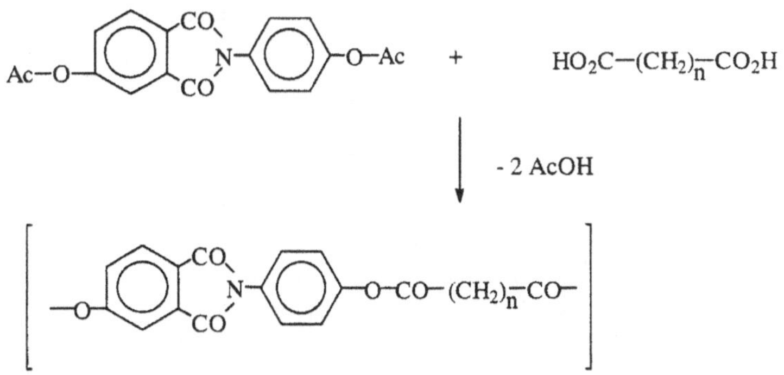

89 a - m	a : n = 3	e : n = 7	i : n = 11
	b : n = 4	f : n = 8	k : n = 12
	c : n = 5	g : n = 9	l : n = 14
	d : n = 6	h : n = 10	m : n = 20

Structure 14 (1)

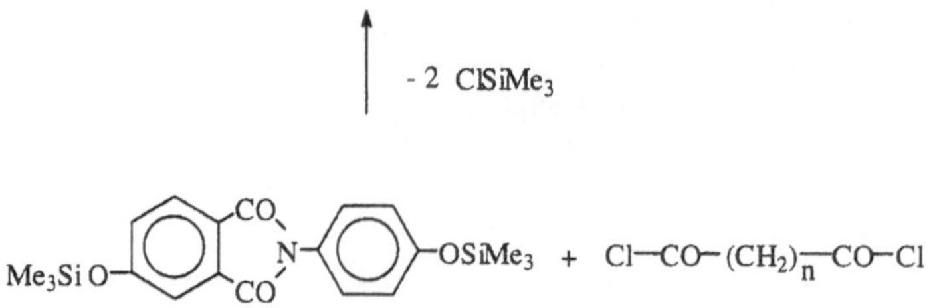

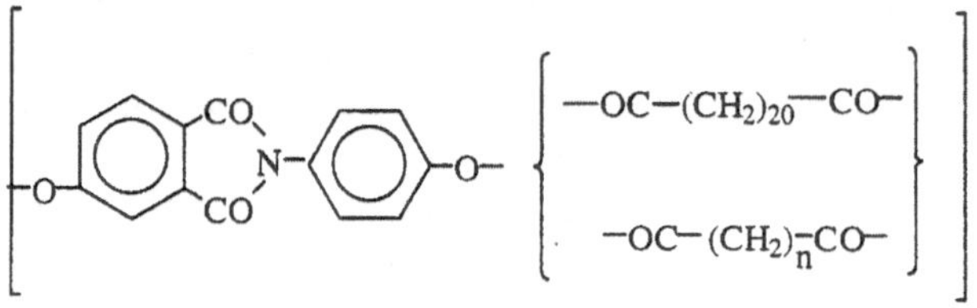

$$-2 \text{ ClSiMe}_3$$

$$\text{Me}_3\text{Si}\,O-\underset{\text{CO}}{\overset{\text{CO}}{\bigcirc}}\!N-\bigcirc\!-\text{OSiMe}_3 \ + \ \text{Cl}-\text{CO}-(\text{CH}_2)_{\overline{n}}\text{CO}-\text{Cl}$$

$$\left[-O-\underset{\text{CO}}{\overset{\text{CO}}{\bigcirc}}\!N-\bigcirc\!-O- \left\{ \begin{array}{c} -\text{OC}-(\text{CH}_2)_{20}-\text{CO}- \\ \\ -\text{OC}-(\text{CH}_2)_{\overline{n}}\text{CO}- \end{array} \right\} \right]$$

90 a, b   **a** : n = 12   **b** : n = 7

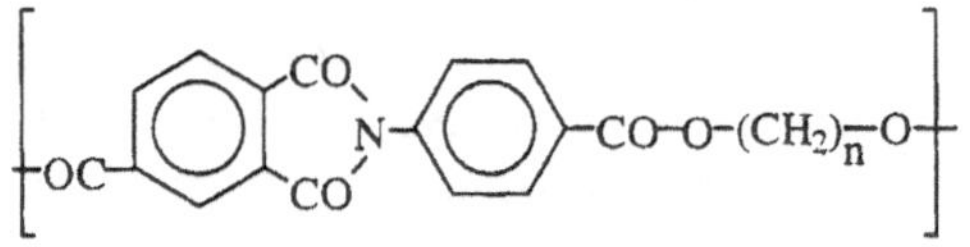

$$\left[-\text{OC}-\underset{\text{CO}}{\overset{\text{CO}}{\bigcirc}}\!N-\bigcirc\!-\text{CO}-\text{O}-(\text{CH}_2)_{\overline{n}}\text{O}- \right]$$

91 a - h

a : n = 4	**d** : n = 7	**g** : n = 10
b : n = 5	**e** : n = 8	**h** : n = 12
c : n = 6	**f** : n = 9	**i** : n = 16
		k : n = 22

$$\left[-\text{OC}-\underset{\text{CO}}{\overset{\text{CO}}{\bigcirc}}\!N-\bigcirc\!-\text{CO}- \left\{ \begin{array}{c} -\text{O}-(\text{CH}_2)_{\overline{n}}\text{O}- \\ \\ -\text{O}-(\text{CH}_2)_{12}-\text{O}- \end{array} \right\} \right]$$

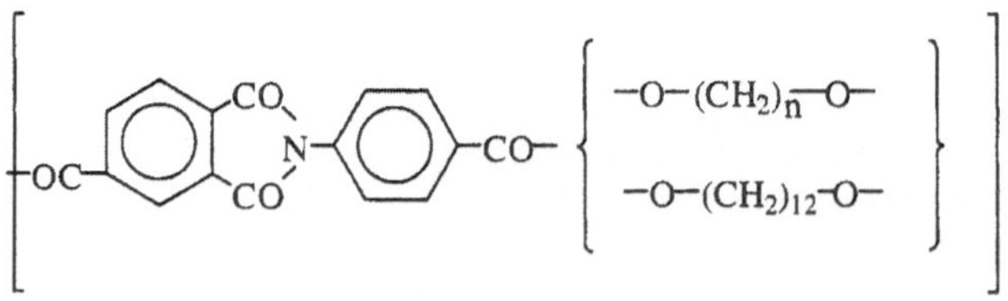

92 a - c   **a** : n = 10   **b** : n = 9   **c** : n = 8

Structure 14 (2)

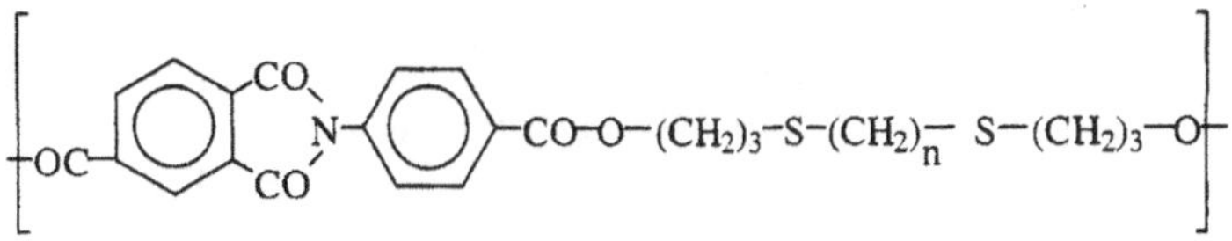

93 a,b   **a:**   n = 4   **b:**   n = 6

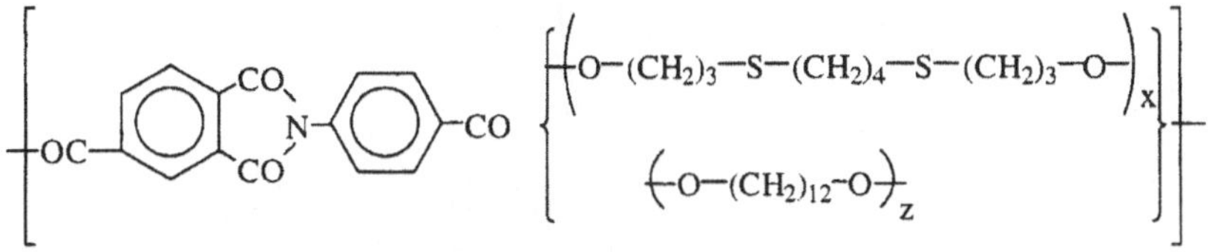

94a - c   **a:**   x/z = 8/2   **b:**   x/z = 5/5   **c : x/z = 2/8**

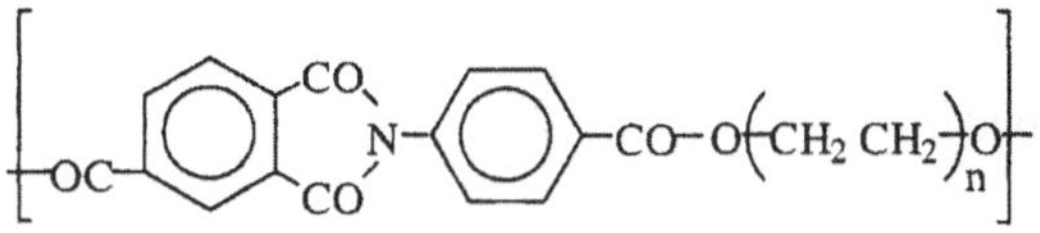

95 a - c   **a:** n = 2   **b:** n = 3   **c:** n = 4

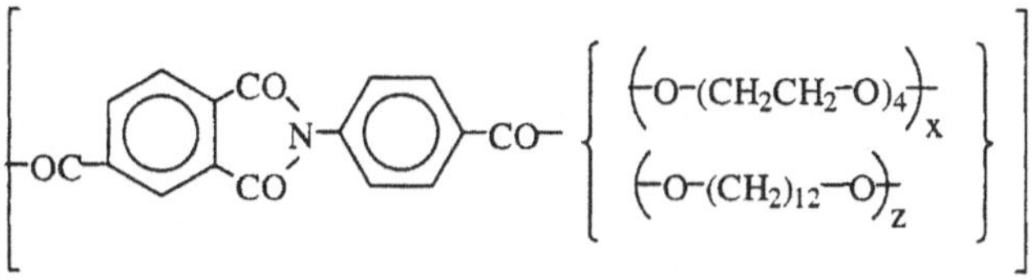

96a - g	**a:** x/z = 9/1	**d:** x/z = 5/5	**f:** x/z = 2/8
	b: x/z = 8/2	**e:** x/z = 3/7	**g:** x/z = 1/9
	c: x/z = 7/3		

Structure 14 (3)

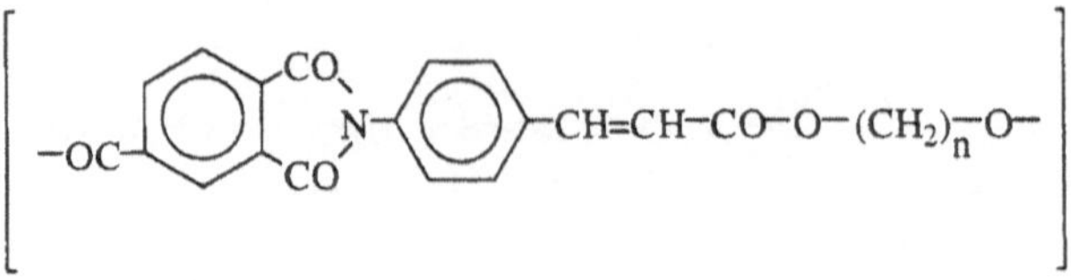

97a - i **a** : n = 5 **b** : n = 6 **c** : n = 7 **d** : n = 8 **e** : n = 9 **f** : n = 10 **g** : n = 12 **h** : n = 16 **i** : n = 22

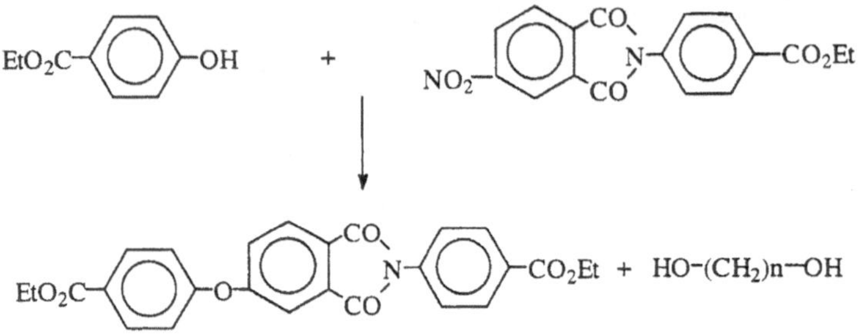

98

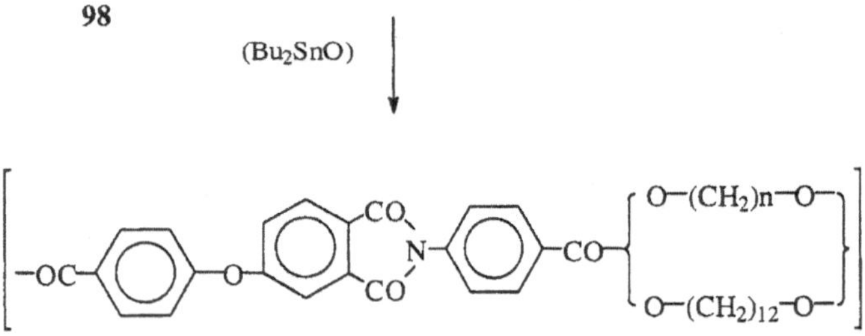

99a - c **a:** n = 12 **b:** n = 10 **c:** n = 8

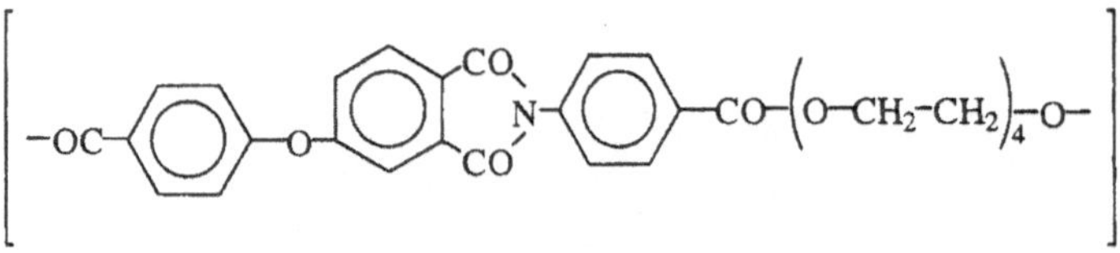

100

Structure 14 (4)

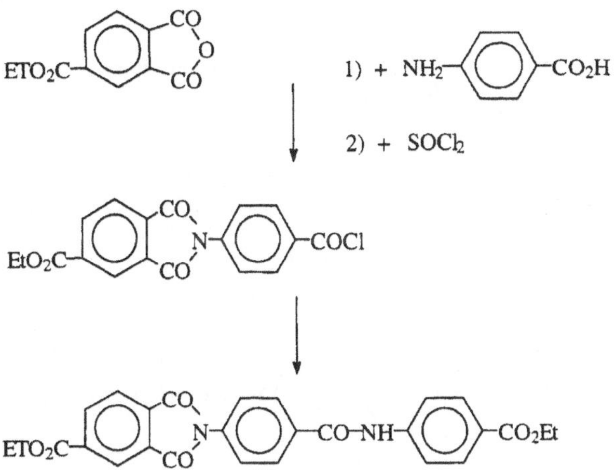

101

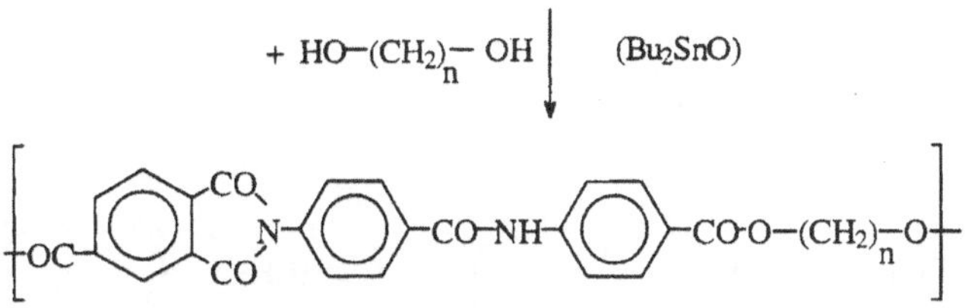

102a - h   **a:** n = 4   **d:** n = 7   **g:** n = 10
              b: n = 5   **e:** n = 8   **h:** n = 12
              c: n = 6   **f:** n = 9

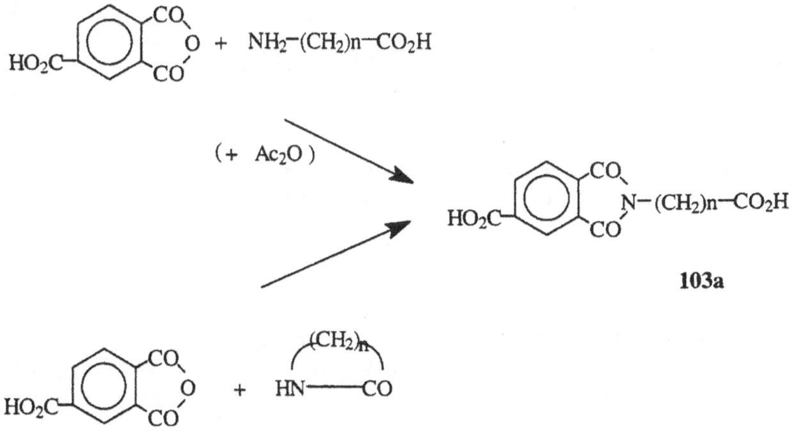

103a

Structure 14 (5)

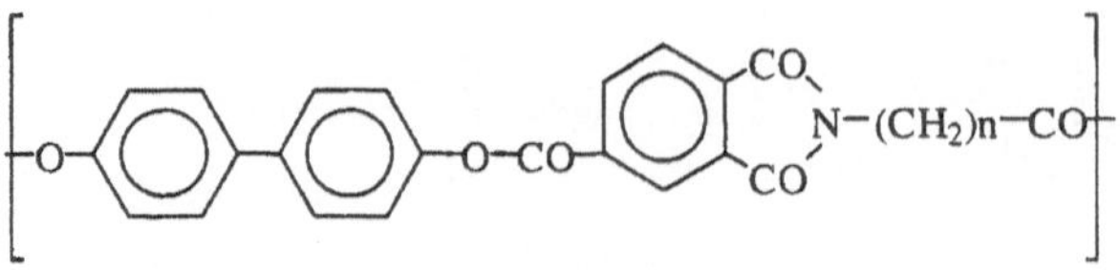

104 a - f

a	: n = 3	c	: n = 5	e	: n = 10
b	: n = 4	d	: n = 6	f	: n = 12

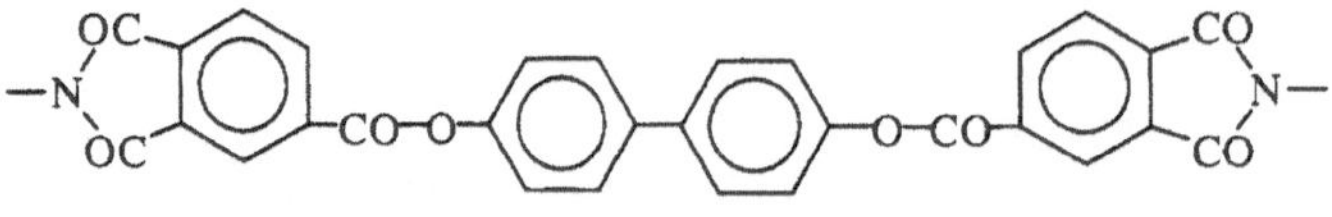

104 A

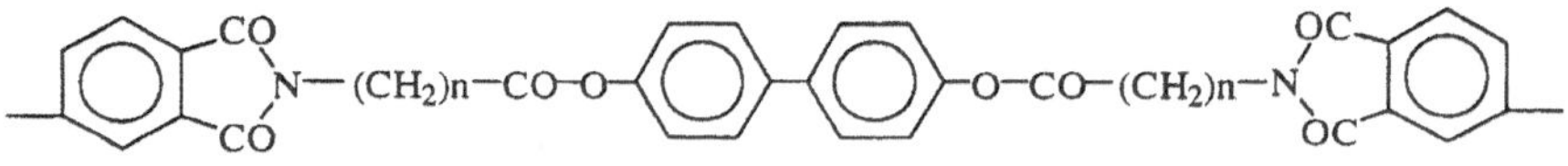

104 B

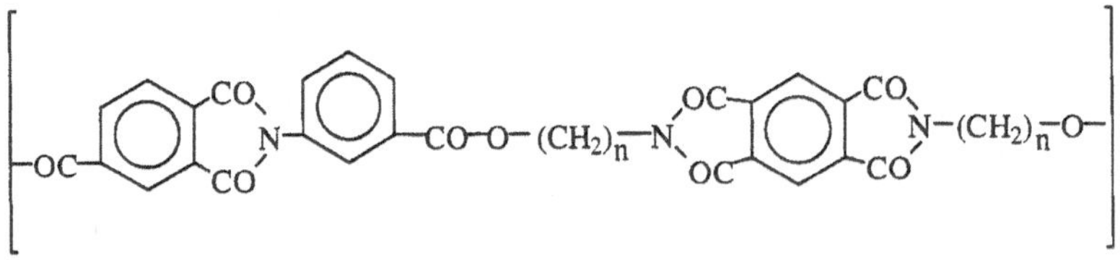

105 a - e

a	: n = 2	c	: n = 4	e	: n = 6
b	: n = 3	d	: n = 5		

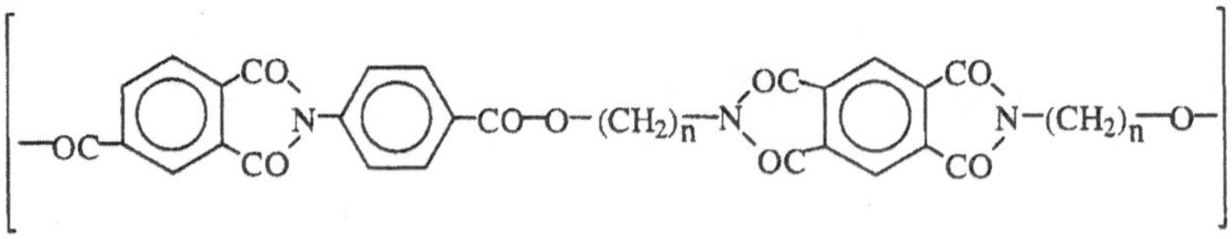

106 a - e

a	: n = 2	c	: n = 4	e	: n = 6
b	: n = 3	d	: n = 5		

Structure 14 (6)

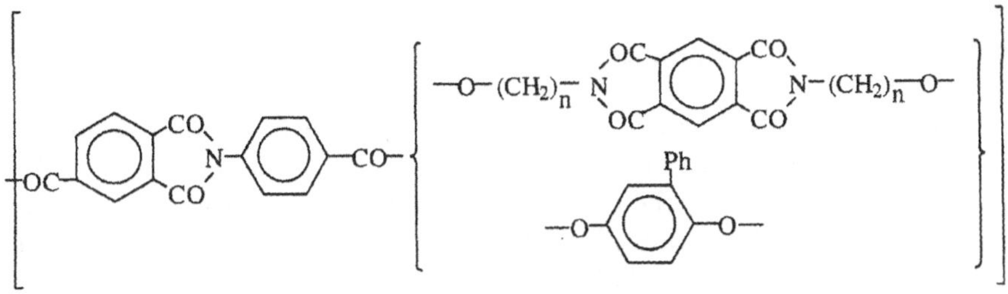

107 a - e

a : n = 2   c : n = 4   e : n = 6
b : n = 3   d : n = 5

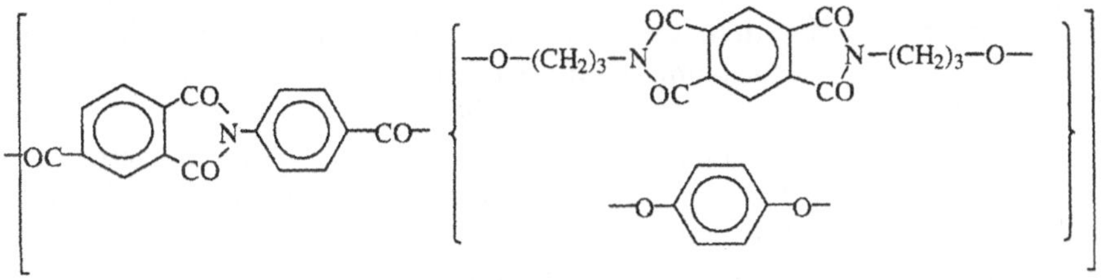

108a - f

a : x/z = 10/0   c : x/z = 5/5   e : x/z = 2,5/7,5
b : x/z = 7,5/2,5   d : x/z = 4/6   f : x/z = 0/10

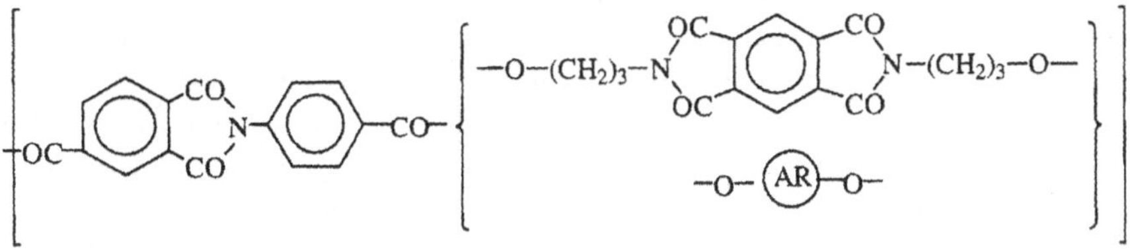

109a - h

a : (AR) =   e : (AR) =

b : (AR) =   f : (AR) =

c : (AR) =   g : (AR) =

d : (AR) =   h : (AR) =

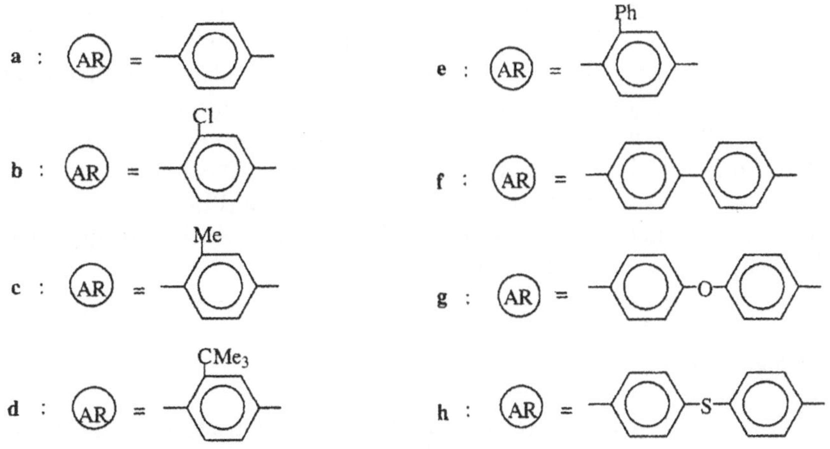

Structure 14 (7)

5
Cholesteric Poly(ester-imide)s

Cholesteric materials have been known for more than 100 years [99] but they still find increasing interest not only for fundamental reasons, but also for potential applications [100]. The cholesteric state is characterized by a helical arrangement of the mesogenic building blocks, and inside one domain the long axes of the helices are more or less parallel to a director. In the ground state of a macroscopic sample the directors of all domains may obey an isotropic distribution. Poling by an electric field or mechanical forces (e.g. shearing) may cause a macroscopic alignment of most or all domains with interesting optical consequences called Grandjean (GJ) texture [101, 102]. When a beam of white light hits a layer having a GJ-texture the wavelength matching the pitch of the helices will be reflected. The transmitted light shows the complimentary colour and no absorption takes place (in the ideal case). Furthermore, the reflected light is circularly polarized. Any potential application of these interesting optical properties requires, of course, a fixation of the GJ-texture. For this purpose three strategies are available. First, a fixation by cooling the GJ-texture below the T_g, second, by photocrosslinking of suitable photoreactive groups, and third, by thermal crosslinking. The latter two strategies have the advantage that the fixation of the GJ-texture is irreversible, but the changes of the chemical structure caused by the crosslinking process are usually unfavourable for the helical order of the mesogens. Fixation by cooling below the T_g requires cholesteric materials with T_gs above 90 °C or better above 120 °C and this property is difficult to achieve by low molar mass or LC-side chain cholesteric materials. Howevere, for this purpose LC-main chain polymers derived from non-symmetrical imide monomers, such as 26b or 27b, are advantageous, inasmuch as crystallinity should be avoided for optical reasons. The presence of crystallites causes diffuse light scattering and reduces the brilliance of the colours.

Three classes of chiral monomers (110–112, 113 and 114, 115 and 116) have recently been used for the preparation of cholesteric PEIs. The "sulfide spacers" 110–112 can easily be synthesized from commercial optically active 3-bromo-2-methyl-1 butanol. Their main advantage is a facile variation of their structure and twisting power, but they are highly expensive. In contrast, isosorbide 113 and isomannide 114 are the least expensive difunctional chiral monomers, they are commercially available and possess a high twisting power. The terephthalic acids 115 and 116 can be prepared in three steps from the inexpensive (S)-2-methyl-1-butanol. The following discussions of cholesteric PEIs are ordered according to the formula numbers of the chiral monomers, which also agrees with the historical sequence of their publication.

5.1
Chiral "Sulfide Spacers"

A first series of chiral PEIs was prepared by polycondensation of the "sulfide spacers" 110–112 with acid chloride of 27b in pyridine [103]. All PEIs 117–119 were prepared from the S-enantiomers of the spacers, but in the case of **119** the R-enantiomer was also prepared. The three PEIs **117–119** proved to possess largely differing properties. PEI **117** was amorphous ($T_g \sim 30$ °C) and isotropic. PEI **118** was semicrystalline ($T_g \sim 78$ °C) with a T_m around 158 °C, but the melt was again isotropic. Only the S- and R-forms of PEI **119** were thermotropic forming a cholesteric melt over a broad temperature range (252–340 °C) (Fig. 18). This PEI (**119**) proved to be particularly interesting because in addition to the cholesteric phase, two chiral smectic phases were found. As indicated by the DSC-measurements of Fig. 18, by optical microscopy and by WAXS measurements with synchrotron radiation PEI **119** forms a true liquid-crystalline smectic-C* phase between 220 and 252 °C and a solid smectic mesophase below 220 °C. The nature of the chiral smectic LC-phase was not clearly identified, but the analogous achiral PEIs **120a–e** formed smectic-A phases suggesting chiral smectic-A phase in the case of PEI **119**. This result is of interest because LC-main chain polymers showing chiral smectic LC-phases are extremely scarce. These results also demonstrate that the mesogeneity of **27b** is not strong enough to yield LC-phases in combination with chiral aliphatic spacers, and thus at least one additional aromatic ring is required in the repeating unit.

On the basis of these results a series of cholesteric copolyesters with phenylene-1,4-bisacrylic acid (**121a–f**) was synthesized [104]. The phenylene-1,4-bisacrylic acid is a highly photoreactive monomer which allows rapid photocrosslinking when its molar fraction (relative to **27b**) is at least 50 mol%. However, it is a poor mesogen, and the homopolyester **121a** does not form a stable GJ-texture upon shearing of the melt in contrast to PEI **119**. The incorporation

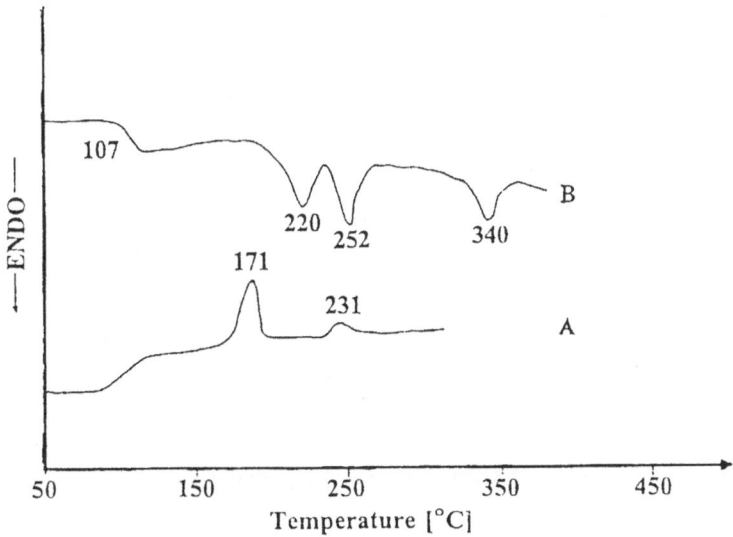

Fig. 18. DSC measurements (heating/cooling rate 20 °C/min) of poly(ester-imide) **119**: **A** 1st heating; **B** 1st cooling

of the imide unit **27b** stabilizes the GJ-texture, so that a fixation by irradiation with light <400 nm is easily feasible. Another interesting aspect concerns the solid state. The lengths of **27b** and phenylene-1,4-bisacrylic acid are similar, and thus, 30 mol% of **27b** suffices to induce the formation of a layer structure in the solid state. However, a chiral smectic LC-phase was never observed regardless of the composition of these co PEIs.

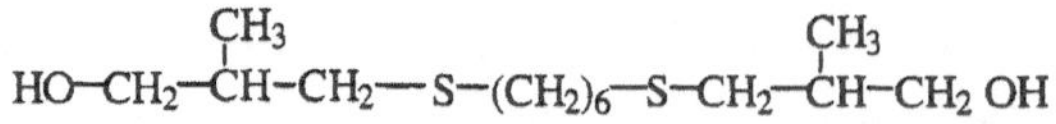

110

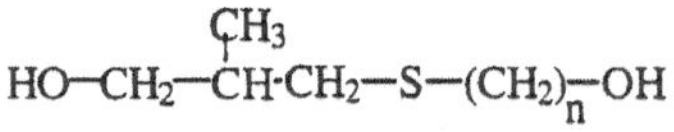

111   a : n = 2

        b : n = 6

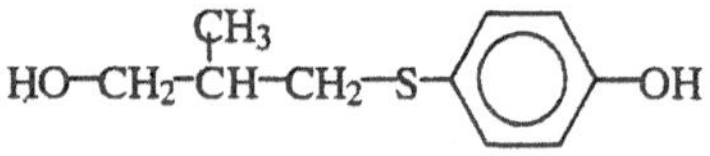

112

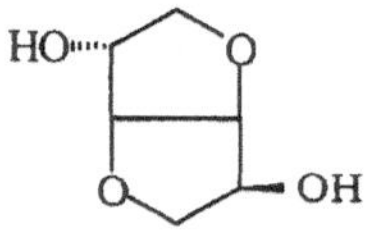

113 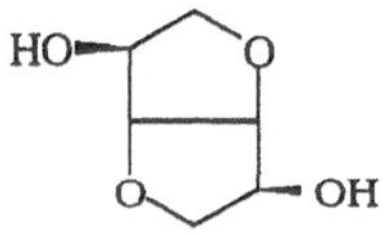   **114**

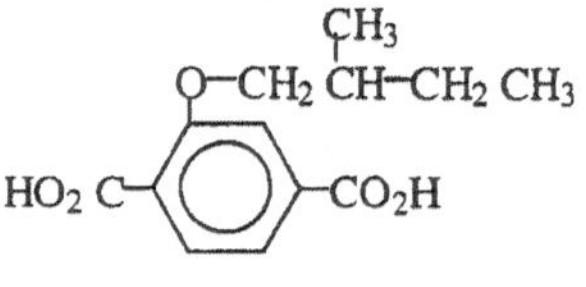

115 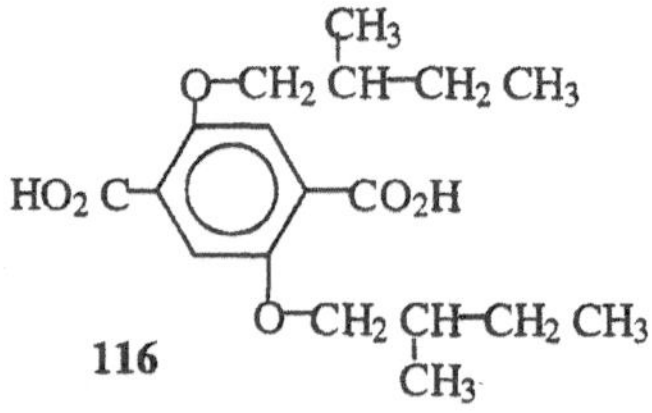   **116**

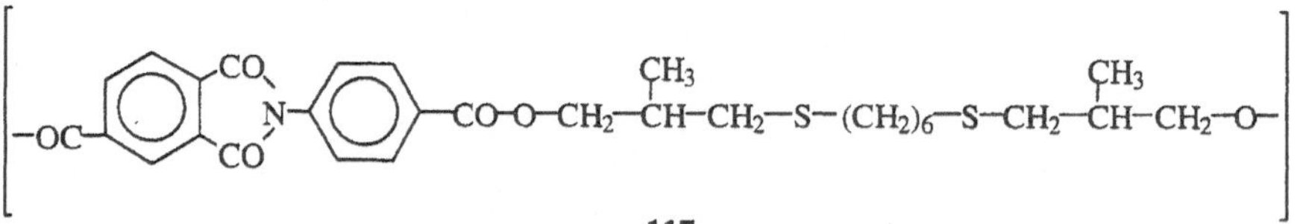

117

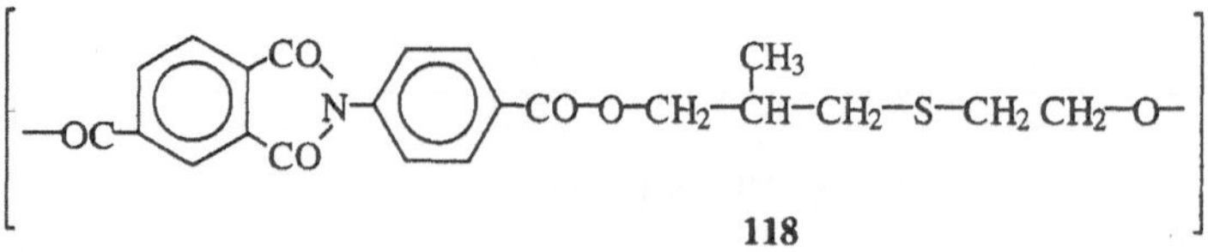

118

Structure 15 (1)

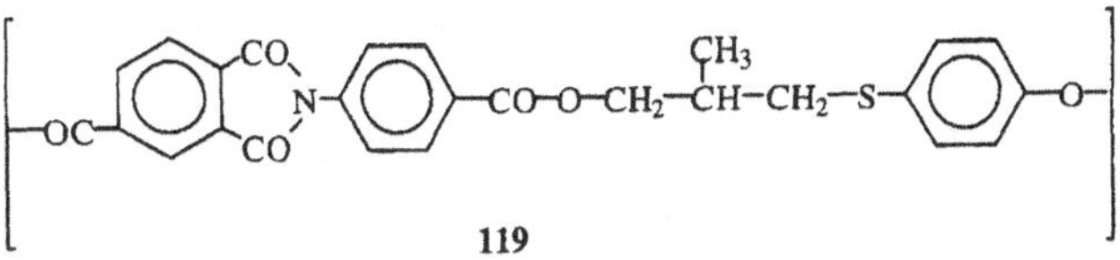

119

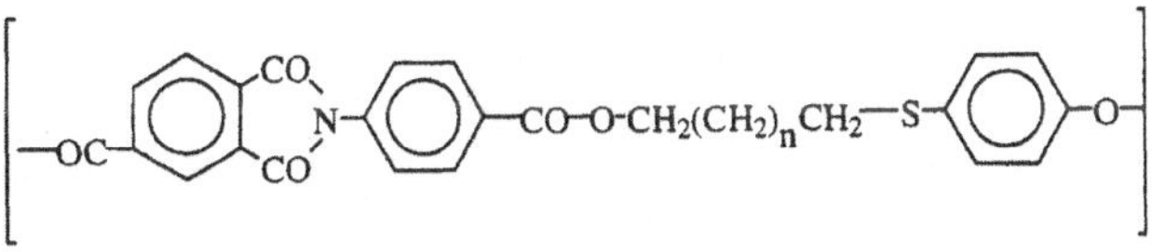

120

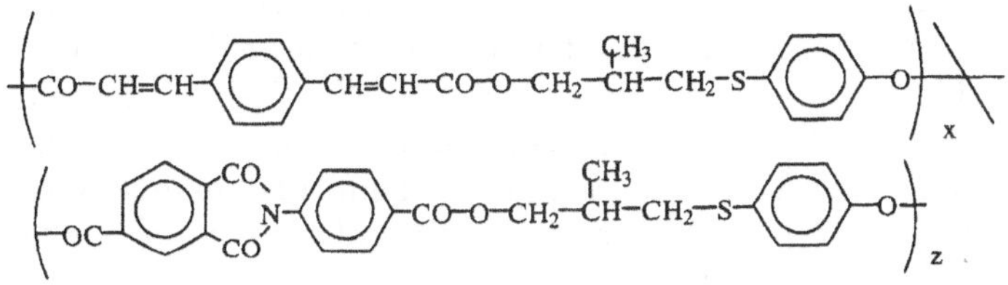

121 a - f

a : x/z = 10/0   d : x/z = 3/7
b : x/z = 7/3   e : x/z = 2/8
c : x/z = 5/5   f : x/z = 1/9

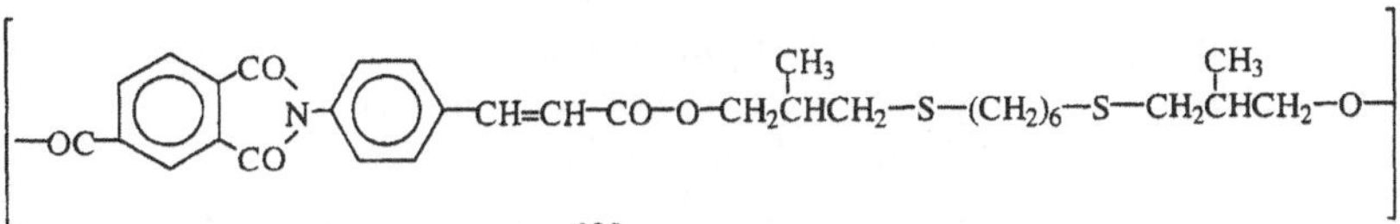

122

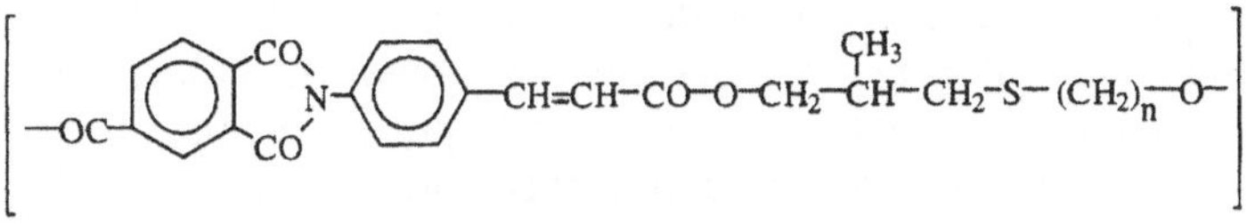

123   a : n = 2   b : n = 6

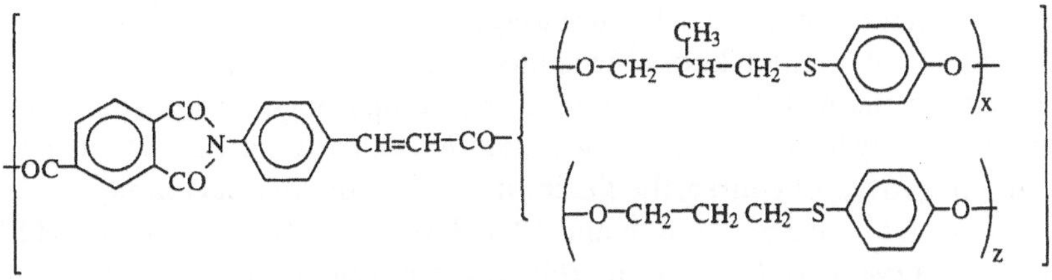

124a - d   a : x/z = 10/0   c : x/z = 5/5
                  b : x/z = 8/2   d : x/z = 0/10

Structure 15 (2)

A third series of chiral PEIs derived from the spacers **110–112** was prepared by polycondensation with 4-aminocinnamic acid trimellitimide (in the form of its dichloride [105]). The trimellitimide of 4-aminocinnamic acid is an interesting new monomer, because it combines good mesogenic properties with photo reactivity. Even this dicarboxylic acid was not mesogenic enough to yield a thermotropic polyester, when polycondensed with the "sulfidespacer" **110**. The PEI **122** proved to be amorphous and isotropic quite analogous to PEI **117**. However, in contrast to PEI **118** the PEIs **123a, b** were cholesteric. These PEIs and **124a–d** have in common that a smectic layer structure is formed in solid state. The layer structure is a smectic glass in the case of **123b** but a smectic crystalline phase in the case of **123a** and **124a–d**. The chiral PEIs **124a–c** form a cholesteric melt over a broad temperature range (Table 6), and which yields stable GJ-textures upon shearing. Particularly interesting are the properties of **124a** because the helical pitch, and thus the colour, changes with the temperature from red (above T_g) to yellowish green, dark green and blueish green (close to T_i). A further change of the colour can be achieved by "dilution of the chiral units" in the case of **124b, c** (with **124d** being nematic [105]. As outlined in a patent claim [106] these colours can be fixed by irradiation of the cholesteric melt with light of wavelengths ≤360 nm.

5.2
Sugar Diols as Chiral Building Blocks

The usefulness of isosorbide (**113**) and isomannide (**114**) for the preparation of cholesteric PEIs was for the first time tested in combination with PEI **33g**. In other words phenylhydroquinone was gradually replaced by one of these sugar diols [107, 108]. All the resulting co PEIs **125a–h** or **126a–h** were non-crystalline. Since the stereochemistry of both sugar diols is unfavourable for the mesogenic character of the PEIs incorporation of more than 50 mol% results in a total loss of the LC-phase. The phase diagram of **125a–h** is illustrated in Fig. 19. A characteristic difference between **125a–h** and **126a–h** concerns the formation of a GJ-texture. In the case of isomannide a GJ-texture was never observed [108]. In the case of isosorbide (**125a–h**) a GJ-texture was obtained upon shearing at temperatures above 250–300 °C, when the molar fraction of isosorbide was low (**125b, c**) [107]. The tendency to form a GJ-texture increased in both cases with higher temperatures. This finding may be interpreted as a consequence of the chain stiffness. The helices underlaying a GJ-texture may be based on mesogenic moieties not by the entire polymer backbone. This means in the case of LC-main chain polymers that the chains have to form bends and loops. Such a conformational behavior is unlikely for a fully aromatic polyester such as **125a**, but increasing temperature improves, of course, the flexibility of the polymer backbone.

In addition to the PEIs **125a–h** and **126a–h** the photoreactive analogs **127a–f**, **128a–d** and **129a–c** were prepared [109]. All polycondensations were conducted in such a way that mixtures of the free diols (e.g. *tert*-butyl hydroquinone and isosorbide) were reacted with the dichloride of 4-aminocinnamic acid trimellitimide in 1-chloronaphthalene at 230 °C. Quite analogous to the properties of

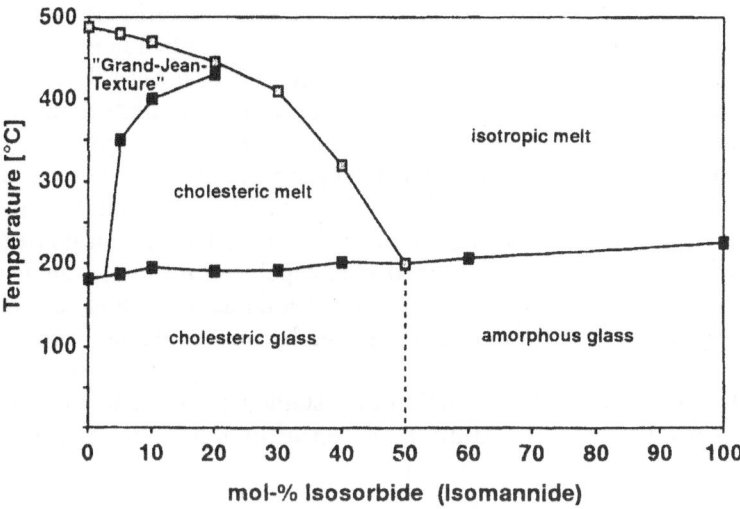

Fig. 19. Phase diagram of copoly(ester-imide)s derived from 4-aminobenzoic acid trimellit-imide and phenylhydroquinone + isosorbide

Table 5. Yields and properties of poly(ester-imide)s prepared from N-(4-carboxy phenyl) trimellitimide and various (acetylated) ortho diphenols

Polym. No.	R	Yield (%)	$\eta_{inh.}{}^a$ (dl/g)	$T_g{}^b$ (°C)	$T_i{}^c$ (°C)
36a	–H	96	0.10	150–155	260–270
36b	–CH$_3$	96	0.17	165–168	370–390
36c	–NO$_2$	94	insol.	98–100	260–270
36d	Condensed naphthalene	97	0.10	178–181	420–430

a   Measured at 20 °C with c=2g/l in CH$_2$Cl$_2$/trifluoroacetic acid (volume ratio 4:1)
b   From DSC measurements with a heating rate of 20 °C/min
c   From optical microsscopy with a heating rate of 10 °C/min.

16a–h and **17a–h**, a broader cholesteric phase was only obtained when the molar fraction of isosorbide was below 50% (x/z>5/5). In the case of **127b, c** a GJ-texture was formed upon shearing above 250 or 300 °C respectively, but these GJ-textures were not stable upon cooling. A flexibilisation of the PEI chain by incorporation of more flexible diphenols (**128a–d** and **129a–c**) did not significantly improve the properties. Incorporation of unsubstituted diphenols rendered these PEIs sensitive to thermal crosslinking above 300 °C. Apparently, the substituent of phenylhydroquinone (**127a–f**) hinders the thermal crosslinking for steric reasons. Taken together, the cholesteric PEIs **125–129** were not well suited for the generation and the fixation of a GJ-texture.

Nonetheless, the concept to produce stable GJ-textures at temperatures <250 °C by a flexibilization of the PEI backbone was studied in more detail and finally with more success. A first approach consisted of the copolycondensation of **27b** (as the dichloride) and adipoyl chloride with a mixture of *tert*-butyl hydroquinone and isosorbide [110]. As expected, the T_gs of the resulting non-crystalline PEIs **130a–h** decreased with higher molar fraction of adipic acid (Table 5). Higher molar ratios of adipic acid also reduced the isotropization temperature and finally destabilized the LC-phase to such an extent that **130h** was completely isotropic. The PEIs **130a–d** were capable of forming a stable GJ-texture which in the case of **130b–d** allowed fixation just by cooling below T_g. The usefulness of this approach was documented by further classes of PEIs having flexibilized main chains.

In the series **130i–o** the formation of a stable GJ-texture was observed for **130l, m**. These greenish GJ-textures persisted again upon cooling below the T_gs.

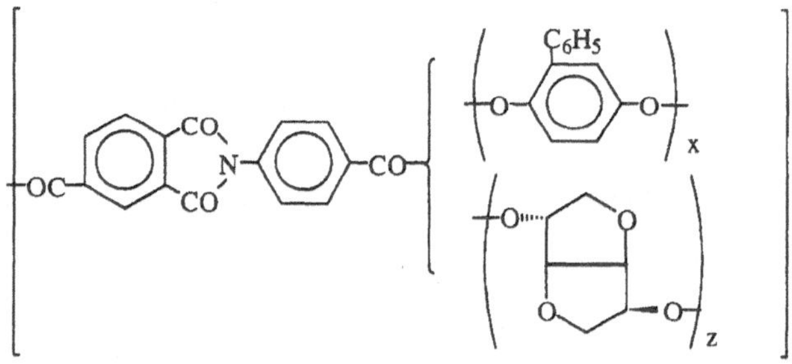

125a - h

a : x/z = 10/0	**e** : x/z = 7/3
b : x/z = 9,5/0,5	**f** : x/z = 6/4
c : x/z = 9/1	**g** : x/z = 5/5
d : x/z = 8/2	**h** : x/z = 0/10

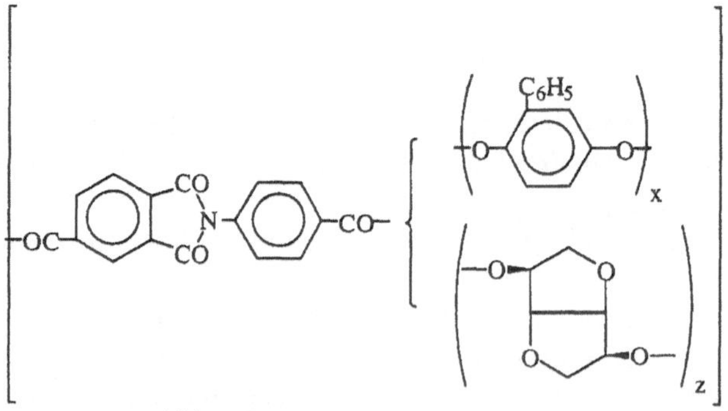

126a - h

Structure 16 (1)

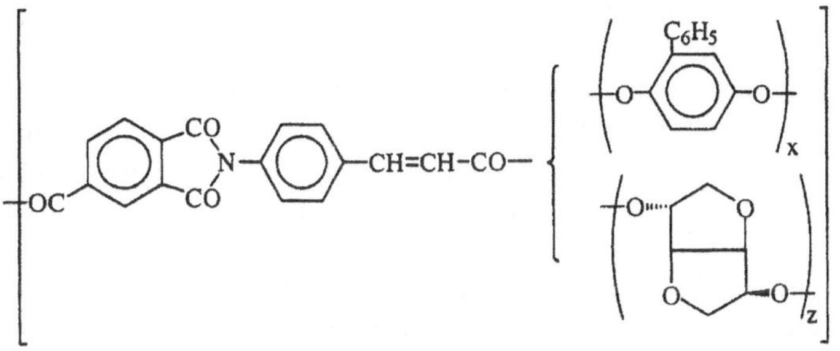

127a - f

a : x/z = 10/0   d : x/z = 7/3

b : x/z = 9,5/0,5   e : x/z = 6/4

c : x/z = 9/1   f : x/z = 5/5

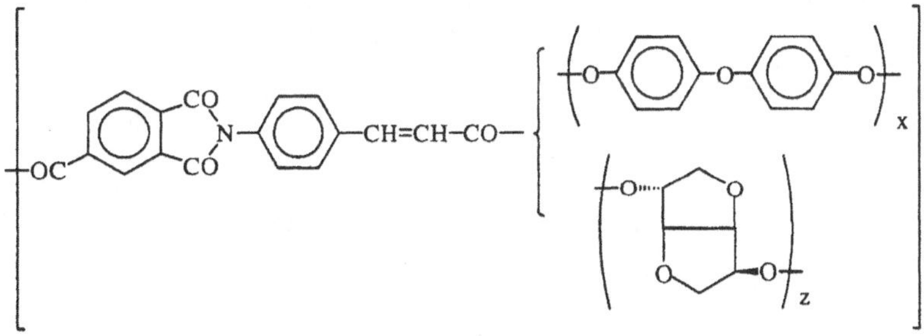

128a - d

a : x/z = 10/1   c : x/z = 7/3

b : x/z = 9,5/0,5   d : x/z = 5/5

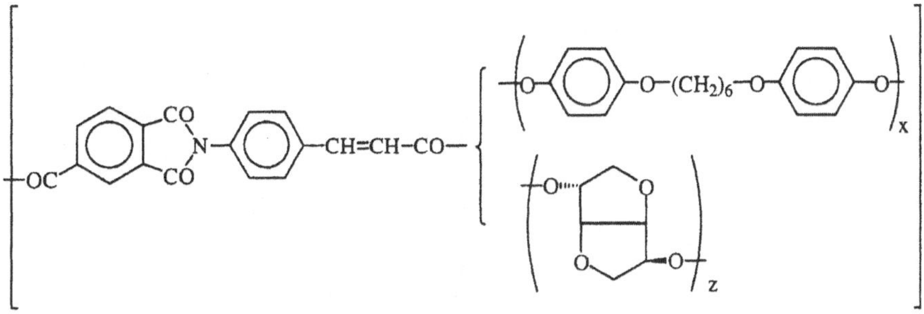

129a - c

a : x/z = 10/0   c : x/z = 5/5

b : x/z = 9,5/0,5

Structure 16 (2)

All other members of this series were cholesteric displaying colourful Schlieren textures. The capability to form GJ-textures was completely lost, when the molar fraction of isosorbide was increased to 20% (or higher). In other words the co PEIs **130a–s** were all cholesteric, but only few of them formed a GJ-texture upon shearing.

When the *tert*-butyl-group was replaced by a methyl group (**131a–d**) a new property emerged. All four co PEIs were cholesteric and capable of forming a GJ-texture, but these co PEIs were semicrystalline. With phenylhydroquinone (**132a–d**) again four non-crystalline and cholesteric polyesters were obtained, but only three of them (**132b–d**) yielded a GJ-texture upon shearing. With 2,7-dihydroxynaphthalene as diphenol two co PEIs (**133a, b**) were obtained which were amorphous and isotropic. In contrast the co PEIs derived from 4,4'-dihy-

Table 6. Yields and properties of the copoly(ester-imide)s containing *tert*-butylhydroquinone and isosorbid

Polym. form	Method	Yield (%)	$\eta_{inh.}^{a}$ (dl/g)	T_g^b (°C)	T_i^c (°C)	Texture
136a	A	99	0.83	194	>400	Grandjean texture blue-green
136a	B	99	0.48	195	>400	Grandjean texture blue
136b	A	99	0.42	174	>400	Grandjean texture green, orange
136b	B	98	0.64	177	>400	Grandjean texture blue
136c	A	98	0.41	143	>400	Grandjean texture blue
136c	B	97	0.54	126	>400	Grandjean textur blue
136d	A	99	0.60	193	>400	Grandjean texture blue-green
136d	B	99	0.47	183	>400	Cholesteric
136e	A	99	0.78	174	>400	Grandjean texture blue
136e	B	98	0.73	175	>400	Grandjean texture blue
136f	A	97	1.14	164	>400	Grandjean texture blue
136f	B	97	0.27	129	>400	Grandjean texture blue
136g	A	99	0.55	190	300	Cholesteric
136g	B	97	0.21	159	250	Cholesteric
136h	A	99	0.80	178	350	Cholesteric
136h	B	96	0.42	162	290–300	Cholesteric
136I	A	98	0.66	149	280	Cholesteric
136I	B	95	0.31	139	270–280	Cholsteric

a Measured at 20 °C with c=2g/l in CH_2Cl_2/trifluoroacetic acid (volume ratio 4:1)
b From DSC measurements with a heating rate of 20 °C/min
c From optical microsscopy with a heating rate of 10 °C/min.

droxydiphenyl ether (134a, b) were cholesteric, but did not show a GJ-texture. However, a GJ-texture was detected in the case of co PEI 135a which contained sebazic acid. Yet, sebazic acid seems to be less favorable than adipic acid, because 135b did not form a GJ-texture in contrast to 130l.

Another group of flexibilized co PEIs containing isosorbide was prepared in such a way that adipic (or sebazic acid) was replaced by an aromatic dicarboxylic acid containing an aliphatic spacer (136a–i, 137a–d, 138a, b and 139a, b) [111]. This strategy proved to be particularly successful, because all these co PEIs were cholesteric. Furthermore, numerous members of these groups formed stable GJ-textures, namely 136a–f and 137a, c (Table 6).

A third variation of the flexibilization concept was realized by the incorporation of an aliphatic α,ω-diol [112]. With 1,6-hexanediol and *tert*-butyl-hydroquinone (140a–h) eight cholesteric co PEIs were obtained, but only three of them (140a–c) were capable of forming a GJ-texture. With methylhydroquinone two co PEIs (141b, c) showed a GJ-texture, but 141a was also thermotropic. When phenylhydroquinone was used as comonomer even all three co PEIs (142a–c) formed a GJ-texture. Also the co PEIs derived from 2,7-dihydroxy naphthalene (143a, b) were cholesteric, but failed to yield a GJ-texture. With 1,10-decanediol and *tert*-butyl hydroquinone again co PEIs forming a GJ-texture were obtained (144a, b).

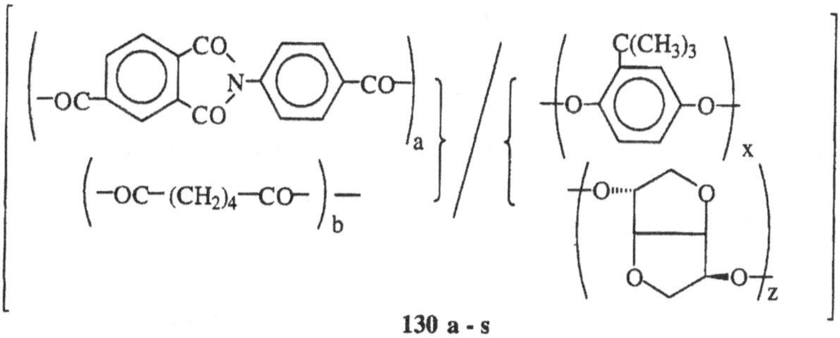

130 a - s

a : a/b = 90/10, x/z = 95/5	e : a/b = 50/50, x/z = 95/5
b : a/b = 80/20, x/z = 95/5	f : a/b = 40/60, x/z = 95/5
c : a/b = 70/30, x/z = 95/5	g : a/b = 30/70, x/z = 95/5
d : a/b = 60/40, x/z = 95/5	h : a/b = 20/80, x/z = 95/5
i : a/b = 90/10, x/z = 90/10	m : a/b = 60/40, x/z = 90/10
k : a/b = 80/20, x/z = 90/10	n : a/b = 50/50, x/z = 90/10
l : a/b = 70/30, x/z = 90/10	o : a/b = 40/60, x/z = 90/10
p : a/b = 90/10, x/z = 80/20	r : a/b = 70/30, x/z = 80/20
q : a/b = 80/20, x/z = 80/20	s : a/b = 60/40, x/z = 80/20

Structure 17 (1)

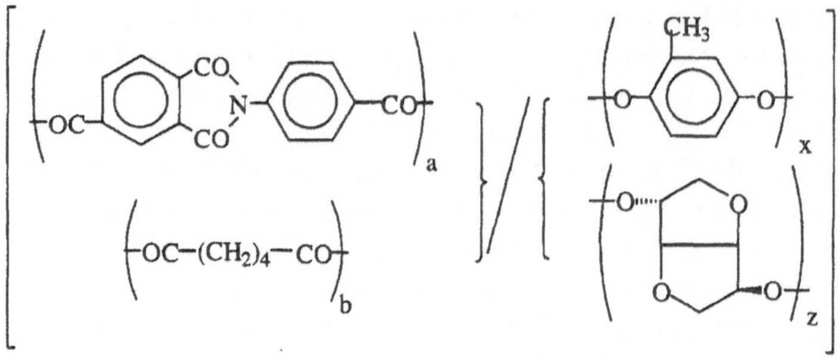

131a - d

a: a/b = 70/30, x/z = 95/5   **c** : a/b = 70/30, x/z = 90/10

b: a/b = 60/40, x/z = 95/5   **d** : a/b = 60/40, x/z = 90/10

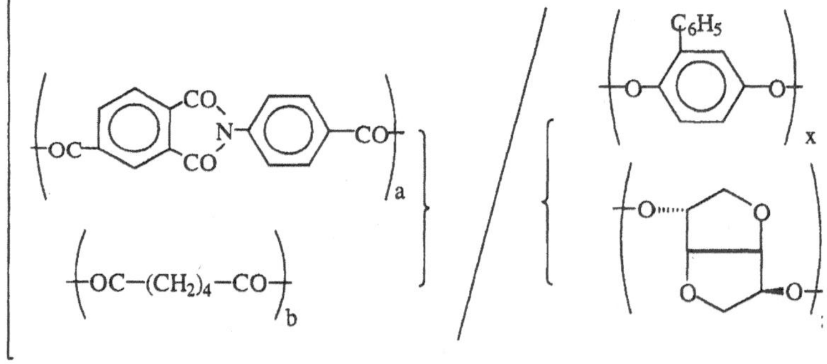

132 a, d   **a** : a/b = 70/30, x/z = 95/5   **c** : a/b = 70/30, x/z = 90/
                  b : a/b = 60/40, x/z = 95/5   **d** : a/b = 60/40, x/z = 90/

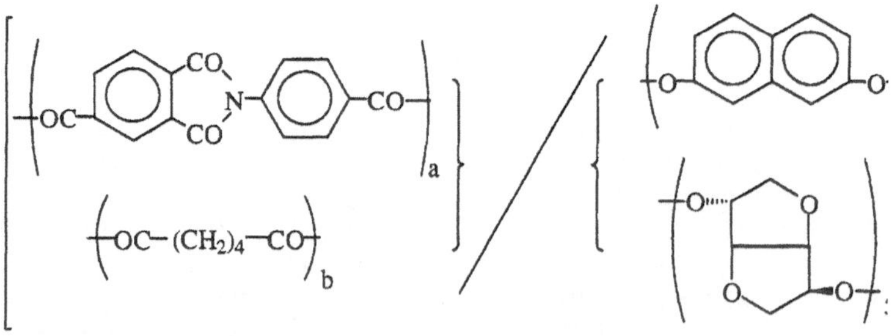

133 a, b   **a:** a/b = 70/30   **b:** a/b= 60/40

Structure 17 (2)

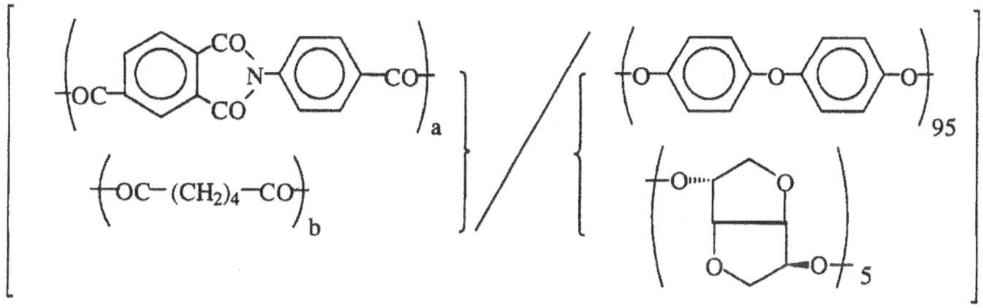

134a, b   **a:** a/b = 70/30   **b:** a/b = 60/40

Structure 17 (3)

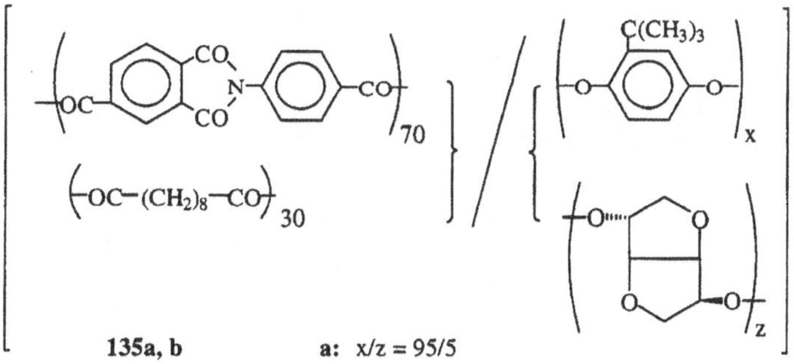

135a, b   **a:** x/z = 95/5

                      b: x/z = 90/10

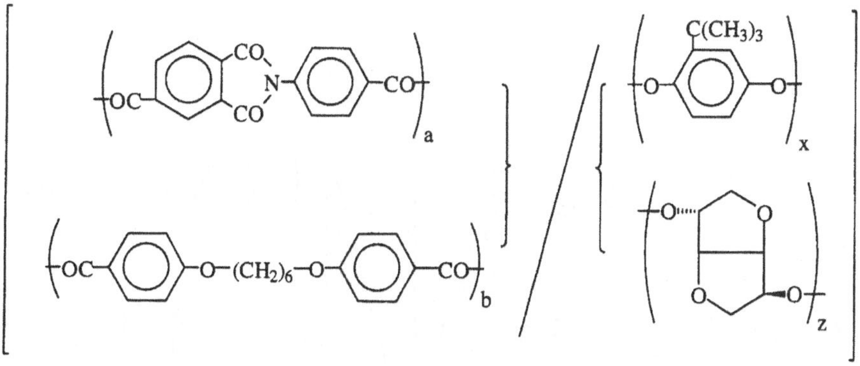

136a - i

a: a/b = 90/10, x/z = 95/5   **d:** a/b = 90/10, x/z = 90/10   **g:** a/b = 90/10, x/z = 80/20
b: a/b = 80/20, x/z = 95/5   **e:** a/b = 80/20, x/z = 90/10   **h:** a/b = 80/20, x/z = 80/20
c: a/b = 60/40, x/z = 95/5   **f:** a/b = 60/40, x/z = 90/10   **i:** a/b = 60/40, x/z = 80/20

Structure 18 (1)

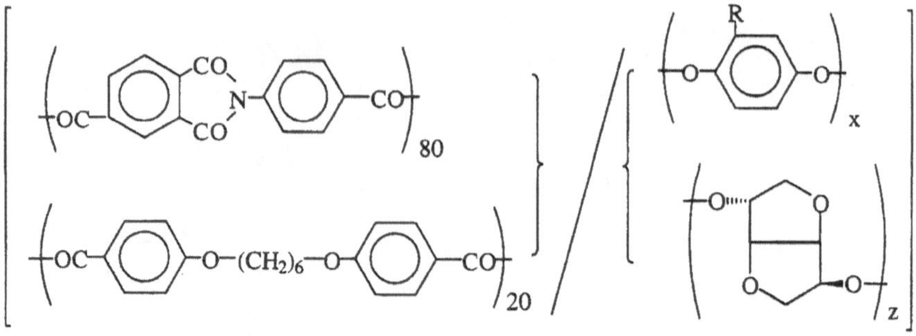

137a - d a: $R = CH_3$, $x/z = 95/5$ c: $R = C_6H_5$, $x/z = 95/5$
b: $R = CH_3$, $x/z = 90/10$ d: $R = C_6H_5$, $x/z = 90/10$

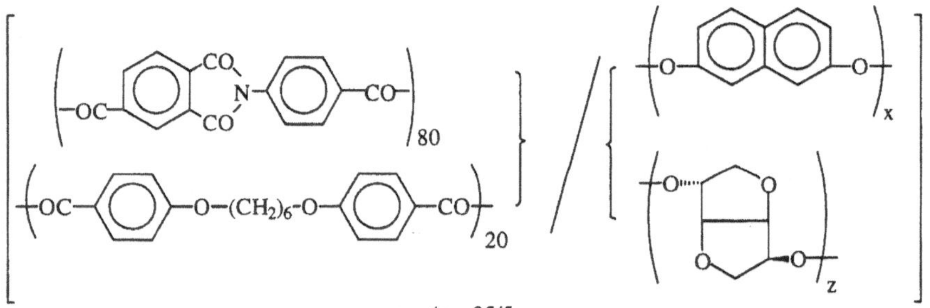

138a, b a: $x/z = 95/5$
b: $x/z = 90/10$

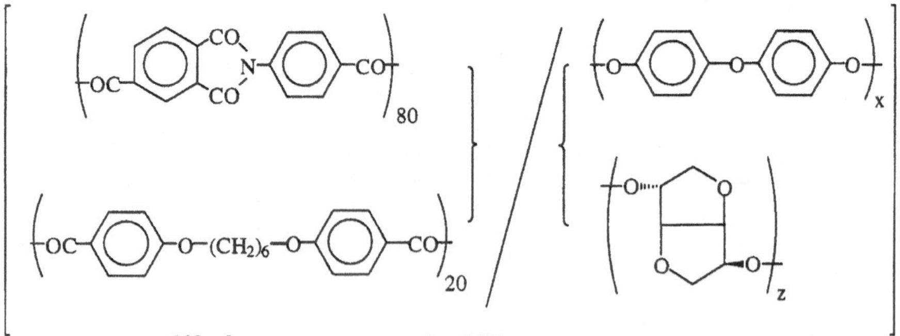

139a, b a: $x/z = 95/5$
b: $x/z = 90/10$

Structure 18 (2)

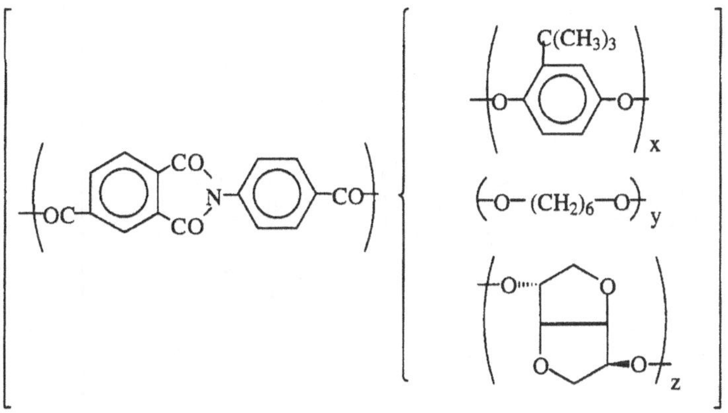

140 a - h

a : x/y/z = 90/5/5
b : x/y/z = 85/10/5
c : x/y/z = 80/15/5
d : x/y/z = 70/25/5

e : x/y/z = 60/35/5
f : x/y/z = 80/10/10
g : x/y/z = 70/20/10
h : x/y/z = 60/30/10

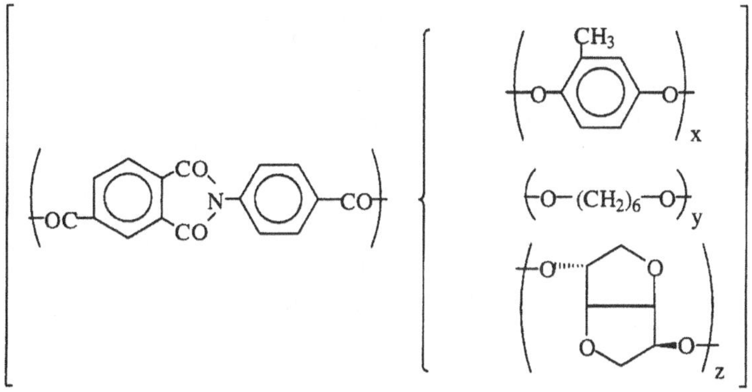

141a - c

a : x/y/z = 85/10/5
b : x/y/z = 80/15/5

c : x/y/z = 80/10/10

Structure 18 (3)

Based on isosorbide as the chiral component a series of cholesteric oligo (ester-imide)s was prepared having either two maleimide endgroups (**145**) or nadimide endgroups. These endgroups allow a chemical or thermal crosslinking, but detailed studies of such crosslinking processes in the cholesteric phase were not reported [113].

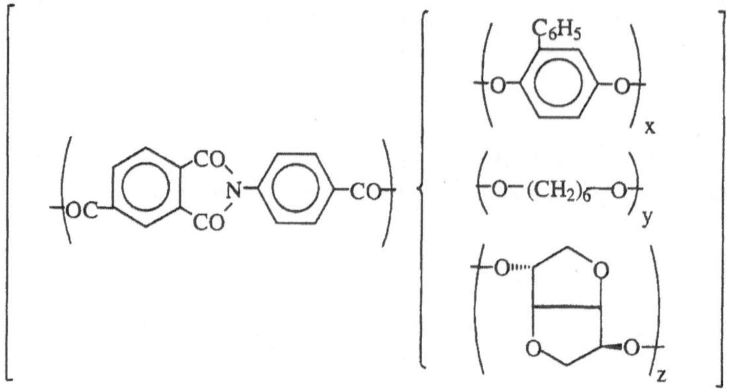

142a - c   **a**: x/y/z = 85/10/5   **c**: x/y/z = 80/10/10
                  b: x/y/z = 80/15/5

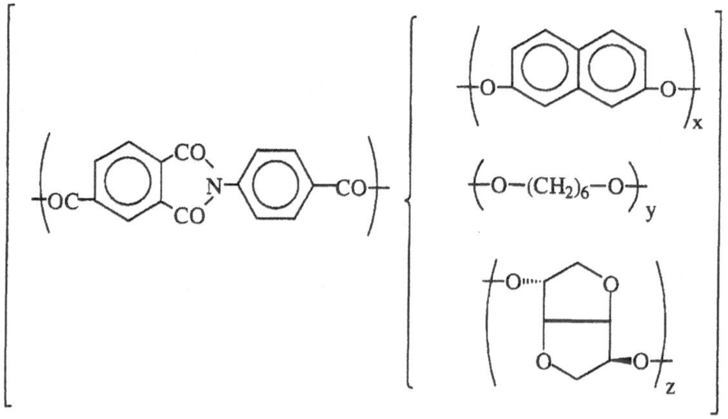

143a, b   **a**: x/y/z = 85/10/5   **b**: x/y/z = 80/10/10

Structure 18 (4)

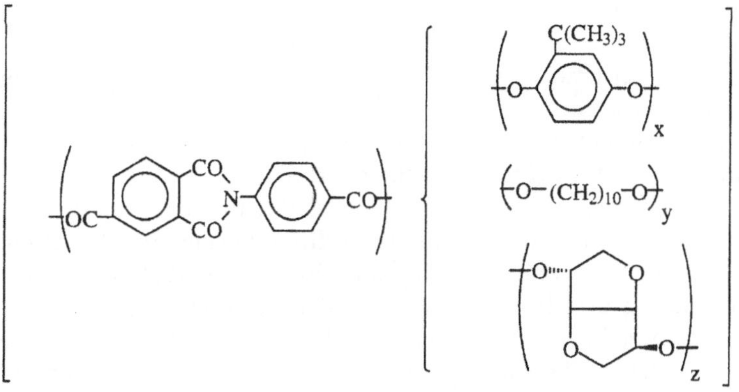

144a,b   **a:** x/y/z = 85/10/5   **b:** x/y/z = 80/10/10

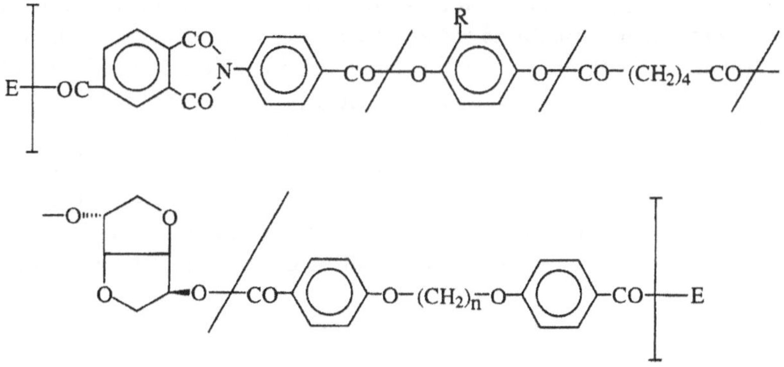

145   R = CH$_3$, C(CH$_3$)$_3$ n = 4, 6, 8, 10

E =

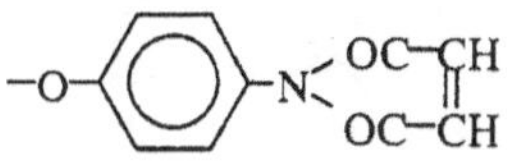

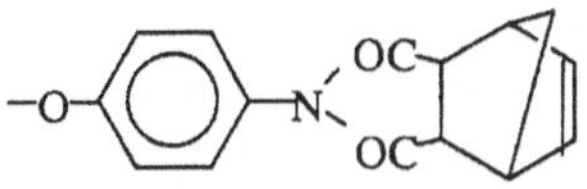

Structure 18 (5)

5.3
Chiral Terephthalic Acids

The syntheses and structures of the cholesteric PEIs **117–145** are characterized by the following two aspects:
1. the chiral comonomers were all incorporated in the form of diols,
2. the chiral centers were all located in the main chain.

A completely different approach was elaborated with the syntheses of the co PEIs **146a–d** and **147a–d**. Although numerous different substituted terephthalic acids were prepared by several research groups in the years 1980–1995, no chiral terephthalic acid has ever been synthesized. The chiral terephthalic acids **115** and **116** were obtained by alkylation of dialkyl mono- or dihydroxy terephthalates with the tosylate of (S)-2-methylbutanol [114]. The PEIs **146a–d** and **147a–d** were than prepared by polycondenzations of the silylated imide diphenol (**26b**) with the substituted terephthaloylchlorides [114]. Unfortunately, the T_m of the homo PEI **146a** was so high (~285 °C) that its melting was immediately followed by thermal degradation of the aliphatic substituents. Therefore, 2-(4-chlorophenoxy) terephthalic acid was used as comonomer which reduced the crystallinity and the T_m. The resulting co PEIs formed indeed a cholesteric melt over a broader temperature range, but a GJ-texture was never obtained. A higher twisting power was expected from the disubstituted terephthalic acid **116**. Furthermore, lower phase transitions were induced by the dodecyl side chains of the comonomer. As demonstrated by the data listed in Table 7, the co PEIs **147a–d** show indeed lower T_gs and broad cholesteric phases with colourful schlieren textures. Unfortunately, stable GJ-textures were never observed.

In conclusion, it may be said that from the viewpoint of fundamental research most syntheses were successful, because LC-PEIs with enantiotropic cholesteric phases were obtained. However, from the viewpoint of a potential application as coloured lacquers or pigments only a handful of these cholesteric PEIs shows promising properties.

Table 7. Yields and properties of poly(ester-imide)s **147a–d**

Polymer formula	Yield (%)	η_{inh}^{a} (dl/g)	$[\alpha]D^{20\,b}$	T_g^{c} (°C)	T_m^{c} (°C)	T_i^{c} (°C)	T_i^{d} (°C)
147a	70	0.38	+14.0	128	290	–	280–290
147b	95	0.56	+ 8.0	–	195	222	225–235
147c	97	0.74	+ 2.9	–	207	220	220–230
147d	98	0.84	0	–	211	228	230–240

a Measured with c=2 g/l in CH_2Cl_2/trifluoroacetic acid (volume ratio 4:1)
b Measured with c=2 g/l
c From DSC measurements with a heating rate of 20 °C/min
d From optical microscopy with a heating rate of 10 °C/min

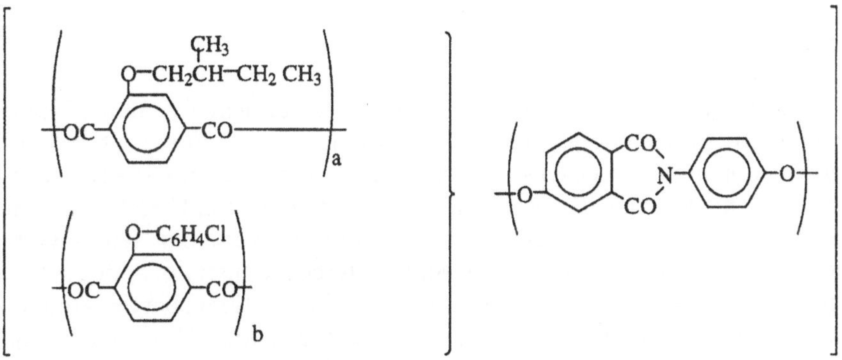

146 a - d

a : a/b = 10/0   **c :** a/b = 1/9

b : a/b = 5/5   **d :** a/b = 0/10

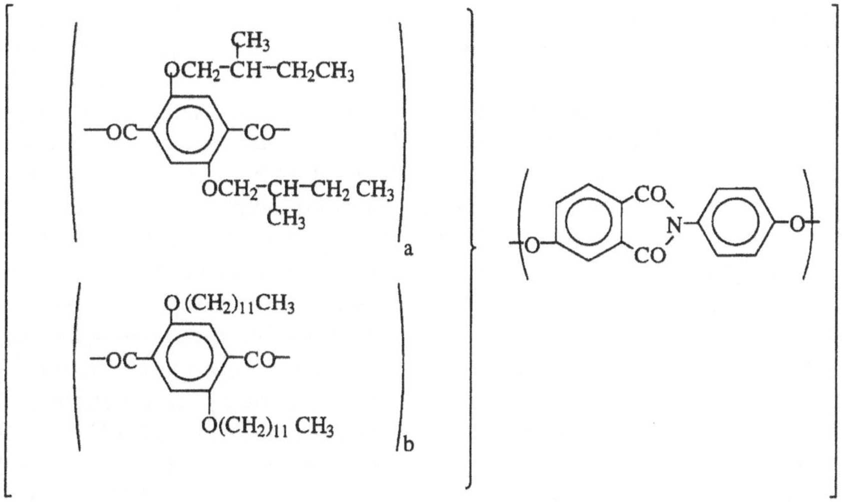

147a - d

a: a/b = 10/0   **c:** a/b = 1/9

b: a/b = 5/5   **d:** a/b = 0/10

Structure 19

6
Various LC-Polyimides

Ester groups favour the formation of LC-phases because they combine flexibility with a high tendency to form linear conformations. In contrast, ether, sulfide, carbonate or urethane groups introduce kinks or bends into a polymer chain which are unfavourable for the stabilization of an LC-phase. Therefore, it is quite obvious that most LC-polyimides contain ester groups and only few polyimides free of ester groups have been reported. The stereochemistry of amide groups is in principle similar to that of ester groups and the rotational barrier around the CO-N bond is even higher favouring a linear conformation. However, the higher dipole moments and the NH-bonds raise both the T_gs and the T_ms, and thus are disadvantageous for a thermotropic character of poly(amide-imide)s (PAIs). Therefore, it was obvious that the first LC-poly(amide-imide)s described in the literature [115] were lyotropic materials.

The PAIs 148a–c and 149 were prepared by polycondensation of the silylated diamines with the dichloride of 27b in N-methylpyrrolidone (NMP) quite analogous to the synthesis of commercial polyaramides. These PAIs proved to be soluble in conc. H_2SO_4 and, in contrast to poly(phenylene terephthalamide) they were also soluble in methane sulfonic acid. A lyotropic solution was observed for 148a at a concentration of 10% w/v in conc. H_2SO_4. Surprisingly, the copolyamides 148b and 148c were not lyotropic despite similar molecular weights and despite the well known lyotropic properties of poly(phenylene-terephthalamide). Obviously, the random sequence of these copolymers is unfavourable for the lyotropic phase. Also the PAI 149 failed to yield a lyotropic phase at 10% in H_2SO_4, whereas the analogous polyamide of a benzoxazole dicarboxylic acid (150) was lyotropic. This was the first, but not the only, example demonstrating that benzoxazole derivatives may be better mesogens than analogous imide moieties. Finally, it may be concluded that the structure/property relationships of potentially lyotropic polymers are not well understood yet(see Structure 20).

Despite high melting temperatures, thermotropic PAIs (152a–d) were obtained from diamines containing long aliphatic spacers (151a–d) by polycondensation with trimellitic anhydride chloride in hot m-cresol. [116]. The bâtonet textures observed by optical microscopy and the synchroton radiations (X-ray) measurements proved that 140a–d form a smectic phase around 300 °C over a temperature range of 50–60 °C. An analogous series of PAIs derived from 3-aminobenzoic acid (153) was, as expected, not thermotropic. Yet, quite interesting is the finding that a polyamide derived from a methyl substituted diamine (154) is not liquid-crystalline. Even a PAI having an alternating sequence of the substituted and unsubstituted diamine (155) proved to be isotropic. This negative influence of the methyl substituents is difficult to understand on the basis of the classical theory of Onsager and Flory, because the lenght/diameter ratio does not significantly change by the methyl groups. However, these substituents successfully hinder the electronic (e.g. DA) interactions between neighbouring chains, and thus, the properties of these PAIs support the hypothesis that en-

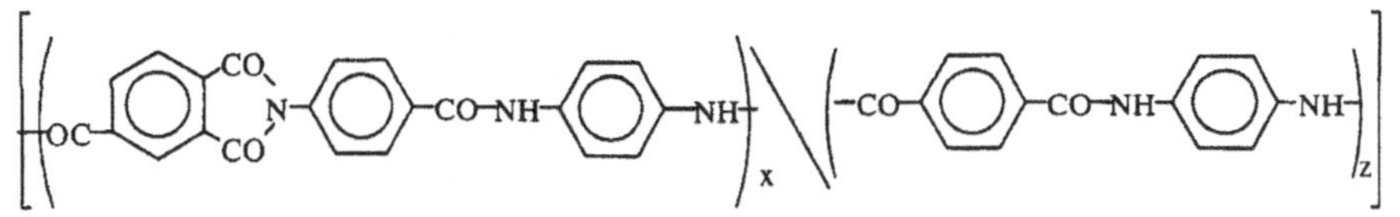

148a - c   **a:** x/z = 10/0   **b:** x/z = 5/5   **c:** x/z = 2/8

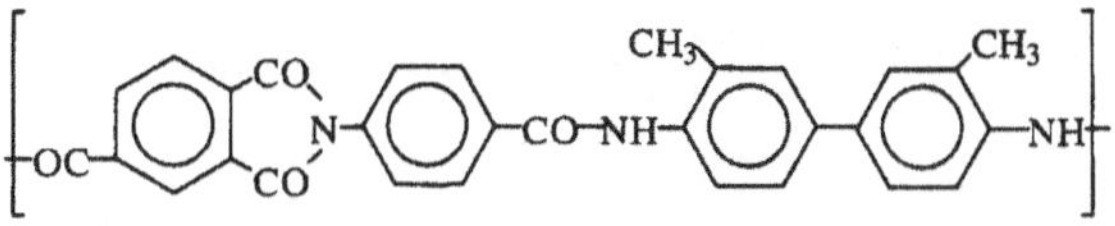

149

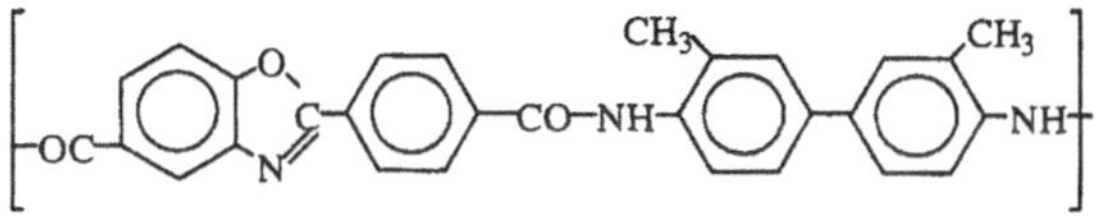

150

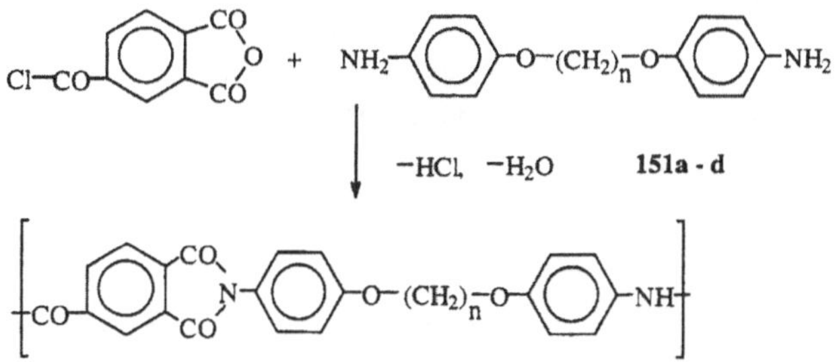

−HCl, −H₂O   **151a - d**

152a - d   **a:** n = 8   **b:** n = 9
                  c: n = 10   **d:** n = 12

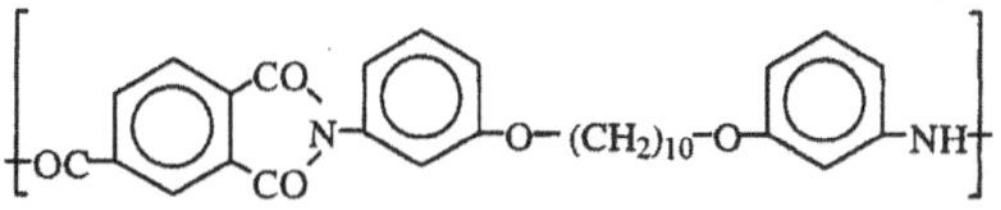

153

Structure 20 (1)

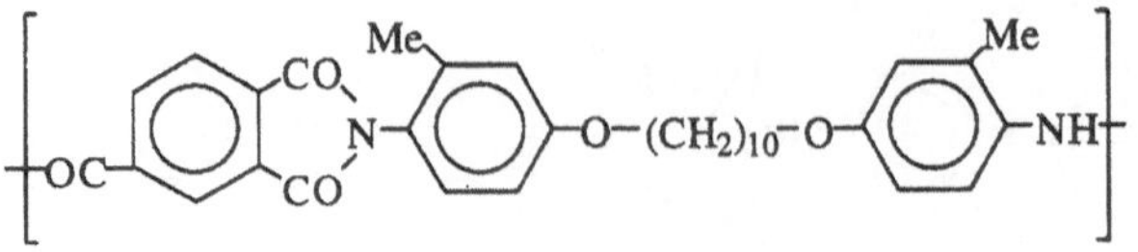

154

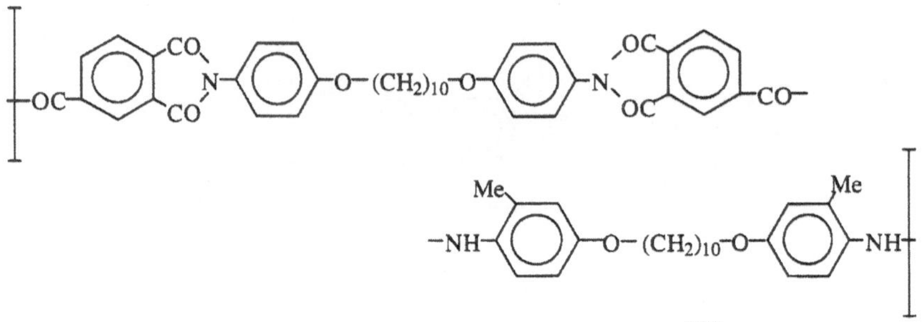

155

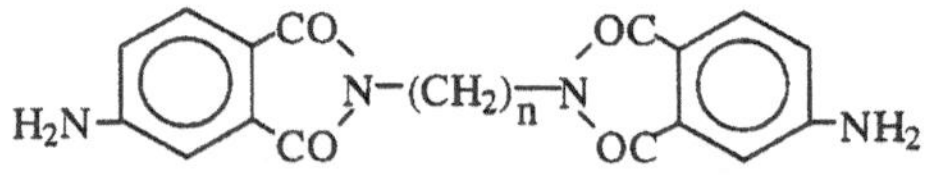

156a - d   **a:** n = 9   **b:** n = 10

                   c: n = 11   **d:** n = 12

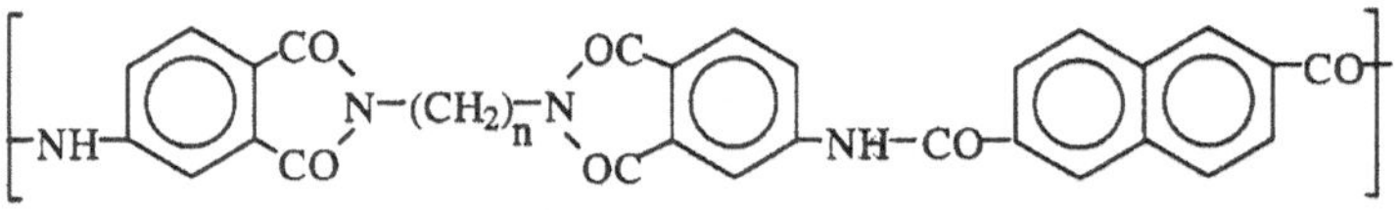

157a - d

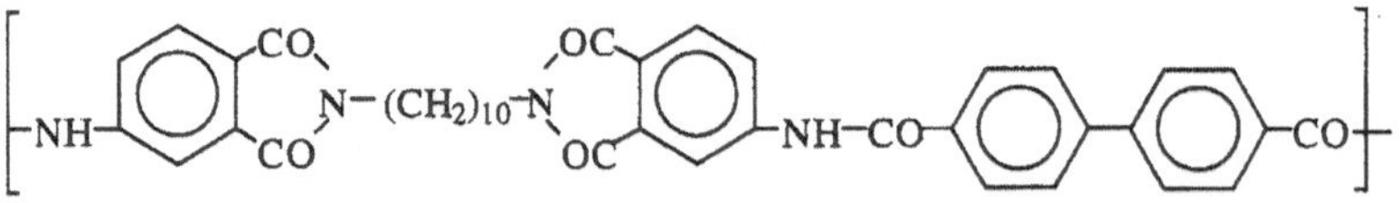

158 a - d

Structure 20 (2)

thalpic attractive interactions between neighbouring mesogens play an important role in the formation of an LC-phase. Finally, it should be mentioned that all the PAIs 152a–d and 153–155 form a smectic type of layer structure in the solid state.

Another class of thermotropic PAIs (157a–d) was prepared by polycondensation of the diamines 156a–d which result from the hydrogenation of the corresponding 4-nitrophthalimides [117]. The comonomers used for the polycondensations were the dichlorides of terephthalic acid, phenylthioterephthalic acid, naphthalene-2,6-dicarboxylic acid (157a–d) and diphenyl-4,4'-dicarboxylic acid (158a–d). Only with the biphenyl dicarboxylic acid as comonomer were thermotropic PAIs obtained. As illustrated by the data summarized in Table 8 three phase transitions were found: a change from smectic crystalline to a smectic-LC-phase (T_{m1}), a transition to the nematic phase (T_{m2}) and finally the isotropization (T_i). The narrow temperature ranges of the smectic LC-phases and the thermal degradation above 350 °C prevented a differentiation between smectic-A and smectic-C textures.

The synthesis of one almost fully aromatic, thermotropic PAI (159) was reported by research groups of Mitsui Toatsu Chem. [118, 119]. Characteristic for this PAI is the long relatively flexible diamine moiety consisting of five aromatic rings. X-ray studies of the author [119] revealed that this unique PAI forms a smectic layer structure in the solid state, but without any lateral order of the main chains and without forming a well defined LC-phase. Quite similar properties were found for the poly(ether-imide) 160 [118, 119]. The dimensions of the extended repeating unit and of the energetically most favourable conformation were calculated by computer modelling and compared to the X-ray patterns obtained by synchrotron radiation up to 320 °C. The results clearly indicate that the polyimide 160 like the PAI 159 are semicrystalline materials after annealing, forming a more or less disordered smectic-B type of layer structure in the solid state. The melt between 282±2 °C and 303±2 °C is a smectic-A phase. Since a thermotropic character was quite unexpected for such fully aromatic polyimides their structure was systematically varied [118, 119]. Both the diamine (161, 162) and the imide groups (163, 164) were varied. However, no other polyimide of this group was thermotropic. Nonetheless, all these polyimides form a smectic type of layer structure in the solid state. This result is interesting enough, be-

Table 8. Phase transitions of the liquid-crystalline poly(ester-imide)s 157a–d

Formula	$T_g{}^a$ (°C)	$T_{m1}{}^a$ (°C)	$T_{m2}{}^a$ (°C)	$T_i{}^b$ (°C)
157a	156	325	351	435–445
157b	173	383	409	440–450
157c	148	322	342	380–390
157d	152	357	383	415–425

a From DSC measurements with a heating rate of 20 °C/min
b From optical microscopy (crossed polarizers) with a heating rate of 10 °C/min

cause main chain LCPs based on fully aromatic polyesters never form smectic layer structures. Therefore, the properties of these polyimides underline the exceptional potential of imide groups to induce the formation of layer structures. In agreement with the pertinent discussion in Sects. 4 and 7, this potential must be attributed to the high polarity of the imide ring.

More recently several poly(ether imide)s containing aliphatic spacers (165–169) were also synthesized and analyzed [13, 120–122]. Most of these polyimides are derived from the diamines 151a–d already mentioned as building blocks of the PAIs 152a–d. However, in contrast to the PAIs with their non-symmetrical mesogen the highly symmetrical polyimide 165 did not melt below 400 °C and decomposed before melting, so that no LC-phase was detectable [13, 120]. Despite lower T_ms (e.g. 237 °C in the case of 166c) even the polyimides 166a–d did not show an LC-phase [122].

For the poly(ether-imide)s 167a–f smectic LC-phases were clearly detectable despite the high temperatures (see Table 2) [122]. Both the bâtonet textures and the X-ray measurements agreed in that the melts were smectic-A phases in all cases. Furthermore, the poly imides 167a–f formed a smectic crystalline state with an odd-even effect of the melting temperatures [124]. As expected, the polyimide derived from a spacer connecting m-amino phenol units (168) was not thermotropic, but proved to possess a reversible solid phase transition [122]. Also the methyl substituted polyimide 169 was not thermotropic [122]. This result is quite analogous to that discussed for the PAIs 153 and 154. Computer modelling showed that the rotational barrier of the C-N bond is far higher due to the methyl groups. Therefore N-phenoxy imide moiety cannot adopt a planar structure which is necessary for electronic (e.g. DA) interactions of neighbouring chains. The lack of such attractive interaction and not the stiffness of the mesogenic building blocks was considered to be responsible for the lack of LC-phases (see Structure 21).

Finally, polyimides bare of any amide, ester or ether groups should be discussed. The polyimides which were studied at first with regard to a potentially liquid-crystalline character were the poly(pyromellit imide)s 170a [13, 123]. When the authors of that study realized that these polyimides were highly crystalline but not thermotropic, they modified the structure systematically to reduce both the crystallinity and the melting temperatures. In this connection the poly(pyromellit imide)s 170b, 171 and 172a–b were synthesized and characterized. However, against all expectations none of these polyimides was thermotropic and thus, that study proved again that the PMDI unit is a poor mesogen in agreement with the properties of poly(ester-imide)s discussed in Sect. 4.

More recently, two research groups reported [122, 124] on the properties of polyimides derived from biphenyltetracarboxylic anhydride (173a–c). Despite the higher length/diameter ratio of the imide moiety even these polyimides were not thermotropic. Also the polyimides derived from diphenylether-3,3',4,4'-tetracarboxylic anhydride (174a), benzophenone-3,3',4,4'-tetracarboxylic anhydride (174b) or diphenylsulfone-3,3',4,4'-tetracarboxylic anhydride (174c) were

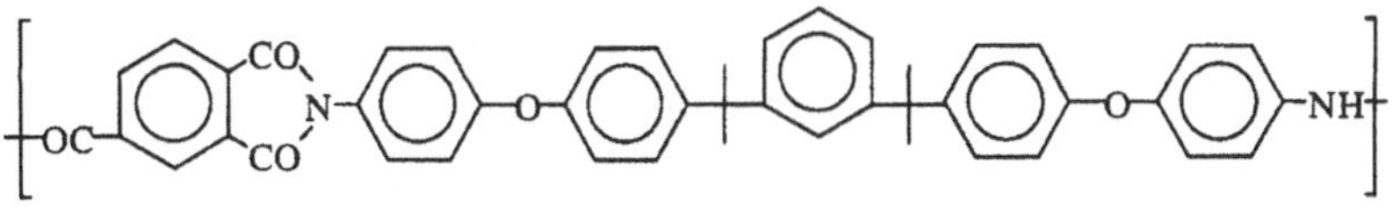

159

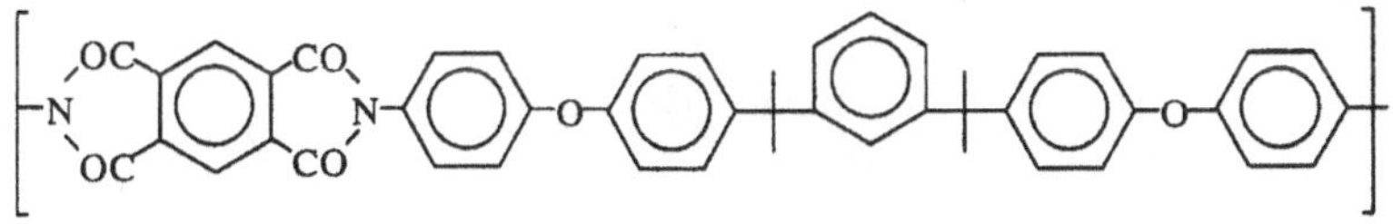

160

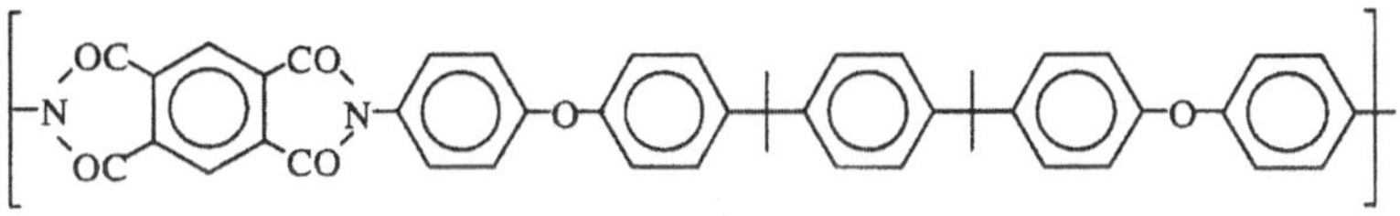

161

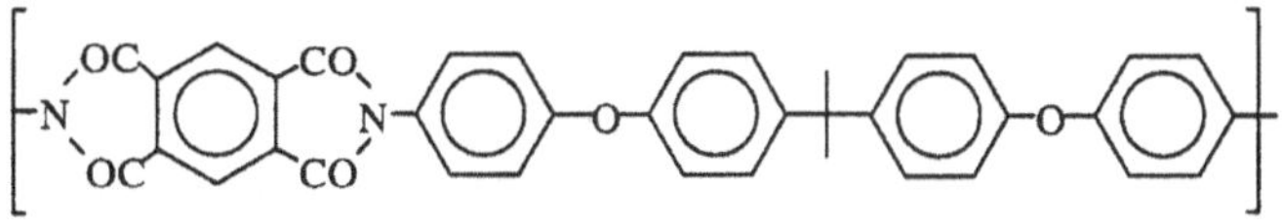

162

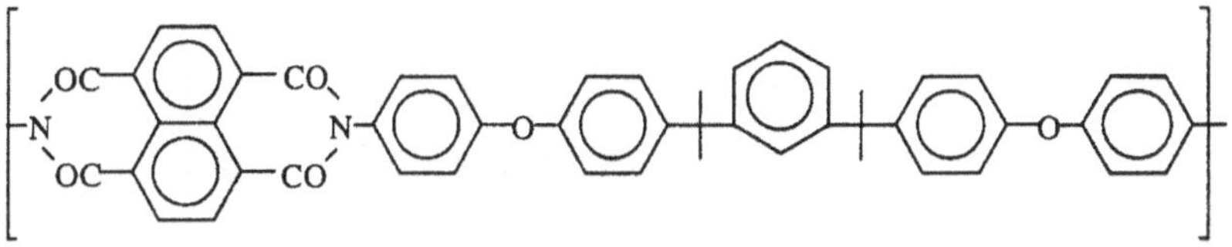

163

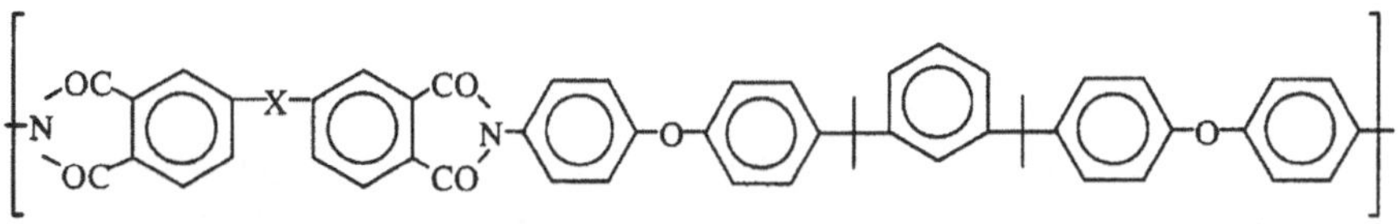

164a -c   **a** : X = σ-bond
                     b : X = CO
                     c : X = O

Structure 21 (1)

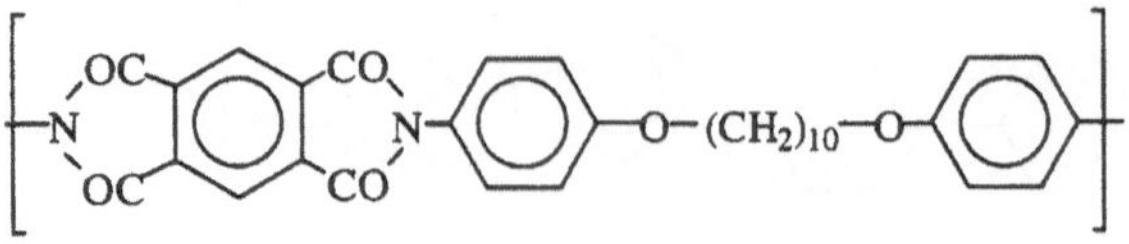

165

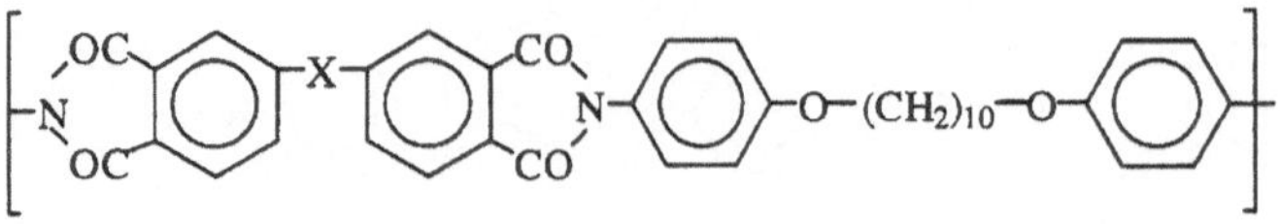

166a - c   **a:** X = O   **b:** X = CO   **c:** X = C(CF$_3$)$_2$

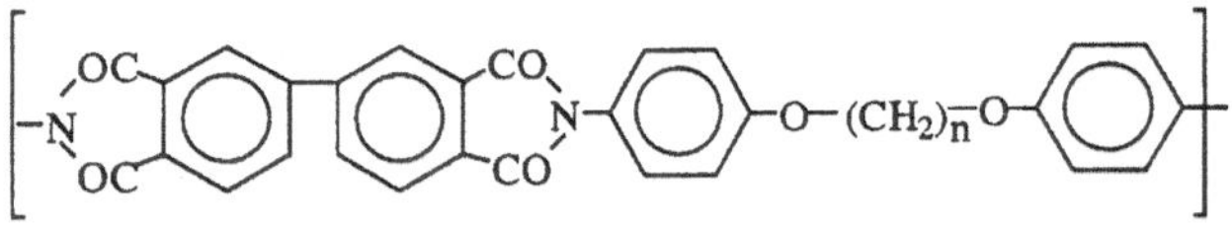

167a - f

a: n = 7	**b:** n = 8	**c:** n = 9
d: n = 10	**e:** n = 11	**f:** n = 12

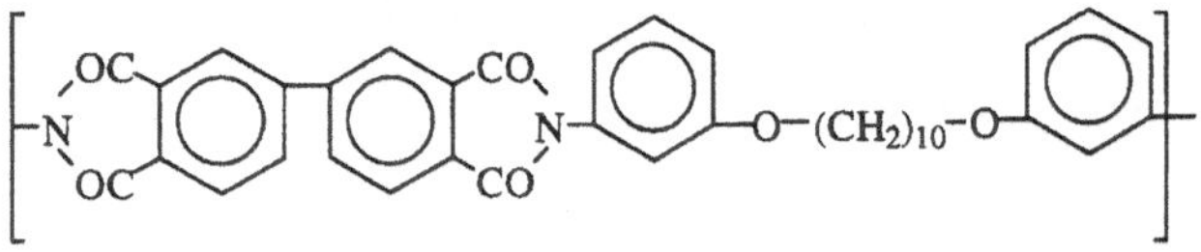

168

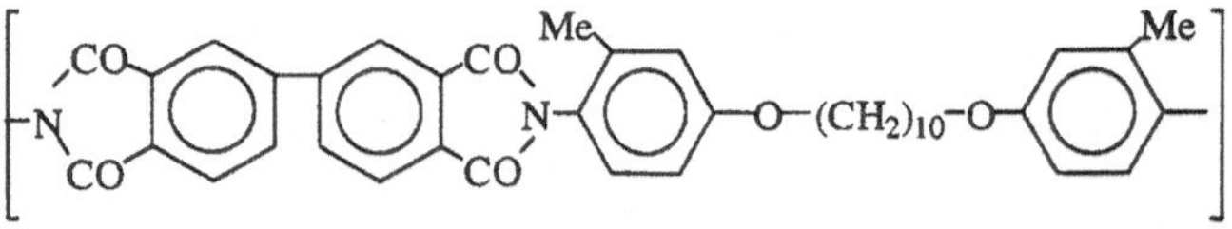

169

Structure 21 (2)

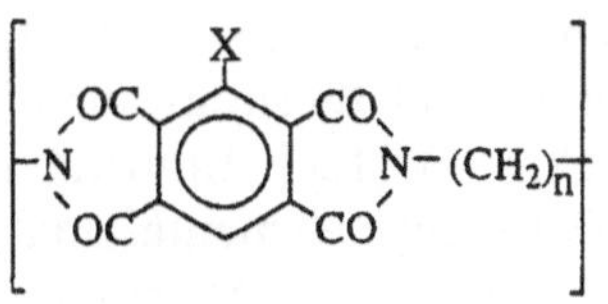

170a, b   a: X = H
                 n = 8, 9, 10, 12

             b: X = Br
                 n = 8, 9, 10, 12

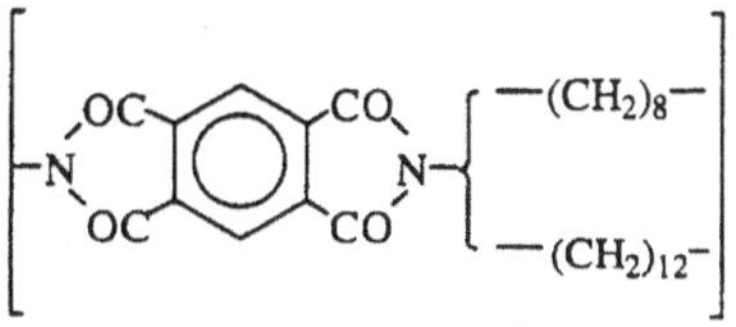

171

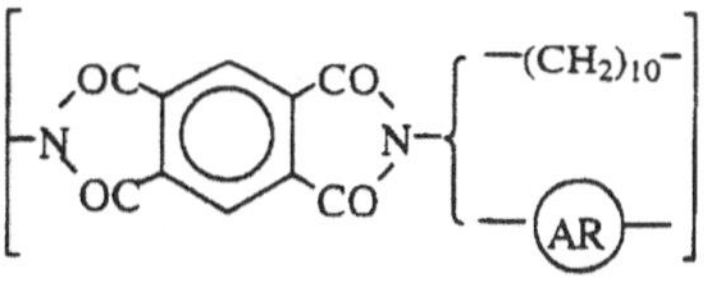

172a - b

a: (AR) =   b: (AR) =

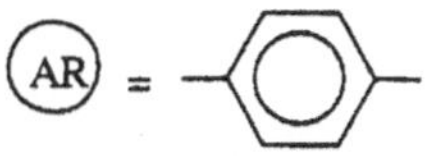

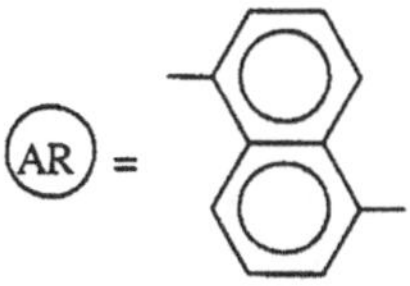

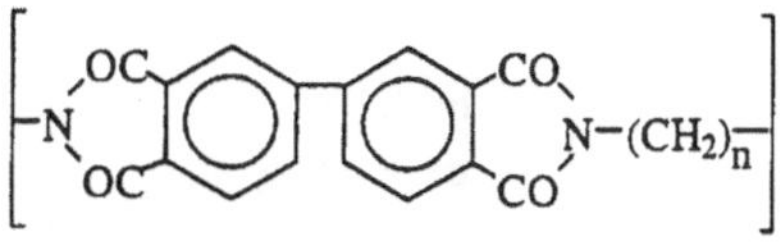

173a - c   a: n = 4   b: n = 9   c: n = 10

Structure 21 (3)

crystalline, but not liquid-crystalline [124]. Only after incorporation of additional aromatic ring an effective mesogen was obtained as illustrated by the thermotropic behavior of polyimide **175** [125]. For this polyimide one phase transition (T_{m1}) was found between two different crystalline smectic phases and a second phase transition (T_{m2}) to a liquid crystalline smectic phase which was not identified (for more details see also the chapter of Y. Imai). Finally, a paper should be mentioned which reports on syntheses of thermotropic oligomers having nadimide endgroups (**176a–d** and **177a–e**) [126]. Unfortunately, the characterization of these materials was poor, because elemental analyses, IR- and NMR spectroscopic data are lacking. Furthermore, no DSC-curves or X-ray patterns were reported and in the case of **177a–e** even the phase transition temperatures are missing. Apparently all these oligoimides undergo complete isotropization before a thermal cure can be achieved.

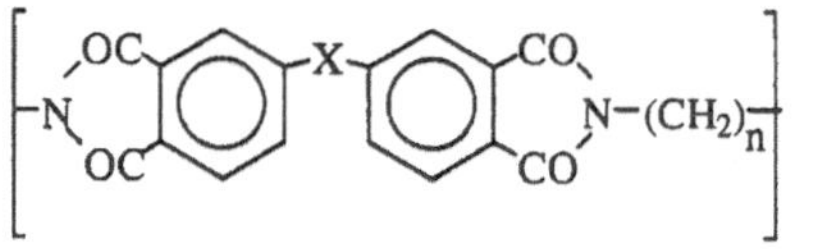

174a - c   **a:** X = O   **b:** X = CO   **c:** X = SO$_2$

(n = 4 - 9)

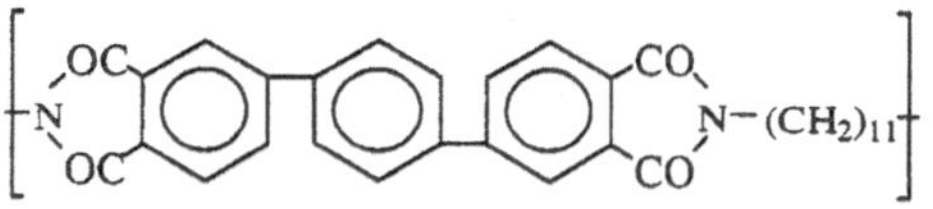

175

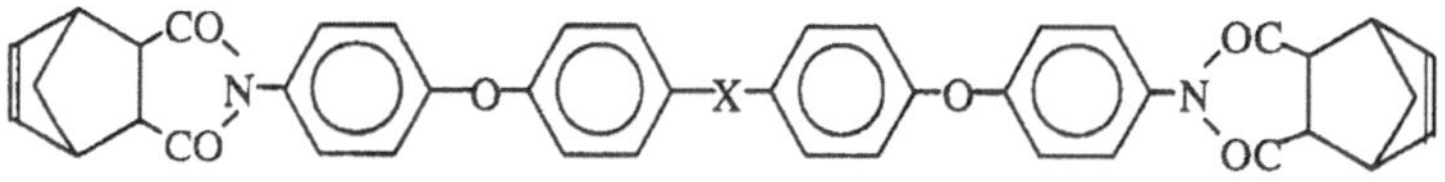

176a - d   **a:** X = σ - bond   **b:** X = O

c: X = C(CH$_3$)$_2$ **d:** X = SO$_2$

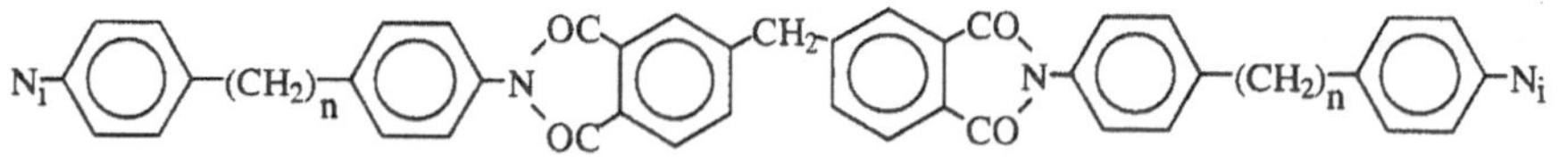

177a - e   **a:** n = 0   **b:** n = 1   **c:** n = 2

(N$_i$ = Nadimide)   **d:** n = 3   **e:** n = 4

Structure 22

7
Characterization of Layer Structures

7.1
General Considerations

The arrangement of molecules or ions in layers is one of the simplest supermolecular structures, and thus widely used by nature for the three dimensional ordering of molecules or their components. Salts like $CdCl_2$, CdI_2 $Mg(OH)_2$ (brucite), silicates and oligopeptides or proteins in the β-sheet (secondary) structure and the arrangement of amphiphilic molecules in the walls of vesicles or micelles may suffice as examples. In the case of synthetic polymers, two classes of layer structures have been found, smectic and sanidic structures. Smectic layers are characterized by layer planes running more or less perpendicular to the chain axes. Their detailed definition, characterisation and nomenclature has been elaborated for low molar mass smectic materials [127, 128]. Figure 20 illustrates the four different smectic layer structures which may be formed by polyimides in the solid state. The sanidic layers are characterised by layer planes running parallel to the axes of the main chains which usually form extended stacks as illustrated by Fig. 21. Depending on the number of substituents (usually *n*-alkyl chains) per repeating unit the substituents form an interdigitated array (Fig. 21A,B) or not (Fig. 21C). Only two papers have appeared [52, 114] containing short descriptions of synthesis and properties of poly(ester-imide)s forming sanidic layers. These PEIs have the chemical structure **33e, f; 34a, b** and **147a–d**, and in agreement with the scheme of Fig. 21 they adopt layer structures of the type A (for **33e, f**) and B (for **34a, b** and **147a–d**) respectively. All other polyimides that have been reported to form layer structures adopt the smectic type of layers, and thus this section concentrates on the discussion of smectic layer structures.

In order to avoid misunderstandings (frequenctly encountered by the author in discussion with colleagues) the following general aspects of smectic layer structures need a short discussion. As already mentioned in Sect. 4, LC-main chain polymers can only form two kinds of smectic LC-phases, namely smectic-A and smectic-C. In these LC-phases the interaction between the mesogens is weak enough to undergo translational motion under the influence of shear forces. When the electronic forces between the mesogens are so strong that they form a perfectly ordered array, they turn immobile, and such a layer of mesogens may be considered as a two-dimensional crystal. The "crosslinking" of the two-dimensional crystals by the numerous spacers yields a hard solid phase. Even a high degree of segmental mobility inside the spacer layers is no contradiction to a hard solid macroscopic appearance. Also in other semicrystalline hard engineering plastics such as poly(butylene terephthalate) or nylon-6,6 the CH_2-groups show a significant mobility in the temperature range between T_g and T_m [129–131]. In this connection it is of interest to see that short chain segments forming a solid smectic phase may play the role of hard segments and physical crosslinks in thermoplastic elastomers [132]. The solid smectic phase replaces the normal crystallites formed for example by PBT segments in PBT-based thermoplastic elastomers (e.g. Arnitel).

The question of whether a solid smectic phase should be labelled mesophase or crystalline phase is easy to answer from the viewpoint of a theoretical definition, but more difficult to answer by experimental techniques. When the mesogens in subsequent layers (i.e. in neighbouring "two dimensional crystals") possess exactly the same orientation in space, they meet the requirement of a three-dimensional long range order and such a layer structure may be called

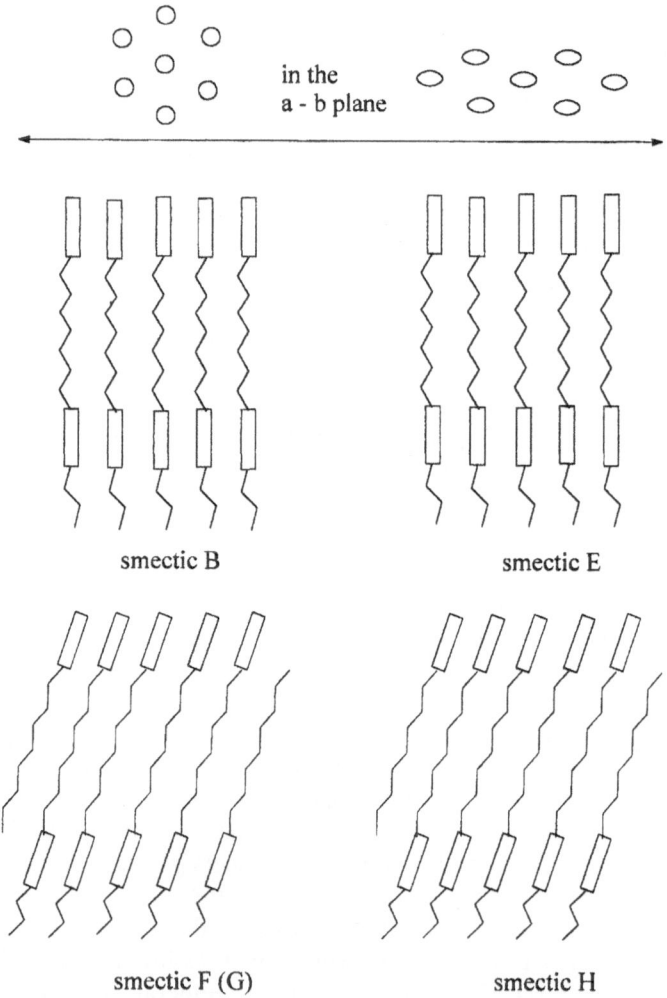

smectic B   smectic E

smectic F (G)   smectic H

Fig. 20. Schematic illustration of the four basically different smectic layer structures which may occur in the solid state of polyimides

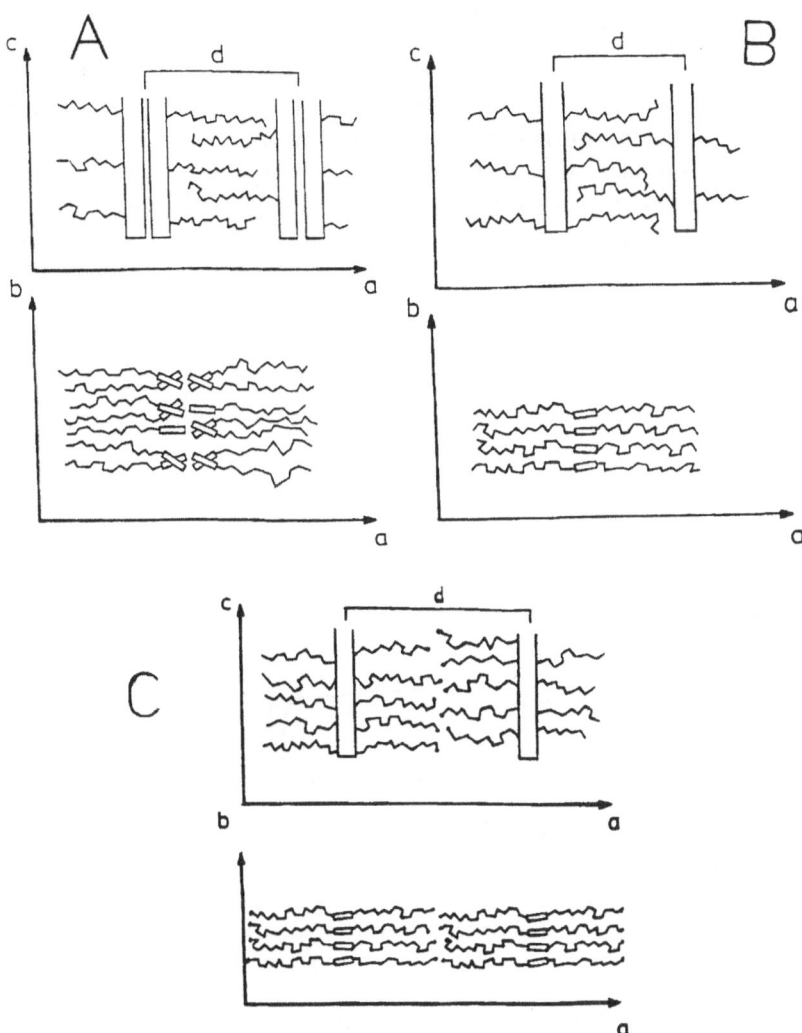

Fig. 21. Schematic illustration of the three most frequently encountered types of sanidic layer structures

smectic crystalline. If the spatial orientation of the mesogens in neighbouring two-dimensional crystals shows a statistical disorder, such as a variation of the tilt direction (e.g. in a smectic F or G phase), this layer structure is a smectic mesophase. A real LC-polyimide may, of course, represent an intermediate situation, with a certain coherence of the spatial order in neighbouring layers of mesogens, modified by a gradual decay of this order over longer distances. An experimental determination of the extent of this three-dimensional order is difficult to achieve, and no study in this direction has been published for LC-polyimides.

Nonetheless, the classification of a solid smectic layer structure as a smectic crystalline phase may be based on the combination of WAXD patterns, DSC measurements and optical microscopy. The following experimental facts suggest the existence of a smectic crystalline phase. First, the observation of sharp middle angle reflections (MARs) indicating the presence of a well ordered layer structure. Second, sharp wide angle reflections (WARs) indicating the existence of a crystalline order inside the layers of the mesogens (i.e. two-dimensional crystals). Third, the detection of a significant supercooling effect of the assumed crystallization process, either by DSC-measurements, optical microscopy or X-

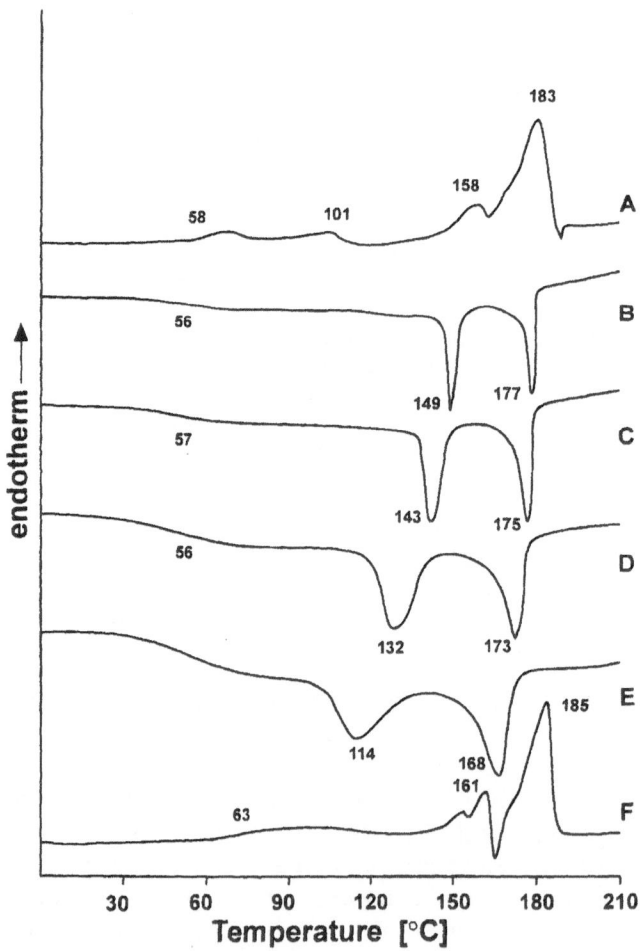

Fig. 22. DSC measurements of the PET 89k: **A** 1st heat, +20 °C/min; **B** 1st cool, –10 °C/min; **C** 2nd cool, –20 °C/min; **D** 3rd cool, –40 °C/min; **E** 4th cool, –80 °C/min; **F** 2nd heat (after cooling with –20 °C/min)

ray measurements with synchrotron radiation. A true crystallization process involves a nucleation step with an activation energy which is significantly higher than that needed for the formation of an LC-phase or any other mesophase. Figure 22 illustrates this relationship for a PEI showing an enantiotropic nematic phase and a smectic crystalline solid state. With increasing cooling rate the temperature of the crystallization (T_c) decreases more rapidly than that of the isotropic/nematic transition (T_{ai}). These different supercooling effects are also useful to detect the existence of short lived monotropic LC-phases as illustrated by Fig. 23 [88]. The existence of such monotropic LC-phases is in turn of interest,

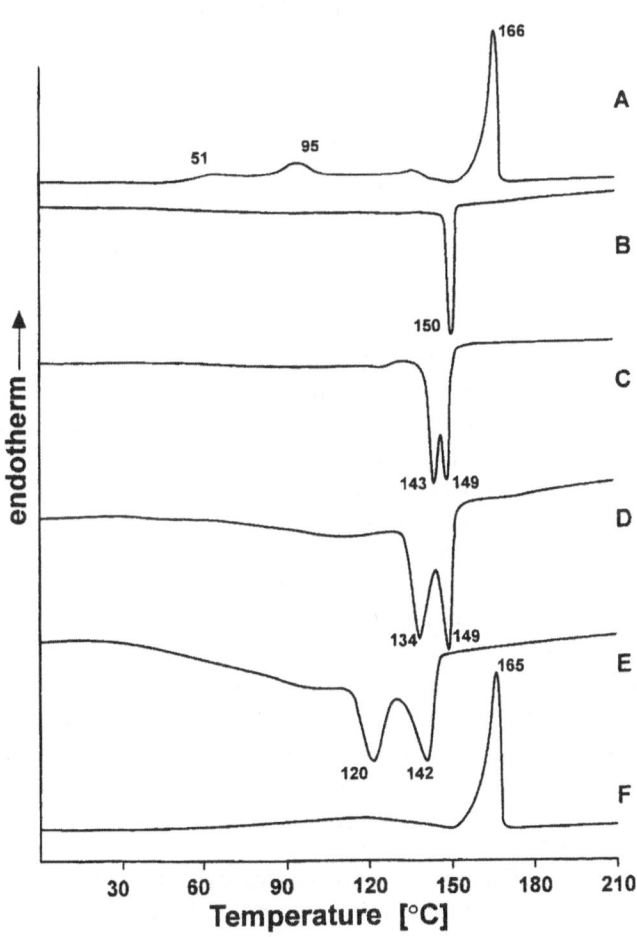

Fig. 23. DSC measurements of the PEI **89m: A** 1st heat +20 °C/min; **B** 1st cool, –10 °C/min; **C** 2nd cool, –20 °C/min; **D** 3rd cool, –40 °C/min; **E** 4th cool, –80 °C/min; **F** 2nd heat (after cooling with –20 °C/min)

when the crystal lattices, morphologies and crystallization kinetics of smectic phases are under investigation. A fourth and rather convincing argument for the identification of a smectic crystalline-phase may come from optical microscopy. Quite recently it has been demonstrated by Wutz [133] that smectic poly(ester-imide)s may crystallize in the form of spherulites. The formation of spherulites is characteristic for the existence of crystallites and absolutely unlikely for a solid mesophase. Taken together, the above-mentioned experimental criteria allow the identification of a smectic-crystalline phase, even when no direct information about the three-dimensional long range order is availabe. Yet detailed X-ray and electron diffration analyses of this interesting aspect are certainly needed to check the validitiy of the above-mentioned criteria and to complete the understanding of solid smectic phases in the case of LC-main chain polymers.

Finally, it should be emphasized that the existence of a smectic solid state does not necessarily entail the formation of a smectic LC-phase above the melting temperature. As outlined in Sects. 4 and 5, numerous poly imides have been synthesized forming smectic layers in the solid state and yielding an isotropic, nematic or cholesteric phase upon melting. Poly(ester-imide)s containing highly symmetrical imide moieties, such as PMDI units, are typical examples of polyimides combining a smectic solid state with an isotropic melt.

7.2
Problems and Experimental Results

The smectic order of low molar mass smectic materials is considered to result exclusively from the interaction between the mesogens. Therefore, several kinds of mesogenic moieties were called smectogens. This simple picture, if it is true at all, is not valid for LC-main chain polymers and systematic studies of polyimides containing aliphatic spacers have significantly contributed to the revision of this simple picture. Particularly informative were the properties of the poly(ester-imide)s **91a–h, 92a–c, 95a–c** and **96a–g**. When fibres of the PEIs **91a–h** were spun from the isotropic melt, so that a rapid cooling from 300–320 °C down to 25 °C occurred, the mesogens had no time for a coordination of the lateral order [89]. Consequently the fibre patterns display merely a broad halo in the wide-angle region (Fig. 24A). In contrast to the total disorder inside the layers of the mesogens a series of perfect layers (with their planes perpendicular to the fibre and chain axes) was formed. A WAXD pattern recorded from a bundle of fibres parallel to the fibre axes displays extremely sharp middle angle reflections up to the eights order! (Fig. 25). After annealing above the T_g the mesogens also acquired a high degree of lateral order corresponding to the orthorhombic order of a smectic-E phase [89]. This result clearly demonstrated that the formation of the layers has nothing to do with the crystallization of the mesogens inside their layers. Therefore, it is obvious the formation of layers is just the consequence of the regular array of highly polar and non-polar groups along the polymer chains, quite analogous to the formation of vesicles or micelles from amphiphilic molecules. This interpretation was supported by the observation that the PEIs **95a–**

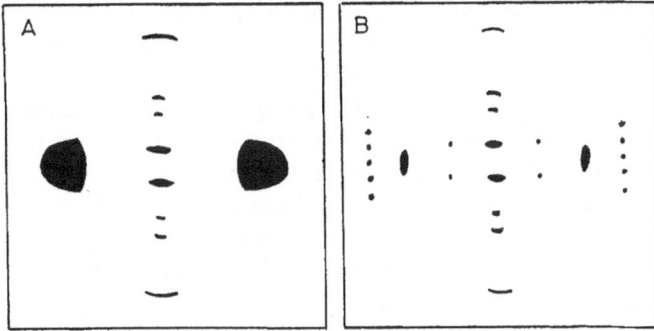

Fig. 24. Fibre pattern of the PEI 91: **A** as spun from the isotropic melt; **B** after annealing

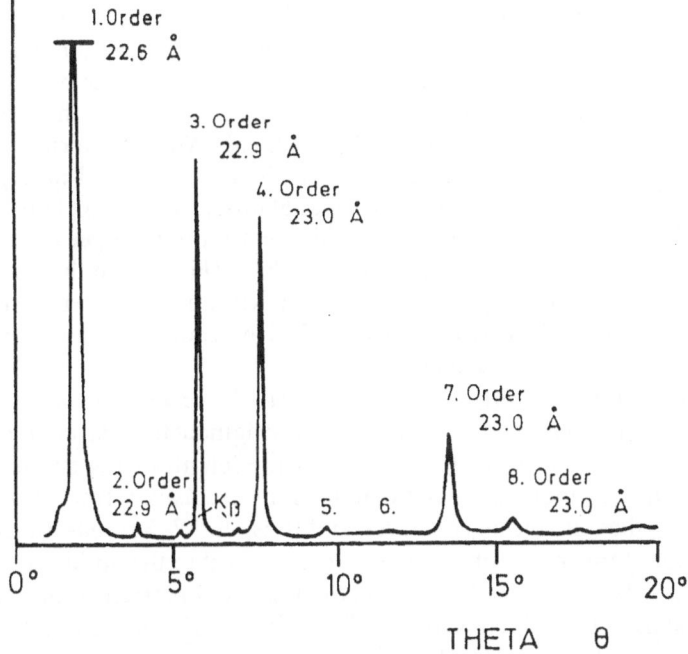

Fig. 25. WAXD pattern of a bundle of PEI 91 fibres (as spun) recorded parallel to the fibre axes

c derived from polar ethylenoxide spacers are amorphous, isotropic and reluctant to form layers [92]. Yet, when the ethylene oxide spacers are combined with a non-polar mesogens as it is true for the polyesters **178** they favour the existence of smectic layers in the melt and in the solid state. From the co PEIs **96a–g** it was learned [92] how many alkane spacers were needed for the stabilization of a layer structure in the solid state (an enantiotropic smectic LC-phase was never observed).

All further research on polyimides containing a regular sequence of the highly polar imide groups and non-polar aliphatic moieties has confirmed that the existence of layers in the liquid or solid state is exclusively due to the regular array of polar and non-polar groups. The most conspicious consequence of this structure-property relationship is the existence of smectic layers by the fully aromatic polyimides of PMDA (**160–163** [119]). In contrast, layer structures have never been observed in fully aromatic polyesters bare of polar heterocycles or polar substituents (e.g. nitro groups). Polyimides containing a regular sequence of alkane spacers are thus excellent model polymers for any kind of chemical or physical studies of layer structures.

In this connection the PEIs **91a–k** are of particular interest not only because of the above-mentioned fibre pattern. It was found that the PEIs **91a–k** can adopt three different kinds of smectic orders in the solid state [89, 92, 134, 135]. The layer distances are nearly identical, but the lateral order of the mesogens can vary. When quenched from the isotropic melt the monotropic smectic-A phase is formed under all circumstances and frozen in. Such a solid state characterized by the fibre pattern of Figs. 24 and 25 and the WAXD powder pattern of Figs. 26A and 27A may also be called a smectic A glass. Upon annealing the "amorphous halo" narrows and forms a sharp tip (Fig. 27B). This WAXD pattern is typical for a smectic-B phase, and means, that the originally disordered mesogens now tend to adopt a more or less perfect hexagonal order in the a–b plane. Upon annealing above the T_m of this smectic-B like phase the mesogens adopt an orthorhombic array analogous to a smectic-E phase (Fig. 22). In the case of short spacers (**91a–c**) even a crystal modification without layer structure may be formed (Fig. 26B). In the case of series **91a–k** also detailed studies of the crystallization kinetics were reported [135].

The WAXD powder pattern of the PEIs **91a–k** (or **89a–m**) varies with the lengths of the spacers, because the reflections originate from layer planes across several layers of mesogens. However, when the length of the spacers reaches 12 CH_2 groups (or more) the WAXD reflections exclusively result from groups and atoms inside one mesogenic layer, and thus, the WAXD reflections become independent of the length of the spacers (in contrast to the middle angle reflections). Figure 27 represents the typical WAXD powder pattern of the two dimensional crystals of **91h** which is identical with the WAXD powder patterns of **91i** and **91k**. Also in the case of the PEIs **89a–m** the border line for spacer independent WAXD reflections is a spacer length of 12 CH_2 groups. This characteristic spacer length may, of course, vary, when the dimensions of the mesogen increase. Finally, it should be mentioned that the textures and morphologies of the PEIs **91a–k** were carefully studied by transmission electron microscopy [134].

The various orientations and conformations mesogenic groups can adopt in a layer has been studied extensively for low molar mass smectic materials, and the classification and terminology of smectic systems is entirely based on these studies. However, low molar mass smectic compound or smectic LC-side chain polymers do, of course, not allow one to elucidate the role spacers play in the layer structures of LC-main chain polymers. Therefore, poly(ester-imide)s, po-

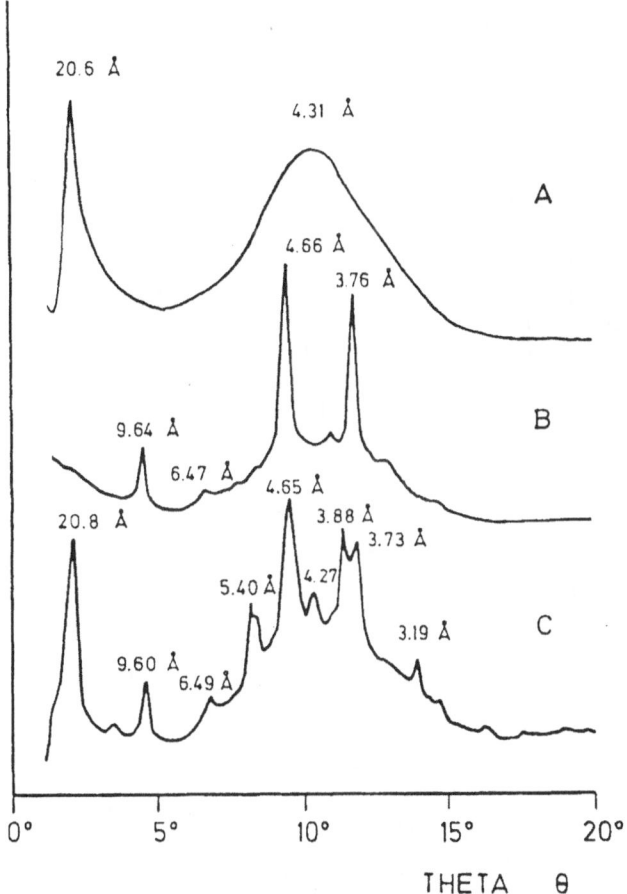

Fig. 26. Fibre pattern of the PEI **91**: **A** as spun from the isotropic melt; **B** after annealing

ly(ether imide)s and poly(amide-imide)s forming smectic layers are useful substrates to study the influence and properties of spacers. The PEIs **89a–m** have demonstrated that increasing length of alkane spacers reduces the isotropization temperature (T_i) to such an extent that the LC-phase may totally fade away. Substituents attached to the alkane spacers (e.g. methyl groups) considerably destabilize smectic LC-phases and may turn them into nematic phases as shown for the PEIs **87a, b** and **88a, b**. Yet neither long spacers nor small substituents hindered the formation of layers in the solid state. Only the ether groups may totally prevent the existence of layer structures as illustrated by the PEIs, **95a–c** and **96a–g**. Another interesting problem arises when alkane spacers of different length are combined in all spacer layers of a sample. Such a structure can easily be prepared when equimolar mixtures of two different diols or equimolar mixtures of two aliphatic dicarboxylic acids are used for the polycondensation. Due

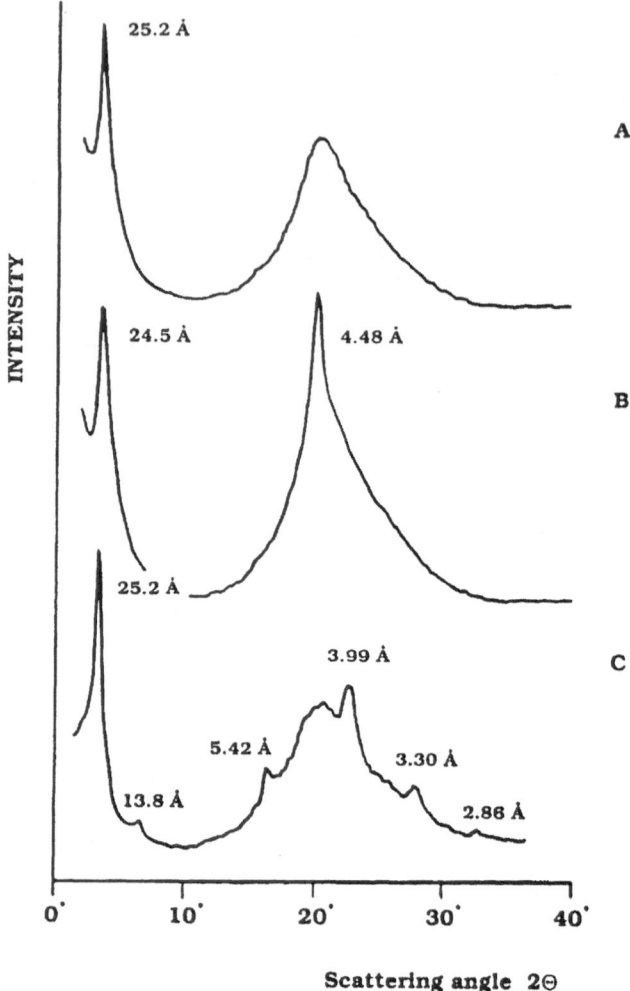

Fig. 27. WAXD powder patterns of PEI **91h**: **A** after quenching from the isotropic melt; **B** after annealing at 70 °C; **C** after annealing at 170 °C

to the identical reactivities of the functional group a random copolycondensation will occur, with the consequence that nearly equal amounts of the two different spacers will be present in most layers.

First examples of this type of copoly(ester-imide)s reported in the literature were the co PEIs **92a–c** [92]. The X-ray measurements indicated the existence of smectic layers, and calculation of the layer distance (d-spacings) via the Bragg equation gave the following result. The d-spacing shrinked with decreasing length of the shorter spacer and matched the length of the fully extended short spacers. Therefore, it was assumed that the short spacers define the layer distance, and the long-spacers are more or less coiled in the space between the extended short spacers and the two dimensional crystals of the mesogens (Fig. 28A). A partial coiling of the long (C=22) spacers is possible because the cross-section of the imide mesogens is nearly twice as large as the cross-section of an alkane chain in the all-trans conformation. When the difference between the alkane spacer became greater, as it was realized in the case of **90a, b** a new

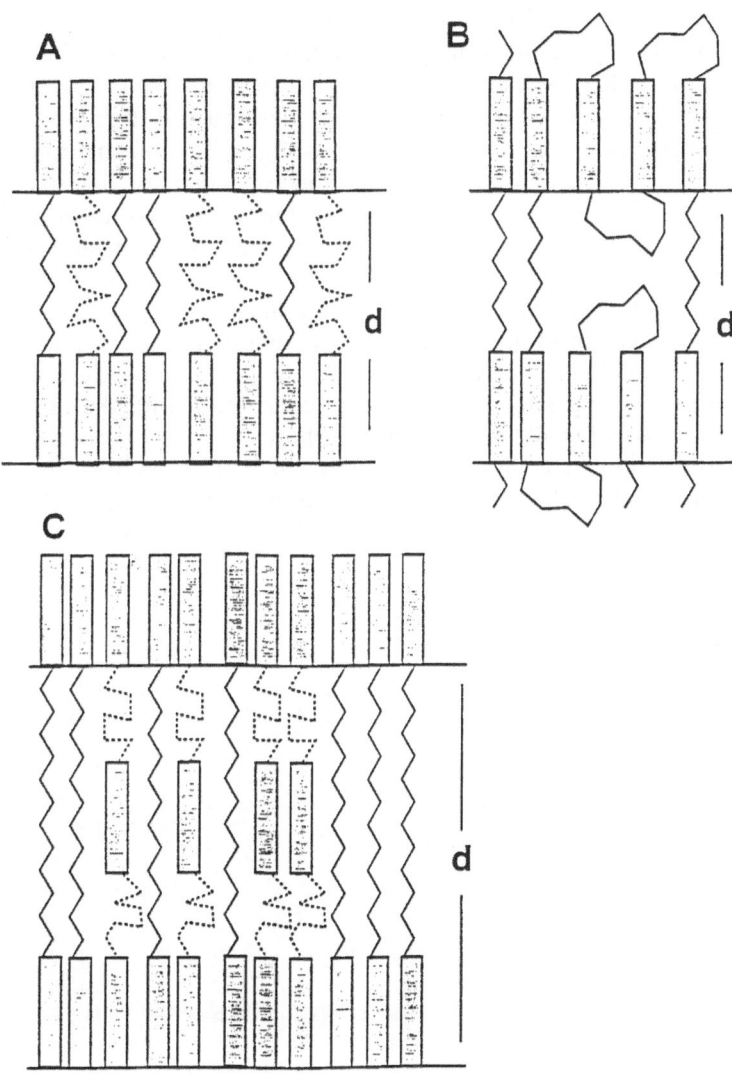

Fig. 28. A Schematic layer structure of the co PEIs **92a–c** (the short spacers are fully extended, the longer spacers are coiled). **B** Schematic layer structure of the co PEI **90b. C** Schematic layer structure of the co PEIs **84a–f**

phenomenon was observed. The X-ray powder pattern of **90b** indicates a d-spacing corresponding to the length of the fully extended long spacer [88]. This result was interpreted by the scheme of Fig. 28C. Obviously the short spacers and some mesogens are embedded in the voluminous soft layer defined by the long spacers. The steric demands of the mesogens dispersed in the spacer layer cause the stretching of the long spacers.

A third variant of the spacer layers came into sight, when the co PEIs **83a–g** were studied [24]. Like their parent homopolyester **82a** [82] they form a smectic crystalline solid state with upright mesogens analogous to a smectic E phase (the homo PEIs **82b, d, f** and **h** show a slight tilt of mesogens [24]). The d-spacing of the homo PEI **82i** corresponds to a nearly extended alkane-spacer. Interestingly, the d-spacings of all co PEIs **83a–g** are identical and slightly shorter than that of **82i** corresponding to a higher fraction of gauche conformations. In other words the layer distances of **83a–g** do not show any indication that spacer shorter than 12 CH_2 groups are present. However, the scheme presented in Fig. 28C cannot serve as explanation in this case, because the mesogens are so long that they cannot be dispersed in the alkane layer. The hypothetical explanation published by the author is based on the assuption that the short spacers form loops inside the spacer layers and on the surface of the smectic crystallites. This scheme presented in Fig. 28B needs, of course, confirmation by further studies.

Another set of problems was studied in connection with the PEIs **179, 180** [136]. It was found that all these PEIs do not show an LC-phase, but form a smectic crystalline solid state with upright mesogens (smectic-E like). However, the PEIs can in principle adopt two different kinds of chain packing and layer structures as illustrated in Fig. 29. The first is that the mesogens pack in an antipar-

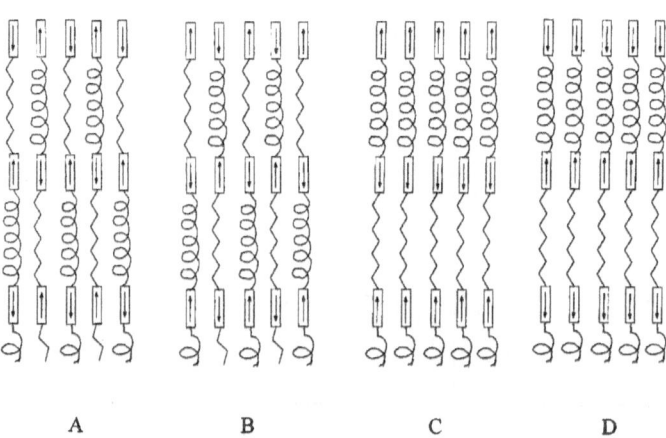

A   B   C   D

Fig. 29. Scheme of alternative layer structures of the PEIs v and w: **A** and **B** antiparallel orientation of the dipoles in the mesogen layers and mixing of alkane and ethylene oxide spacers; **C** and **D** parallel array of the mesogens and separate layers of alkanes and ethylene oxide

allel fashion, and the spacer layers contain an equimolar mixture of alkane and ethylene oxide spacers (models A and B). This arrangement may be energetically more favourable for the mesogens and has the consequence that the properties of the PEIs 179 and 180 are almost identical. The second is that the mesogens pack in a parallel fashion and the alkane spacers form a homogeneous layer in an alternating sequence with homogeneous oligoethylene-oxide layers (models C and D). This layer structure is energetically more favourable for the spacers. It turned out that the properties of both isomeric PEIs are different, and only the PEI 179 showed a reversible first order solid phase transition; thus supporting the chain packing outlined in Fig. 9C,D. This hypothesis was supported by the properties of the analogous poly(benzoxazole ester)s 182 and 183 [137]. Their properties were again quite different, and only one of them formed a smectic LC-phase. These results suggest in turn that also the PEIs 181a, b crystallize in such a way that homogeneous spacer layers are formed. In this connection it was found that the X-ray beam did not see the full periodicity of the six layers. Only the reflections of sublayers were detectable [136, 137].

The X-ray analyses discussed above only provide a static picture of the smectic layer structures. However, it is obvious that at least the spacers posses a certain degree of segmental mobility at room temperature, which will be even higher at elevated temperature. Therefore, it was of great interest to find out how the conformation and segmental motions of the spacers depend on the neighbouring rigid layers of the mesogens. ^{13}C NMR cross-polarization/magic angle (CP/MAS) spectra allow a crude quantification of the tt, gt and gg conformations of long alkane spacers, an analytical approach originally elaborated for polyethylene and other alkanes. As illustrated in Fig. 10, the long alkane spacers of PEIs may contain a high fraction of tt conformations even at temperatures up to 100 °C. Two quite different kinds of "chain packing" and segmental motions may account for this finding. Either the middle segments of the alkane spacers form

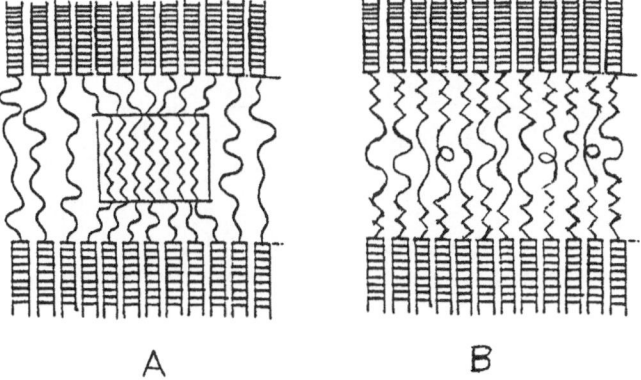

A   B

Fig. 30. A Scheme of paraffin domains in the spacer layers. **B** Scheme of of layer structures containing conformationally disordered spacers "paraffin domains" based on the all-trans conformation (Fig. 30A), or the middle segment are highly mobile and the CH_2-groups attached to the mesogens are responsible for the tt-conformations and are less mobile (Fig. 30B).

In order to discriminate between both hypotheses PEIs having selectively deuterated spacers were synthesized (**184–186**). The deuterated spacers were prepared by Pt catalyzed deuteration of spacers containing a diacetylene unit in their midst. The deuterated PEIs were examined by 2H NMR spectroscopy in the

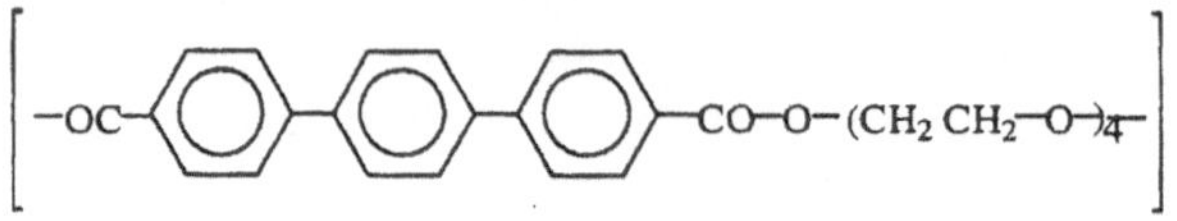

178

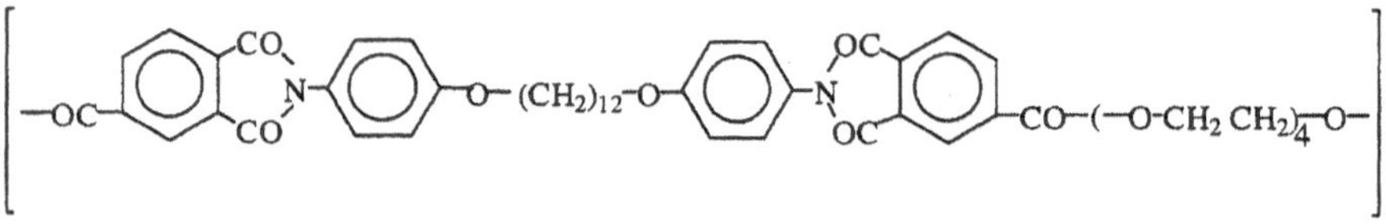

179

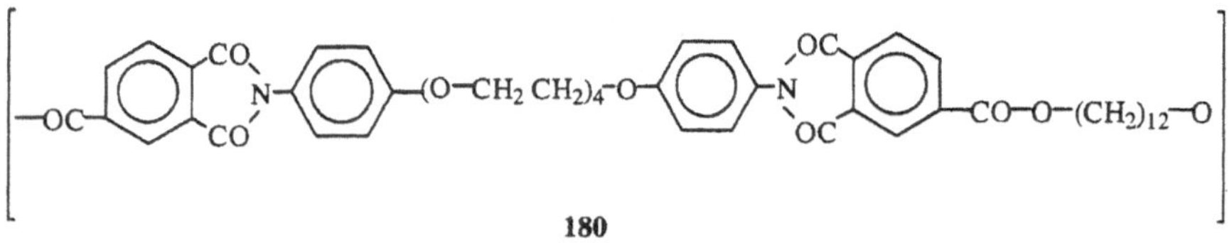

180

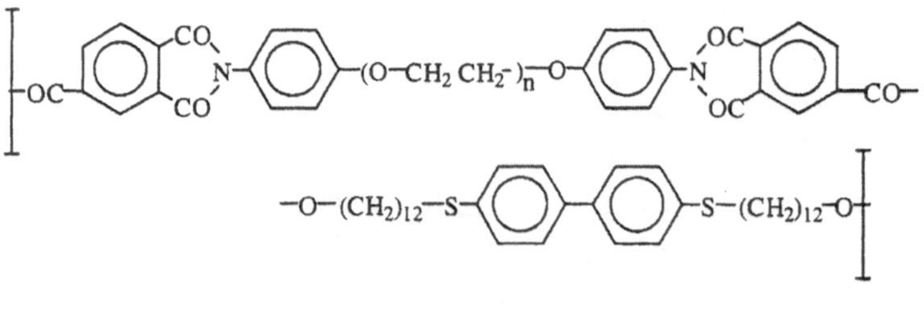

181a, b   **a:** n = 2   **b:** n = 4

Structure 23 (1)

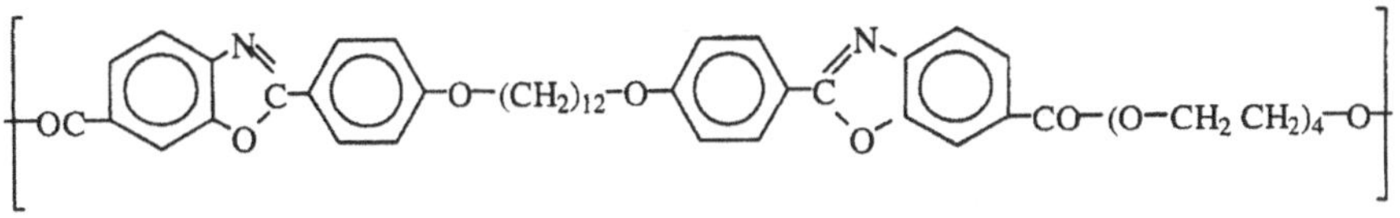

182

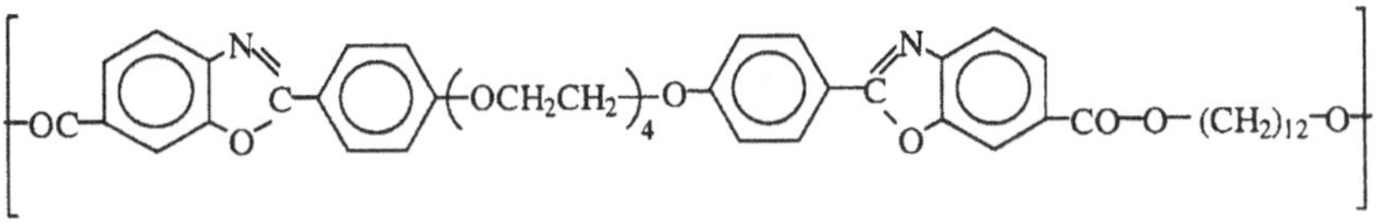

183

Structure 23 (2)

temperature range from –100 to 180 °C. These measurements proved that the middle part of the spacers is highly mobile above room temperature. Solid state 13C NMR spectroscopy with quadropol suppression allowed to eliminate the signals of the deuterated carbons. In this way it was found that the CH_2-CH_2 groups neighbouring the mesogen layers prefer the tt-conformation. In other words, both NMR methods confirmed the validity of the hypothesis outlined in Fig. 10B. Finally it should be mentioned that quite recently two papers have appeared [138, 139] reporting on dielectric relaxation measurements of the PEIs **89a–m** and **91a–k**. Three relaxation regions were found in the temperature range 100–400 K which were identified as α-, β- and γ-relaxations. For the α-process a WFL-like temperature dependence was found and for the β-relaxation an Arrheniuses-type temperature dependence. Furthermore, both series of PEIs showed similar β-processes. The motions detected by the dielectric relaxations concern vibrations and reorientations of strong dipoles, and thus result from the mesogens and not from the alkane spacers. Therefore, all these measurements together gradually provide a complete picture of the chain dynamics in solid PEIs forming smectic layer structures (see structure 24)

Finally, it should be emphasized that the tendency of polyimides containing aliphatic spacers to form stable layer structures of various dimensions and degrees of order can be utilized for an even wider variety of studies. A recent example is a study of the epitaxial growth of polyethylene on smectic crystallites of PEI [140]. Most likely smectic crystallites adopt a lamellar form with the large surface "covered" by loops of the aliphatic spacers. This hypothesis still need detailed studies and confirmation. Anyway, smectic crystalline polymers are interesting substrates for studies of epitaxial crystallizations. Furthermore, layer structures derived from long aliphatic spacers (alkanes or oligoethers) may play

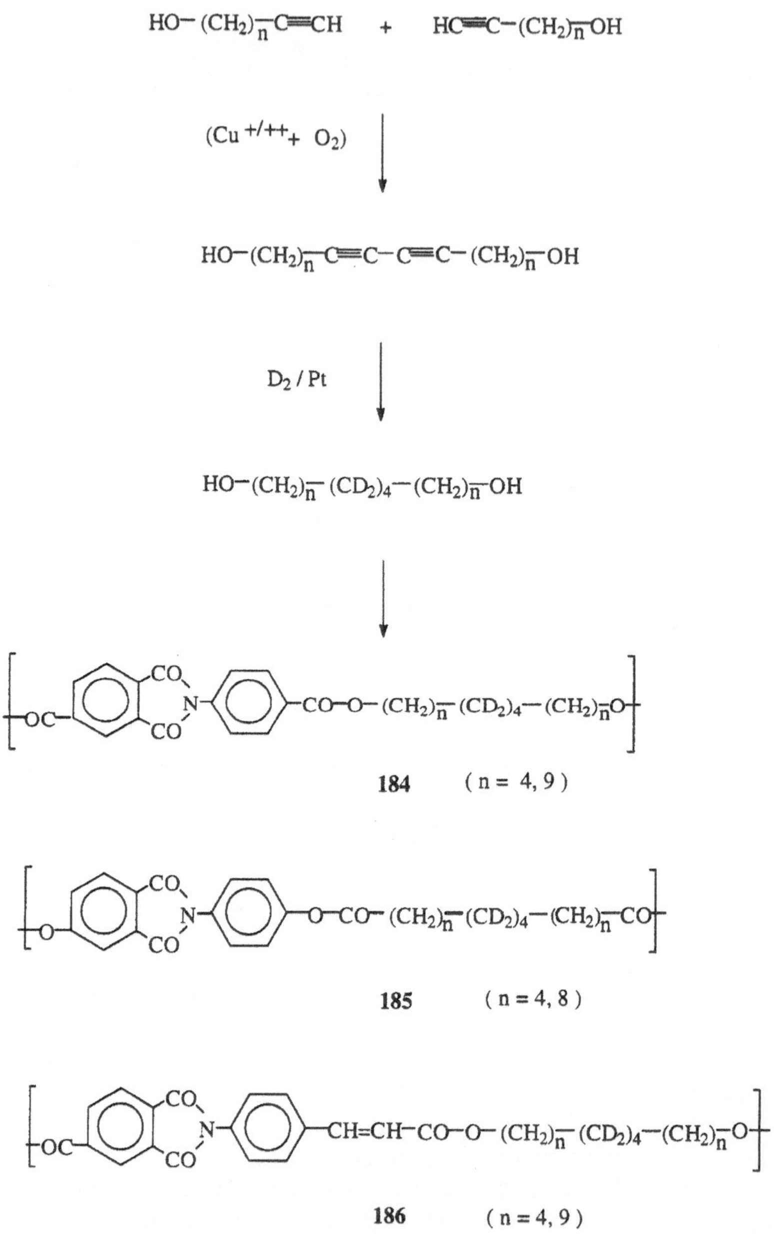

$$HO-(CH_2)_{\overline{n}}C\equiv CH \quad + \quad HC\equiv C-(CH_2)_{\overline{n}}OH$$

$$(Cu^{+/++} + O_2) \quad \downarrow$$

$$HO-(CH_2)_{\overline{n}}C\equiv C-C\equiv C-(CH_2)_{\overline{n}}OH$$

$$D_2/Pt \quad \downarrow$$

$$HO-(CH_2)_{\overline{n}}(CD_2)_4-(CH_2)_{\overline{n}}OH$$

$$\downarrow$$

184   (n = 4, 9)

185   (n = 4, 8)

186   (n = 4, 9)

Structure 24

the role of hosts for the embedding of guest molecules. Such guest molecules may be ions, deuterated photoreactive or chiral components enabling a broad variety of physico-chemical studies. However, syntheses and investigations of such guest-host layers have not been published yet.

8
References

1. Cassidy PE (1982) Thermally stable polymers, chap 5. Marcel Dekker, New York
2. Elias HG, Vohwinkel F (1983) Neue Polymere Werkstoffe, chap 11, 2nd edn. Hanser, München, p 257
3. Mittal KL (ed) (1982) Polymides, vol 1. Plenum Press, New York London
4. Verbicky W Jr (1988) In: Bikales M, Mark HF, Overger CG, Menges G (eds) Encyclopedia of polymer science and engineering, vol 12, 2nd edn. Wiley, New York, p 363
5. Sillion B (1989) In: Allen G, Bevington JC (eds) Comprehesive polymer science, chap 30, vol 5. Pergamon, Oxford
6. Alam S, Kandpal LD, Varma IK (1933) J Macromol Sci Rev Macromol Chem Phys 33:291
7. de Abajo J (1992) In: Kricheldorf HR (ed) Handbook of polymer synthesis, chap 15. Marcel Dekker, New York
8. Sroog CE (1991) Prog Polym Sci 16:561
9. Feger C, Khohasteh MM, McGrath J (eds) (1989) Polyimides chemistry, materials and characterization. Elsevier, Amsterdam
10. Gosh MK, Mittal KL (eds) (1996) Polyimides – fundamentals and applications. Marcel Dekker, New York
11. Irwin RS (1979) US Pat 4,176,223 to EI Du Pont; Chem Abstr 92:95,539u (1980)
12. Irwin RS (1983) US Pat 4,383,105 to EI Du Pont; ChemAbstr 99:72113f (1983)
13. Evans JR, Orwoll RA, Tang SS (1985) J Polym Sci Polym Chem Ed 23:971
14. Kricheldorf HR, Pakull R (1985) J Polym Sci Polym Lett Ed 23:413
15. Onsager LA (1949) Ann NY Acad Sci 51:627
16. Flory PJ (1953) Principles of polymer chemistry. Cornell Univ Press, Ithaca, New York
17. Flory PJ (1976) R Soc London A 234:73
18. Flory PJ (1984) Adv Polym Sci 59:1
19. Bialecka-Florjanczyk E, Drzeszko A (1993) Liq Cryst 15:255
20. Kricheldorf HR, Linzer V (1995) JMS Pure Appl Chem A 32:311
21. Kricheldorf HR, Domschke A, Schwarz G (1991) Macromolecules 24:1011
22. Kricheldorf HR, Linzer V (1995) Polymer 36:1893
23. Kricheldorf HR, Gurau M (1995) J Polym Sci Part Polym Chem 33:2241
24. Kricheldorf HR, Linzer V, Leland M, Cheng SZD (1997) Macromolecules 30:4828
25. Mosher WA, Chlystek SJ (1972) J Heterocycl Chem 9:319
26. Gitis SS, Ivanova VM, Nomleva SA, Seina IN (1967) Zh Org Khim 2:1265; (1966) Chem Abstr 66:85,592e
27. Serra A, Cadiz V, Mantecon A, Martinez PA (1985) Tetrahedron 41:763
28. Dorogova NK, Soloveva LM, Mironov GS, Filipova TP (1982) Izv Kyssh Uchebn Zaved Khim Khim Technol 25:415; ChemAbstr 97:109,842
29. Mulvaney IE, Figeroa FR, Wu SJ (1986) J Polym Sci Part A Polym Chem 24:613
30. Hisgen B, Kock HI (1987) Ger Offen 3,542,798 to BASF AG; Chem Abstr 107:218,264 (1987)
31. Hisgen B, Portugell M, Blinne G (1987) Ger Offen 3,542,833 to BASF AG; Chem Abstr 107:199,170m (1987)
32. Hisgen B, Portugell M, Reiter V (1987) Ger Offen 3,542,857 to BASF AG; Chem Abstr 107:199,173g (1987)
33. Wang S, Yang Z, Mo Z, Zhang H, Feng Z (1996) Polymer 37:4397
34. Kricheldorf HR, Krawinkel ThM, Schwoltz G (1998) JMS-Pure Appl Chem (in press)
35. Sidorovich AV, Baklagina YG, Kenarov AV, Nadezhin YS, Adrova NA, Florinsky FS (1977) J Polym Sci Polym Symp 58:359
36. Kricheldorf HR, Linzer V, Bruhn Ch (1994) JMS – Pure ApplChem A31:1315
37. Fujiwara H, Ozake H, Isaba T, Tayama T (1990) Jap Pat 0,433,925 to Mitsubishi Chem; Chem Abstr 117:27,491g (1992)

38. Sasaki S, Hasuda Y (1987) J Polym Sci Part C Polym Lett 25:377
39. Kricheldorf HR, Schwarz G; Nowatzky W (1989) Polymer 30:935
40. Kricheldorf HR, Schwarz G, Adebahr T, Wilson DI (1993) Macromolecules 26:6622
41. Culbertson BM (1970) US Pat 3,542,731 to Ashlaud Oil; Chem Abstr 74:55430m (1971)
42. Kurita K, Matsuda S (1983) Makromol Chem 184:1223
43. Kurita K, Mikawa N, Koyama Y, Nishimara S-I (1990) Macromolecules 23:2605
44. Tomioka T, Takeya T (1987) Jap Pat 62,161,832 to Idemitsu Petrochem Co; Chem Abstr 108:39,533 (1988)
45. Wakabayashi M, Fujiwara K, Hayashi H (1989) Eur Pat 0,314,173 to Idemitsu Petrochem Co; Chem Abstr 111:154,637k (1989)
46. Fujiwara K, Hayashi H, Wakabayashi M (1990) Eur Pat 0,375,960 to Idemitsu Petrochem Co
47. Kricheldorf HR, Linzer V, Bruhn Ch (1994) Eur PolymJ 30:549
48. Dicke HR, Genz I, Eckhardt V, Bottenbruch L (1989) Ger Offen 3,737,067 to Bayer AG; Chem Abstr 111:215,120k (1989)
49. Smolka MG, Jehnichen D, Komber H, Voigt D, Böhme F, Rätzsch M (1995) Angew Makromol Chemie 229:159
50. Yamanaka T, Inoue S (1991) Jap Pat 0,333,125 to Toray Ind Inc; Chem Abstr 116:21,700 (1992)
51. Kricheldorf HR, Hüner R (1990) Makromol Chem Rapid Commun 11:211
52. Kricheldorf HR, Schwarz G, Domschke A, Linzer V (1993) Macromolecules 26:5161
53. de Abajo I, de la Campa J, Kricheldorf HR, Schwarz G (1992) Eur Polym J 28:261
54. Kricheldorf HR, Bruhn Ch, Rusanov A, Komarova L (1993) J Polym Sci Part A Polym Chem 31:279
55. Kricheldorf HR, Hüner R (1992) J Polym Sci Part A Polym Chem 30:337
56. Bonfanti C, Lezzi A, Pedrelti V, Roggero A, La Mantia FP (1993) Eur Pat 0565 195; Chem Abstr 120:246,615r (1994)
57. Kricheldorf HR, Gerken A, Alanko H (1998) J Polym Sci Part A Polym Chem (in press)
58. Alanko H, Kricheldorf HR, Salmela A (1994) Fin Pat Appl 943,188 to NESTE Oy; ChemAbstr 124:P234,016r (1996)
59. Pospiech D, Häußler L, Eckstein K, Komber H, Voit B, Jehnichen D, Meyer E, Jahnke A, Kricheldorf HR (1997) ACS Polym Prepr Div Polym Chem 38(2):398
60. Pospiech D, Häußler L, Eckstein K, Komber H, Voit B, Jehnichen D, Meyer E, Jahnke A, Kricheldorf HR (1998) In: Designed monomers and polymers 1 (in press)
61. Pospiech D, Eckstein K, Shaik A, Kricheldorf HR
62. Eck Th, Gruber HF (1994) Macromol Chem Phys 195:3543
63. Kricheldorf HR, Pakull R, Buchner S (1988) Macromolecules 21:1929
64. Aducci JM, Nie F, Lenz RW (1990) ACS Polym Prep Div Polym Chem 31:63
65. Kricheldorf HR, Pakull R, Schwarz G (1995) Makromol Chem 194:1209
66. Sato M, Hirata T, Mukaida K (1992) Makromol Chem 193:1724
67. Hirata T, Sato M, Mukaida K (1993) Makromol Chem 194:2861
68. Hirata T, Sato M, Mukaida K (1994) Macromol Chem Phys 195:2267
69. Sun S-J, Chang Th C (1994) J Polym Sci Part A Polym Chem 32:3039
70. Orzeszko A, Mirowski K (1990) Makromol Chem 191:701
71. Mustafa IF, Dujaili AH Al, Alto AT (1990) Acta Polymerica 41:310
72. Sato M, Hirata T, Mukaida K (1994) Macromol Rapid Commun 15:203
73. Hirata T, Sato M, Mukaida K (1994) Macromol Chem Phys 195:1611
74. Sato M, Ujiie S (1996) Macromol Phys 197:2765
75. Kricheldorf HR, Jahnke P (1990) Eur Polym J 26:1009
76. Karayannidis G, Standos D, Bikiaridis D (1993) Makromol Chem 194:2789
77. Sato M, Hirata F, Kamita T, Makaida K-I (1996) Eur Polym J 32:639
78. Kricheldorf HR, Rabenstein M (manuscript in preparation)
79. Kricheldorf HR, Rabenstein M (manuscript in preparation)
80. Kricheldorf HR, Rabenstein M (manuscript in preparation)

81. Kricheldorf HR, Pakull R (1987) Polymer 28:1772
82. Kricheldorf HR, Pakull R (1988) Macromolecules 21:551
83. Leland M, Wu Z, Chhajer M, Ho R-M, Cheng SZD, Keller A, Kricheldorf HR (1997) Macromolecules 30:5249
84. Kricheldorf HR, Pakull R (1989) New Polym Mater 1:165
85. Leland M, Wu Z, Ho R-M, Cheng SZD, Kricheldorf HR (1998) Macromolecules 31:22
86. De Abajo J, De la Campa J, Schwarz G, Kricheldorf HR (1997) Polymer 38:5677
87. De Abajo J, De la Campa J, Kricheldorf HR, Schwarz G (1990) Makromol Chem 191:537
88. Kricheldorf HR, Probst N, Schwarz G, Wutz Ch (1996) Macromolecules 29:4234
89. Kricheldorf HR, Schwarz G, De Abajo J, De la Campa J (1991) Polymer 32:942
90. Aducci IM, Facinelli IV, Lenz RW (1994) J Polym Sci Part A Polym Chem 32:2931
91. Pardey R, Zhang A, Gabori PA, Harris FW, Cheng SZD, Aducci I, Facinelli IV, Lenz RW (1992) Macromolecules 25:5060
92. Kricheldorf HR, Schwarz G, Berghahn M, De Abajo J, De la Campa J (1994) Macromolecules 27:2540
93. Kricheldorf HR, Probst N, Wutz Ch; (1995) Macromolecules 28:7990
94. Kricheldorf HR, Domschke A, Böhme S (1992) Eur Polym J 28:1253
95. Reinecke H, De la Campa J, De Abajo J, Kricheldorf HR, Schwarz G (1996) Polymer 37:3101
96. Kricheldorf HR, Pakull R, Buchner S (1989) J Polym Sci Part A Polym Chem 27:431
97. Hung T-Ch, Chang T-Ch (1996) J Polym Sci Part A Polym Chem 34:2455
98. Hung T-Ch, Sun S-J, Chang T-Ch (1996) J Polym Sci Part A Polym Chem 34:2465
99. Reinitzer F (1988) Monatsh Chemie 9:421
100. Buning TI, Kreuzer F-H (1995) Trend in Polym Sci 3:318
101. Bouligand Y (1973) J Phys 34:603
102. de Jeu HW, Vertogen G (1977) In: Thermotropic liquid crystals – fundamentals. Springer, Berlin Heidelberg New York
103. Kricheldorf HR, Berghahn M (1995) J Polym Sci Part A Polym Chem 33:427
104. Kricheldorf HR, Probst N (1995) Macromol Chem Phys 196:3511
105. Kricheldorf HR, Probst N, Gurau M, Berghahn M (1995) Macromolecules 28:6565
106. Kricheldorf HR, Berghahn M, Gurau M, Gailberger M, Barth A, Vill V (1995) Ger Pat Appl 4,461,993 to Daimler Benz AG; Chem Abstr 124, P 120,282 (1996)
107. Kricheldorf HR, Probst N (1995) Macromol Rapid Commun 16:231
108. Probst N, Kricheldorf HR (1995) High Perform Polym 7:461
109. Kricheldorf HR, Probst N (1995) High Perform Polym 7:471
110. Kricheldorf HR, Krawinkel T (1997) High Performance Polym 9:91
111. Kricheldorf HR, Krawinkel T (1997) High Performance Polym 9:105
112. Kricheldorf HR, Krawinkel T (1997) High Performance Polym 9:121
113. Kricheldorf HR, Krawinkel T (1998) Macromol Chem Phys (in press)
114. Kricheldorf HR, Wulff D (1996) J Polym Sci Part A Polym Chem 34:3511
115. Kricheldorf HR, Thomsen SA (1993) Makromol Chem Rapid Commun 14:395
116. Kricheldorf HR, Gurau M (1995) J Polym Sci Part A Polym Chem 33:2241
117. Kricheldorf HR, Gurau M (1996) JMS-Pure Appl Chem A32:1831
118. Asanuna T, Ockawa H, Ookawa Y, Yamashita W, Matsuo M, Yamaguchi A (1994) J Polym Sci Part A Polym Chem 32:2111
119. Schwarz G, Sun S-J, Kricheldorf HR, Ohta M, Oikawa H, Yamaguchi A (1997) Macromol Chem Phys 198:3123
120. Koton MM, Zhukova TI, Florinsky FS, Bessonor MI, Kuznetsov NR, Laius LA, Prikl Zh (1997) Khim 50:2354; Chem Abstr 88:7448c (1978)
121. Marek M Jr, Labsky J, Schneider B, Stokr J, Bednar B, Kralicek J (1991) Eur Polym J 27:487
122. Kricheldorf HR, Linzer V (1995) Polymer 36:1893
123. Edwards WM, Robinson IM (1955) US Pat 2,710,853
124. Cor Koning E, Teuwen L, Meijer EW, Moonen I (1994) Polymer 35:4889

125. Inocu T, Kakimoto M, Imai Y, Watanabe J (1995) Macromolecules 28:6368
126. Alder P, Dolden JG, Smith P (1995) High Perform Polym 7:421
127. Demus D, Richter L (1978) The textures of liquid crystals. VEB, Leipzig
128. Gray GW, Goodby JW (1984) Smectic liquid crystals. Leonhard Hill/Blackie, Glasgow
129. Hirschinger J, Micera H, Gardener KH, English AD (1990) Macromolecules 23:2153
130. Dorlitz H, Zachmann HG (1997) J Macromol Sci-Phys B36:205
131. Leisen J, Boeffel C, Spiess HW, Yoon DY, Sherwood MH, Kawasumi M, Percec V (1995) Macromolecules 28:6937
132. Kricheldorf HR, Linzer V (1996) Macromol Chem Phys 197:4183
133. Wutz Ch (1998) Polymer 39(1):1
134. Pardey R, Shen D, Gabori PA, Harris FW, Cheng SZD, Aducci J, Facinelli JV, Lenz RW (1993) Macromolecules 26:3687
135. Pardey R, Wu SS, Chen J, Harris FW, Cheng SZD, Keller A, Aducci J, Facinelli JV, Lenz RW (1994) Macromolecules 27:5794
136. Wutz Ch, Thomsen S, Schwarz G, Kricheldorf HR (1997) Macromolecules 30:6127
137. Schwarz G, Thomsen S, Wutz Ch, Kricheldorf HR (1998) Acta Polymerica 49:162
138. de Abajo J, de la Campa J, Alegria A, Echave JM (1997) J Polym Sci Part B Polym Physics 35:203
139. Abdallah M, Groothues H, Kremer F, Kricheldorf HR (1997) Macromol Chem Phys 198:2817
140. Huang Y, Petermann J (1996) Polymer Bull 36:517

Received: April 1998

Calculation of a Mesogenic Index with Emphasis Upon LC-Polyimides

John G. Dolden

The Thatched Cottage, Upper Ifold, Dunsfold, Surrey, GU8 4NX, UK

The Mesogenic Index is an empirical method based on functional group contributions, in which the basic unit corresponds to a double bond. The method has been applied to a wide range of condensation polymer and copolymer types containing ester, amide, imide or carbonate groups. A second empirical concept, termed the "Mesogenic Length" has been introduced which has the effect of correcting the Index for random copolymers so that mesophases occur at approximately the same score (ca. 10) irrespective of the types or ratios of different linking groups in copolymers. It has been shown that an MI value of ca.10±0.5 is the borderline condition to predict whether it will contain a mesophase in the melt or in solution (subject to it being fusible or soluble). It has been demonstrated in this chapter that the condition for the mesophase applies to imide-containing polymers as well as esters, amide and carbonate polymers. The mesogenic length of the imide group was set at two, the same as a polyester, but overall polyimides are much less mesogenic than polyesters because aromatic linked rings are slightly kinked and score midway between the para- and meta- orientations. More specifically, MI threshold values around 9.5 seem to characterise PEIs, especially those containing NCPT or BPTI units, whereas a threshold value of 10 more closely characterises poly(amide-imide)s. Lower threshold values down to 9.3 have occasionally been observed with PEIs and as low as 9.1 for poly(carbonate-imide)s (PCI). Exceptionally, values below 9 have been observed with poly(carbonate-imide)s containing pyromellitic anhydride units. The low values are attributed to donor-acceptor interactions between neighbouring chains. In the case of PCIs the effect seems to be particularly large due to the electron rich nature of the carbonate group. It is anticipated that MI threshold values lower than 10 will be observed in other copolymers containing mixed groups, but that in the majority of cases the deviations are not likely to be as large.

Keywords. Liquid crystallinity, Mesogenic Index, Mesophase prediction, Poly(amide-imide)s, Poly(esterimide)s, Poly(etherimide)s, Polyimides

Advances in Polymer Science, Vol.141
© Springer-Verlag Berlin Heidelberg 1999

1
Introduction

There are few actual reports of either liquid crystal low molar mass imides or polyimides in the literature. Thus for example "The Handbook of Liquid Crystals" [1] does not contain any examples of mesogenic imides. Polyamides, on the other hand, can form lyotropic phases and Aharoni [2] has explored the structural requirements of mesogenic units, indicating that a minimum of three amide-linked aromatic units are needed for mesophase formation in solution. The possibilities for thermotropic polyamides are restricted by interchain hydrogen bonding, but this restriction does not apply to polyimides. Polyimides tend to be very high melting or infusible, hence it is difficult to design polymers with stable thermotropic phases. Aromatic polyimides also tend to be very insoluble thereby precluding the formation of lyotropic materials.

Whang and Wu [3] have described the liquid crystalline state of polyimide precursors and shown that certain polyamic acids derived from pyromellitic anhydride exhibit lyotropic behaviour. Liquid crystal phases have also been observed by Wenzel et al. [4] in polyimides derived from pyromellitic anhydride and 2,5-di-n-alkoxy-1,4-phenylene diisocyanate. Dezern [5] has disclosed a synthesis for linear polyamide-imides derived from benzophenone dianhydride but the occurrence or otherwise of mesophases is not mentioned.

The main classes of mesogenic copolymers containing imide groups which are reported in the literature have been classified as poly(esterimide)s (PEIs), either containing flexible spacer units or as wholly rigid copolymers; poly(amide-imide)s; flexible polyimides (also containing ether groups) and wholly aromatic polyimides. The majority of these liquid crystalline polyimides contain ester groups to confer meltability and mesogenicity, and have been synthesised in the last few years. A comprehensive review and discussion of LCPs containing imide groups is given in Chap. 8. A short review of the main types is given in the following section.

The author worked for many years at BP Research on the synthesis of LCPs and devised an empirical method called the Mesogenic Index, which employs functional group contributions on an additive score basis to predict whether a particular random copolymer is likely to exhibit a mesophase (subject to the polymer being soluble or fusible). This chapter explores the general features and theoretical aspects of the chemical structures of main chain LCPs and describes the Mesogenic Index and how it was successfully applied to polyesters, polyamides and polycarbonates. The final section describes the extension of the MI empirical method to the various types of LC polyimides reported in recent years.

2
Review of LCPs Containing Imide Groups

By far the largest group of liquid crystalline polymers containing imide groups are copoly(esterimide)s (PEIs). These have been sub-divided into flexible copolymers containing aliphatic spacers and wholly aromatic copolymers. There are several reports of copoly(amide-imide)s and copoly(ether-imide)s (both thermotropic and lyotropic) and these are treated separately. There are only a few examples of wholly aromatic imide containing LCPs, as polyimides frequently melt well above their decomposition temperature. Successful attempts have been made to incorporate flexible spacers, but in most instances the spacers are based on ether units, and the polymer is more strictly classified as a poly(etherimide).

2.1
Poly(esterimide)s (PEIs) with Aliphatic Spacers

Polyimides, and derivative copolymers such as poly(ester-imide)s have been known for decades but none were reported before 1987 to contain lyotropic or thermotropic mesophases, except certain copoly(esterimide)s with three or more monomers [6, 7].

As most polyimides are extremely high melting (>500 °C) with aromatic symmetric spacers, the introduction of linear aliphatic spacers is one successful strategy for producing meltable and soluble polymers. PEIs derived from highly symmetrical imide units (e.g. pyromellitic imide, naphthalene 1,4,5,8 imide units) give rise to isotropic and not mesophasic melts [8–11]. However, PEIs containing N, N'-dihydroxypyromellitimide (1) have been reported to contain broad nematic phases in the melt [12].

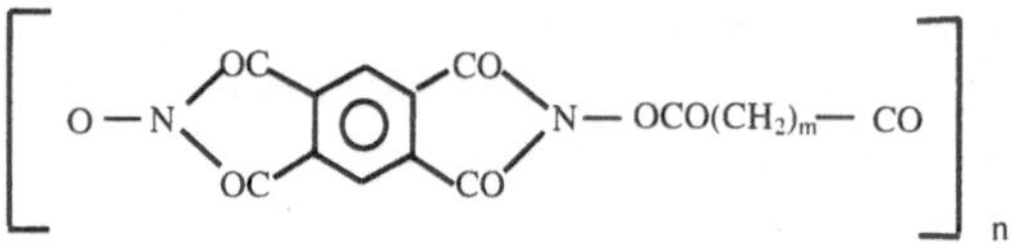

$$(1)$$

Thermotropic PEIs can be formed by polycondensation of dicarboxylic acids with diphenols such as hydroquinone, bisphenol or 2,6-dihydroxy naphthalene. Alkylene bis-trimellitimide units form PEIs with a characteristic smectic A melt. There is an odd-even effect with melting points, with even spacers yielding higher Tm [13, 14]. The aliphatic spacers prefer gauche formation, according to spectral studies. Copolymerisation of the PEI of hydroquinone and alkylene trimellitimide with p-hydroxy benzoic acid caused the solid state layered structure to convert gradually to a columnar mesophase with almost perfect hexagonal packing (2). The LC phase shifts from smectic to nematic [15].

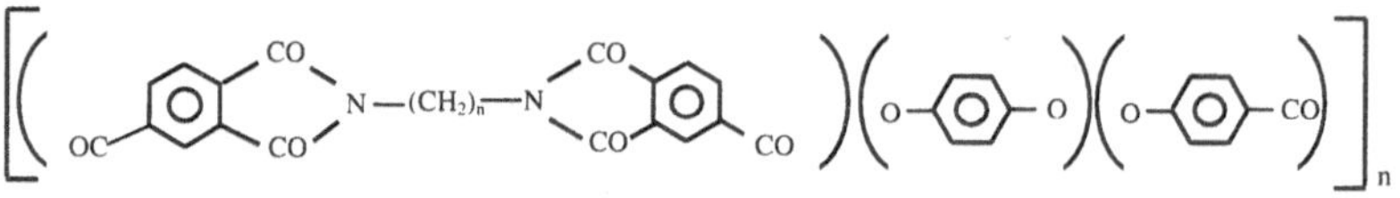

$$(2)$$

Nematic melts are also formed by PEIs based on imide diphenol alkylene diester (3). WAXD patterns and optical microscopy indicate a layered structure in the crystalline state melting to a nematic phase, once more with a strong odd-even effect of the isotropisation temperature [16].

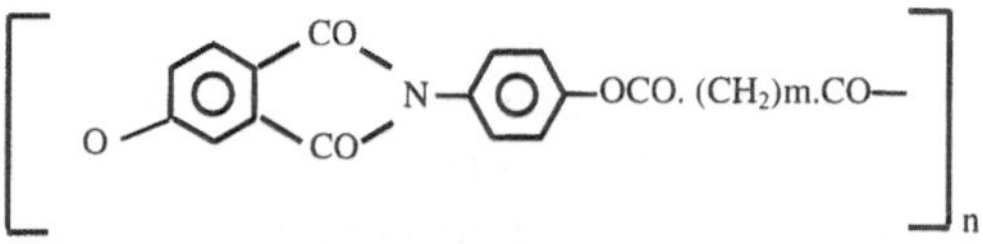

$$(3)$$

Similar polymers formed from the isomeric trimellitimide of p-amino benzoic acid (4) do not however form an enantiotropic phase but rather a short-lived monotropic phase upon cooling from an isotropic melt [17, 18].

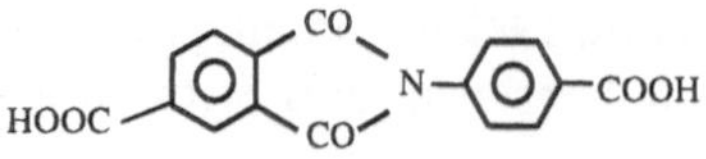

$$(4)$$

Extension of the isomeric trimellitimide of p-amino benzoic acid with a hydroxy alkyl thiophenol (p-HO-Ph-S-$(CH_2)_n$-OH) does yield nematic PEIs (5) [19].

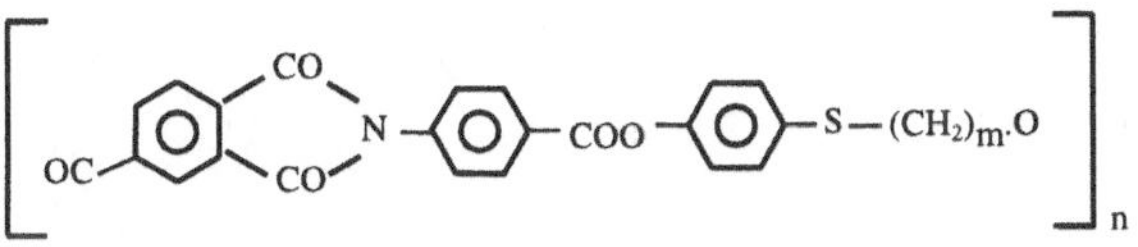

(5)

Cholesteric PEIs may also be obtained by extension with chiral spacers – a methyl substituent placed on the alkyl chain of the aforementioned thiophenol on the β carbon [20]. Photo-reactive cholesteric PEIs with this type of structure have subsequently been reported by the same author [23] formed from the condensation of 1,4-phenylene di-acrylic acid and N-(4-carboxyphenyl) trimellitimide with a chiral spacer prepared from (S)-3-bromo-2-methylpropanol and 4-mercaptophenol. The resulting polymer had enantriotropic cholesteric melt with grandjean texture upon slight shearing (6).

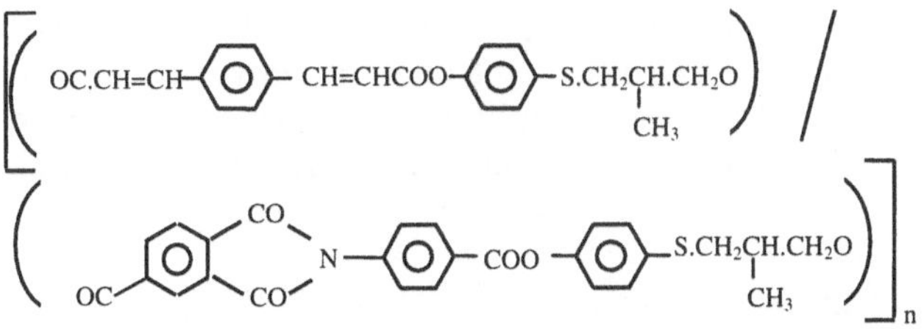

(6)

Extension of the above isomeric mesogen (4) with linear ether or thioether structures, e.g. HO-$(CH_2$-CH_2-O)n-H where n=2,3 or 4 does not however form mesophasic polymers. The active mesogen structure appears to require three aromatic rings. If one of the rings is attached through a non-linear ether group then once again no mesophase is observed (e.g. 7).

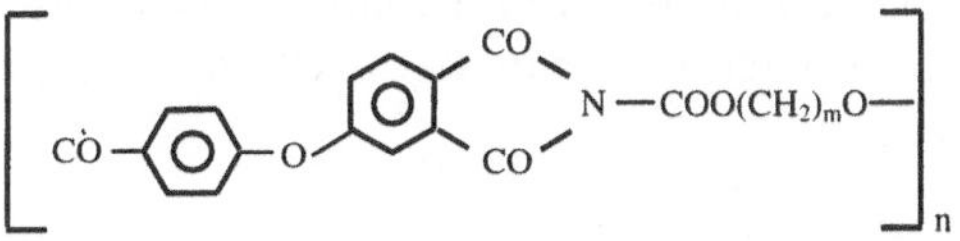

(7)

PEIs of 4,4'-bisphenol and trimellitic anhydride linked by a linear aliphatic aminoacid (8) form enantiotropic LC-phases but it is the bisphenol unit which predominates to give this behaviour. For n=10 or 11 both smectic and nematic phases are observed [21].

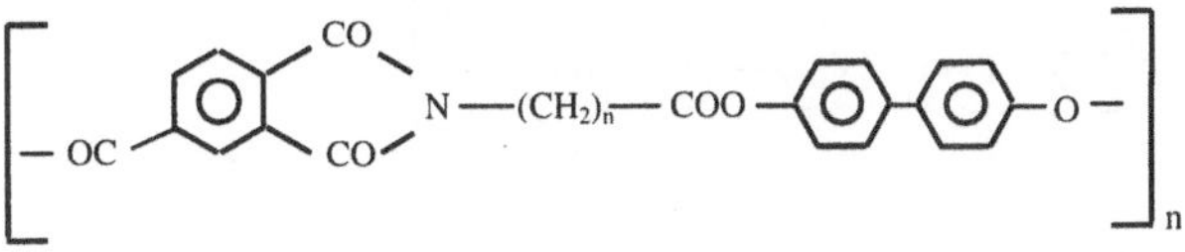

(8)

PEIs derived from benzophenone dianhydride and linear amino acids (9) generally give isotropic melts with most aromatic phenols, excepting 4,4'-bisphenol
[22].

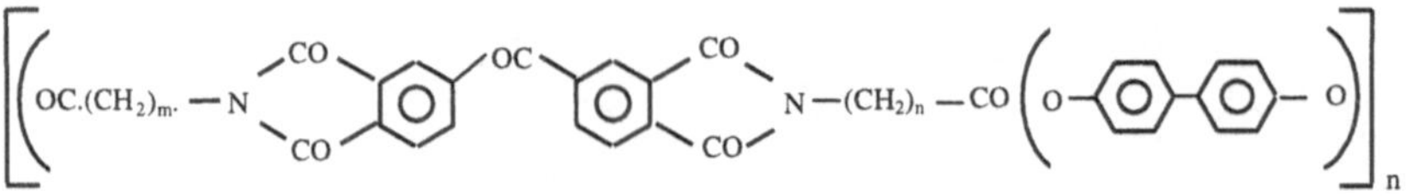

$$(9)$$

A photo-reactive and mesogenic dicarboxylic acid was prepared from 4-ami-
no-cinnamic acid and trimellitic anhydride (10), and condensed via its dichloride with isosorbide, methylhydroquinone, phenylhydroquinone, 4,4'-dihydrox-
ydiphenylether or 1,6-(4'-dihydroxyphenoxy)hexane. WAXD patterns indicated
that all four homopoly(esterimide)s were semi-crystalline, but incorporation of
isosorbide gradually reduced and eliminated crystallinity to form a cholesteric
glass. Grandjean textures were observed above 250 °C in copolymers containing
5–10% isosorbide and phenylhydroquinone. All poly(esterimide)s were sensitive to photocross-linking for wavelengths <360 nm [24]. The presence of the
additional double bond in the cinnamic group made this dicarboxylic acid a better mesogenic unit than the prior one based on trimellitic anhydride condensed
with p-aminobenzoic acid [25], in that mesophases were formed by direct condensation with aliphatic diols and an additional aromatic ring was not required.

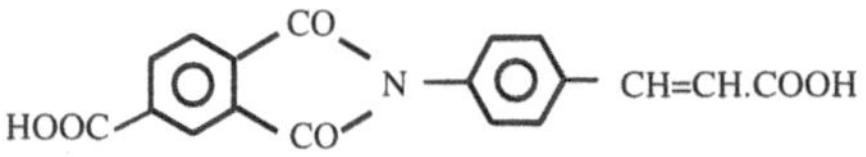

$$(10)$$

PEIs derived from N-(4-carboxyphenyl) trimellitimide and aliphatic spacers
(11) are not thermotropic, irrespective of whether the spacer is chiral or not. If
the spacers are semi-aliphatic and contain a benzene ring, then thermotropic
PEIs may be formed with both a smectic and nematic phase. If chiral spacers are
then used a chiral smectic A* or C* phase may additionally be obtained. Such
phases may be ferroelectric, which is extremely rare for main-chain polymers
[26]. Examples of chiral and non-chiral spacers used in the copolymers (12) are:
1) p-O-Ph-S-CH$_2$-CH(CH$_3$)-CH$_2$-O- and
2) p-O-Ph-S-(CH$_2$)n-O-

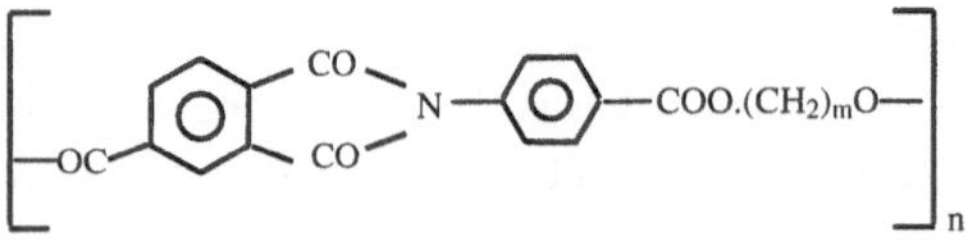

$$(11)$$

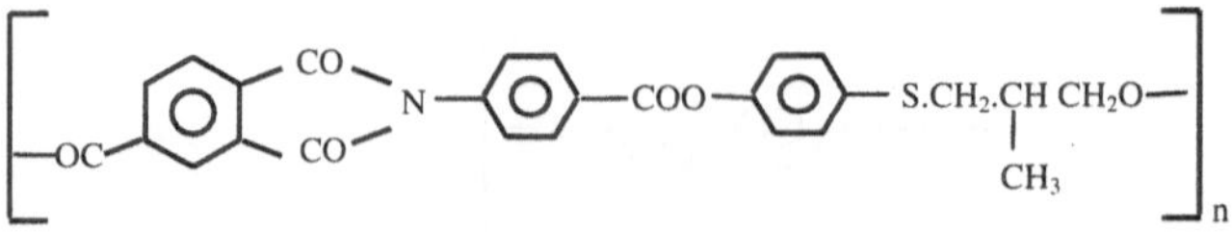

$$(12)$$

PEIs were synthesised from pyromellitic anhydride and ω-aminoacids, condensed first to form a di-acid (13) and subsequently reacted with a range of mesogen forming diols, e.g. 4,4'-biphenyldiol, 2,6-dihydroxynaphthalene, methyl-, chloro- and phenylhydroquinone. None of the polymers were found to contain a liquid crystal phase [27].

$$\text{HOOC.(CH}_2)_m\text{—N} \underset{\text{CO}}{\overset{\text{CO}}{\diagdown}} \underset{\text{CO}}{\overset{\text{CO}}{\diagup}} \text{N—(CH}_2)_m\text{COOH} \tag{13}$$

An imide monomer was synthesised from 4-nitrophthalic anhydride: N-(4'-carboxyphenyl), 4-(4'-carboxyphenoxy)phthalimide. (14). This was reacted (via its dimethyl ester) with chloro- or phenylhydroquinone to form wholly aromatic co-poly(esterimide)s which were found to exhibit a nematic phase in the melt. Similar PEIs derived from aliphatic diols produced only isotropic polymers.Thus, PEI derived from 4-(4'-carboxyphenoxy)phthalimide and n-dodecane diol melted at 162 °C, Tg=66 °C, and was isotropic [28]. The above imide monomer based on phthalic anhydride was therefore shown to be a poor mesogen compared with its trimellitic anhydride analogue (4).

$$\text{HOOC}–\langle O \rangle–\text{O}– \underset{\text{CO}}{\overset{\text{CO}}{\diagup}} \text{N—COOH} \tag{14}$$

A further study of the influence of chain packing was made by these authors [21] with regard to the influence of spacers and chain packing on the phase transitions of PEIs derived from N-(4-carboxyphenyl)trimellitimide (4). The latter is the cheapest mesogenic dicarboxylic acid known so far and polymers derived from it may be of commercial interest, according to the authors. Polyalkylene, polyglycol ether and polyglycolthioether spacers were used to link the imide units together via ester links. None of the copoly(esterimide)s were mesogenic. However, three different kinds of solid states were observed which were interpreted as frozen smectic-A, B and E phases [29].

Mesogenic PEIs have also been synthesised from benzophenone dicarboxyl-icimide in conjunction with 4,4'-dihydroxybiphenyl [30] (15).

$$\left[–\text{O}–\langle O \rangle–\langle O \rangle–\text{O}–\text{CO(CH}_2)_m \text{ N} \underset{\text{CO}}{\overset{\text{CO}}{\diagdown}} \underset{\text{CO}}{\overset{\text{OC}}{\diagup}} \overset{\text{CO}}{\underset{\text{CO}}{\diagdown}} \text{N—(CH}_2)_m\text{—CO—} \right]_n \tag{15}$$

2.2
Wholly Rigid Aromatic PEIs

Wholly aromatic polyesterimides which exhibit nematic phases were first described in patents to Du Pont [6] and Idemsitsu Chem [7]. These were essentially copolyesters of three monomers, one of which was a dicarboxlic acid imide.

Thermotropic homopoly(ester-imides) have subsequently been described by Kricheldorf in recent publications [31–33]. These were synthesised from an imide diphenol preconstructed from phenol 3,4-dianhydride and p-aminophenol, reacted with a substituted terephthalic acid. Those derived from monothioalkyl or dithioalkyl substituted acids were especially interesting as these contained the side chains $S(CH_2)_nCH_3$ (**16**).

Where X=S-(CH₂)ₘCH₃
Where Y= H or S-(CH₂)ₘCH₃
m = 7 or 15   (16)

The mono-substituted PEIs melt from a saniditic layered structure into a normal nematic phase whereas the disubstituted presumably yield a biaxially oriented nematic melt. The disubstituted PEIs isotropise from the melt around 200 °C lower than the monosubstituted of equal chain length. As the second substituent should increase chain stiffness rather than reduce it, Kricheldorf suggests that interactions between temporarily coplanar aromatic π systems, particularly donor-acceptor interactions, can make an efficient contribution to the stabilisation of nematic phases. If this hypothesis is correct then a greater variety of non-linear monomers may be regarded as building blocks for liquid crystal polymers.

Similarly constructed PEIs, in which the imide unit was preconstructed from trimellitic anhydride and p-aminobenzoic acid and reacted with substituted hydroquinones, have also been reported [32] by Kricheldorf and co-workers. Monosubstituents of methyl, chloro or phenyl groups resulted in nematic PEIs, but disubstituents again destabilised the nematic phase. By contrast PEIs of o-diphenols formed thermostable enantiotropic nematic melts, despite a linear conformation being energetically unfavourable It was again suggested that interchain forces between temporarily coplanar chain segments play a much more significant role in stabilising the nematic phase of these polymers than geometric factors. Polycondensation of the preformed diacid derived from trimellitic anhydride reacted with equimolar p-aminobenzoic acid (**17**) and monosubstituted hydroquinones produces meltable or amorphous PEIs which form a nematic melt over a broad temperature range [31, 33].

Where X = Cl, C(Me)₃,(-C₆H₅)
(-O C₆H₅),, (-S C₆H₅).   (17)

Although a higher tendency to crystallise and form LC-phases was expected for disubstituted hydroquinones reacted with the same diacid, in fact amorphous isotropic PEIs were generally obtained. Since there is no reason why a second bulky substituent should reduce chain stiffness it seemed more likely that extra bulky side substituents interfere with electronic interactions between neighbouring chain segments.

Since 1991 Kricheldorf has published many other articles on liquid crystal polyesterimides. These include two series of thermotropic copoly(esterimides) derived from N-(3'-hydroxyphenyl)trimellitimide and 4-hydroxybenzoic acid or 6-hydroxy-2-naphthoic acid, in which the critical compositions for obtaining mesophases were discovered by varying monomer levels [34]. A series of homopoly(esterimide)s were synthesised from N-(4-carboxyphenyl)trimelliti-mide and various diacetylated linked diphenols. When the diphenol contained an ether or carbonyl linking group, mobile nematic phases were observed. Various naphthalene diols were also employed instead of the diphenol, but only the 2,7-dihydroxy naphthalene produced a nematic phase [35].

Fully aromatic liquid crystalline poly(esterimide)s have also been derived from N-(4-carboxyphenyl)trimellitimide and diphenylether-3,3', 4,4'- tetracarboxylic imide [36] (18).

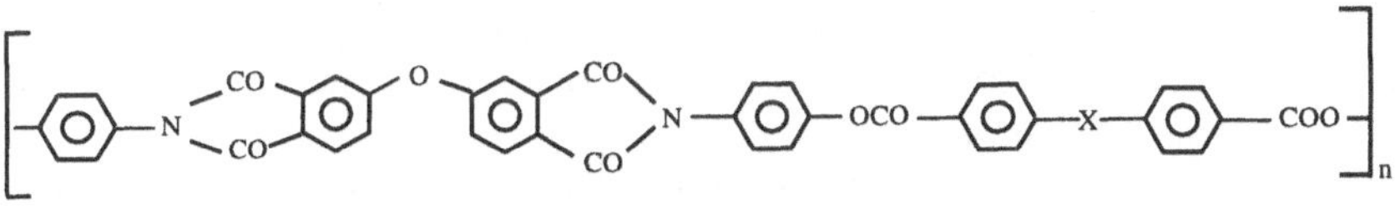

$$ \text{(18)} $$

All PEIs derived from 3-aminophenol exclusively formed only isotropic melts. Thermotropic character was however observed in PEIs derived from 4-aminophenol and diphenylether 3,3',4,4'-tetracarboxylic acid, but not when diphenyl sulphone or isopropylidene diphenyl units were employed. PEIs formed from 4-aminophenol and biphenyl-3,3',4,4'-tetracarboxylic anhydride have also been found to form nematic melts [37].

2.3
Poly(amide-imide)s

Kricheldorf has reported the synthesis of lyotropic poly(amide-imide)s and poly(benzoxazole-amide)s . These were prepared by the polycondensation of N,N'-bis(trimethylsilyl)-p-phenylenediamine or N,N'-bis(trimethylsilyl)-3,3'-dim-ethylbenzidine with the diacyl chloride of trimellitimide of p-aminobenzoic acid, or the imide formed from p-amino benzoic acid and terephthalic acid . Lyotropic behaviour was observed in conc. sulphuric acid solution [38]. A series of thermo-tropic poly(imide-amide)s was prepared based on trimellitimides formed from trimellitic anhydride and an α, ω-bis(4-aminophenoxy) alkane with carbon chain lengths 9–12. Melting points were in the range 250–300 °C. They formed smectic A phases and tended to degrade around the isotropisation temperatures (around 350 °C). Pendant methyl groups or occupied meta- groups tended to pre-vent mesophase formation [39]. Novel LC poly(imide-amides) have also been synthesised from new diamine spacers derived from linear diaminoalkanes and 4-nitrophthalic anhydride. A smectic and nematic phase were observed when 4,4'-biphenyl dicarboxylic acid was used as co-monomer [40].

2.4
Polyimides

In 1994 Kricheldorf reviewed the field and showed that it was difficult to obtain LC-polyimides free of ester groups. Only three classes were then cited, with ether or amide supplementary linkages, which contained either aliphatic spacers in the main chain with a high tendency to form layer structures in the solid state, or a fully aromatic chain sufficiently substituted to stabilise the formation of nematic phases [41].

The first reports of actual liquid crystal polyimides first appeared in 1995. A series of oligomeric polyimides were synthesised at Lancaster University [42] and later some of these were shown by Dolden et al. [43] to contain transient nematic phases, which were observed during melting but disappeared rapidly due to onset of thermal degradation. The materials contained nadic anhydride end-groups which tended to decompose below the melting point, hence rapid heating was required to spot the phase on a hot stage polarising microscope (structures given in Table 7).

Neither polyimides based on pyromellitic dianhydride nor on biphenyltetracarboxylic anhydride with aliphatic spacers have yielded LC polyimides [77–79]. The need for an extra ring to obtain mesogenic properties was evident. Employment of a terphenyl tetracarboxylic anhydride has been reported to produce a LC polyimide free of other functional groups (19).

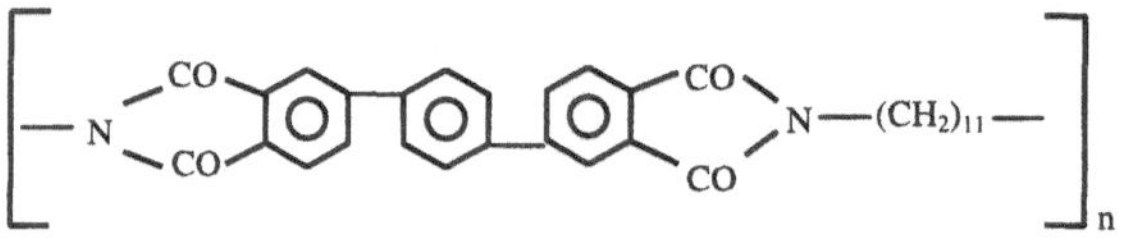

$$\left[-N \underset{CO}{\overset{CO}{\diagup}} \bigcirc\bigcirc\bigcirc \underset{CO}{\overset{CO}{\diagdown}} N-(CH_2)_{11}- \right]_n \qquad (19)$$

2.5
Poly(etherimide)s

Dolden et al. [43] successfully synthesised a thermotropic polyetherimide from biphenylene dianhydride and 1,12-bis(4-aminophenoxy) dodecane. This polyimide was mobile in the melt at 300 °C and exhibited a highly viscous anisotropic phase. Several LC oligoimides were also synthesised by the same workers who showed that more than two aromatic rings per repeat unit seem to be required in order for anisotropic behaviour to be observed.

Kricheldorf also reported the synthesis of a thermotropic polyetherimide based on biphenyl-3,3',4,4'- tetracarboxylic anhydride. and α,ω-bis(4-aminophenoxy) alkane. He similarly observed the need for aromatic rings additional to the two in the anhydride. Batonet and fan-shaped textures indicative of smectic A phases were observed [44] (20).

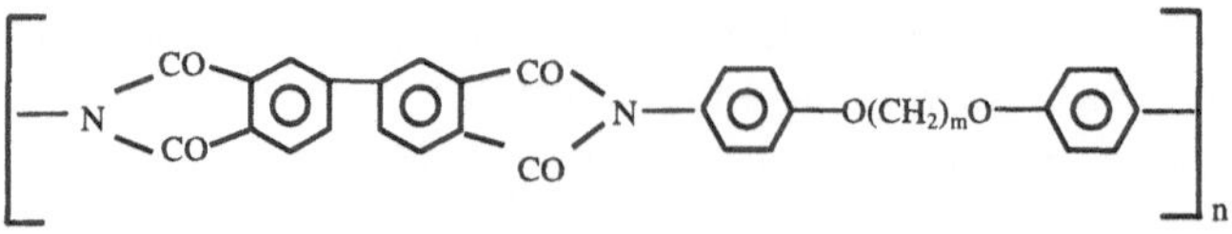

$$\left[-N \underset{CO}{\overset{CO}{\diagup}} \bigcirc\bigcirc \underset{CO}{\overset{CO}{\diagdown}} N-\bigcirc-O(CH_2)_mO-\bigcirc- \right]_n \qquad (20)$$

Some examples of wholly aromatic polyetherimides were synthesised by workers at Liverpool University [42] and shown to be liquid crystalline [43]. The structures are given in Table 8.

A series of wholly aromatic poly(etherimide)s were synthesised [80, 81], one of which was found to contain a smectic A phase in the melt. Its structure is given in Chap. 8 (number 160) and consists of pyromellitic anhydride co-reacted with a five ringed diamine containing a central meta- unit linked via an isopropylene unit to two para- linked ether diphenylamine units. If the central ring is made para-oriented, then an LC phase is not observed.

3
Structure and Theoretical Aspects of Main-Chain LCPs

Liquid crystal polymers have created a great deal of interest in recent years finding a number of commercial applications ranging from high-strength engineering plastics to optical display devices. A liquid crystal molecule possesses anisotropy and, as a mobile fluid, can spontaneously order. It therefore exhibits some of the properties of a liquid (mobility, flow) as well as a degree of order usually associated with a crystalline structure.

The hierarchy of liquid crystals can be characterised in a number of ways. The major division is between liquid crystals that gain mobility through the addition of a solvent, termed *lyotropic*, and those which form a fluid by the action of heat, termed *thermotropic*. Within each type there are numerous subdivisions and classes, many of which are related by symmetry.

Most LCPs of commercial interest are thermotropic main-chain nematic materials. The other main structural type of interest are the side chain LCPs which are generally more complex to synthesise and have not yet found significant commercial applications. The author's Mesogenic Index model, which is useful for predicting whether a given polymer has a mesophase, has been applied only to main-chain LCPs to date.

Within the extensive literature on this subject, there are many examples of the synthesis of thermotropic polyesters, polyesteramides, polycarbonates, polyethers, polyurethanes and polyester-imides. Until recently, the main omissions had been thermotropic polyamides and polyimides; however, many examples of polyamides that show lyotropic behaviour have been known for a long time.

Liquid crystalline aromatic polyamides were the first chemical class to be commercialised. The best known example is Kevlar fibre which is spun from liquid crystalline solution to obtain the benefit of the high orientation present in the nematic phase. Subsequently, melt-processable main-chain polyesters were developed and brought to the market (Amoco with Xydar, Hoechst-Celanese with Vectra).

3.1
Effects of Chemical Structure on Liquid Crystallinity

Structural effects are best illustrated with respect to polyesters. The vast majority of aromatic polyesters which are fusible and contain a substantial number of para links tend to be liquid crystalline in the melt. Their low solubility means that such materials are not normally found to be lyotropic.

Wholly para substituted polyesters are normally infusible, unless disruptive groups are substituted onto the aromatic rings, e.g. methyl or chlorine. Commercially, it has proved more attractive to use symmetrical monomers which are able to disrupt crystallinity because they are significantly different in size. Sequences therefore match at infrequent intervals along the chain. Xydar is an example of a commercial LCP using this principle.

The degree of alignment and reinforcement achieved with these polymers is much better than LCPs based upon the meta- substituted isophthalic acid, which kinks the backbone severely to destroy crystallinity, but consequently also destabilises the mesophase, reduces thermal stability and detracts from modulus.

The incorporation of aliphatic spacers between rigid rods with as few as two aromatic rings can still give a liquid crystalline phase (especially if the linking group is ester); even lengths tend to give smectics and odd lengths nematic phases [47]; mesogens are reported to confer a trans configuration to the chain, the effect weakening with chain length.

The liquid crystalline phase can however disappear if branches are present on the alkylene chains too close to the mesogen [48]. Bridging groups also adversely affect liquid crystal behaviour. Single atom links between two benzene rings reduce the tendency to form a mesophase due to chain kinking. On the other hand two methylene or a methyleneoxy linkage between rings can adopt a trans conformation, which facilitates mesophase formation [49].

Several examples are known, in which one or even both monomers in a liquid crystalline [A-B]n condensation polymer contain cyclo-aliphatic rings [50]. The trans chair form makes a larger contribution than the cis boat form.

Bond angles formed by a central group X between two aromatic rings can have a pronounced effect upon mesogenic properties. The bond angle may not be taken in isolation [51] for whereas an ether or carbonyl link distort by 55–60°, sulphide or diphenylpropane links cause a relative small additional distortion (75°) One might expect these groups to behave similarly, but in fact the former groups tend to support mesogenic properties more readily.

The nature of ring substituents can play an important part in determining mesogenicity. Nematic phases are often observed with single substituents, e.g methyl, phenyl, chloro-, but lead to isotropic polymers when the number of substituents on the ring is increased to two or more. Kricheldorf [52] gives examples of polyesters and polyesterimides which show this effect.

3.2
Theoretical Aspects of Main-Chain LCPs

The main features which describe the characteristics of a mesogen are:
1. high rigidity with high length to diameter (aspect) ratio;
2. good molecular parallelism;
3. substantially larger polarisability along the chain relative to the transverse direction.

The most commonly quoted theoretical parameters that are considered to be of importance are:
1. aspect ratio of domain;
2. Mark Houwink exponent of equation relating solution viscosity to molecular weight of the polymer (1.8=ideal value for mesophase to be present, compared to 0.5 for a flexible isotropic polymer);
3. persistence length of chain (no. of units affected by liquid crystal alignment);
4. mean square end to end distance of chain;
5. radius of gyration;
6. chain flexibility.

The original theoretical model has evolved from Flory's lattice theory [53], which demonstrated that rigid rods with aspect ratio greater than 6.4 align automatically under shear conditions in dilute solutions (hard interactions which are entropy driven). Using virial theory, a value of 3.5 was obtained for energy driven systems (soft interactions applicable to thermotropes). This approach considered both entropy and energy driven transitions and allowed for the effect of temperature upon the aspect ratio of the mesogen. In general, aspect ratios calculated by this theory at the nematic to isotropic transition fall between 3.5 and 6.4. Krigbaum et al. [54] discuss measurements made using the Flory-Ronca theory [55] involving Kuhn semi-rigid chain segments.

Ronca-Yoon theory [56] predicts an aspect ratio >4.45 for the persistence length of semi-flexible polymers. The persistence length is defined as the number of units along the chain which are subject to the orienting power of the mesogen. Coulter and Windle [57] have calculated values for about ten polymers and have found semiquantitative agreement with the Ronca-Yoon theory. They have developed their own software package (Cerius) which allows the systematic estimation of "persistence length", orientation parameter, radius of gyration etc. to be computed for a polymer or copolymers with fixed or statistical sequences. The ability to quantify these quantities as a function of composition, temperature, sequence or configuration provides a powerful tool for assessing the liquid crystalline nature of a polymer.

New theories for predicting the occurrence and transition temperatures in backbone and side-chain LCPs have been described by Dowell [58]. The need for wholly rigid structures has also been questioned. Percec and Yourd [59] have produced thermotropic polyethers using mesogens containing two rings separated by two methylene groups. Percec has called this type "conformational isomers". Other examples [60] exist in which longer spacers separate two aromatic rings.

There is some doubt about the relevance of aspect ratio considered in isolation and as applied just to the empirically identified rigid mesogen. It can only really be applied to in-chain (or nematic unidirectional phases) but this is not applicable to discotic phases [61]. The examples quoted above also make it clear that flexible elements can be included in the mesogen. Strict definition of the mesogen is therefore difficult.

Moreover, coplanar conformations of entire repeat units are not always energetically favourable, perhaps for entropic reasons. Steric hindrance may play a part in destabilising liquid crystal phases. Thus, Kricheldorf has observed this effect comparing mono-substituted rings to disubstituted . He posed the question whether the second substituent reduces the persistence length or whether it hinders specific electronic interactions between neighbouring chain segments. He argued that interactions between temporarily coplanar aromatic π-systems, particularly donor-acceptor interactions, can have a considerable stabilising effect upon nematic phase formation [52].

4
Prediction of Mesophases in Copolymers

An empirical method for predicting the chemical compositions of random or partially ordered condensation copolymers which are capable of exhibiting mesophases (either in solution or in the melt) was devised by the author in 1989, while working on liquid crystal copolymer synthesis for BP Chemicals. A brief description of the method and its application to the chemical synthesis of amorphous thermotropic polyamides has been given in a previous paper [45] and a further more detailed description of the method is to be published shortly [46]. Subsequently, the method has been updated and applied to polycarbonates and polyimides. Thermotropic polyimides have also been synthesised by the author resulting from the use of the predictive method [43].

The new approach has been termed "The Mesogenic Index" (MI) and has successfully been applied to 23 copolymer systems in which the critical compositions for mesophase formation have been established by means of varying constituent monomer concentrations. It is also consistent in predicting liquid crystalline behaviour in several hundred main-chain polymer systems containing amide, ester, carbonate, ether and urethane groups.

It is inherent in the MI system that the mesogenic length is defined in terms of the number of monomer units for a given polymer class. From published work in the literature, it was first established that ester and amide groups need at least two and three aromatic rings in the mesogen respectively. Using these values to define *mesogen length* in polyester and polyamide copolymers, the condition MI>10 has been determined for mesophase formation. This rule has been applied to other linking groups, such as carbonate or imide, with the surprising result that the corresponding mesogen lengths for these condensation polymers are simple numbers. Moreover, the rule has been applied successfully to mixed systems, by simply averaging the contributions of the different groups to the *mesogen length* in strictly molar proportions.

The Mesogenic Index is based upon summing functional group contributions towards rigidity and/or resonance stabilisation of the mesogen. To the best of our knowledge, this is the first simple method to be published that successfully correlates chemical composition with mesophase formation.

In this chapter the application of the MI to polyimides and mixed copolyimide polymers will be examined in detail.

4.1
The Mesogenic Index (MI)

The Mesogenic Index (MI) is the sum of empirically derived contributions from constituent groups in a condensation polymer with alternating backbone structure of type [(A)-(B)]n. It accounts for the positive contribution of trans oriented rings or unsaturated groupings, and their contribution by anisotropy along the chain towards rigidity. It is an overall measure of a monomer unit's ability to 'stiffen' the backbone, after accounting for the negative effects of in-chain flexible groups, exo-cyclic substituents and other groups pendant to the backbone. Similarly, the effects created by hinged rings or meta- orientation are also reflected in group contributions. A comprehensive list of our estimated group contributions to the Mesogenic Index of polymers is given in Tables 1a–f for mono-, di- and multi-ring systems, linking groups, spacers and miscellaneous.

The MI is first calculated for the empirical monomer structure, on a molar basis, for all [A]+[B] units. It is then multiplied by "the mesogenic length factor" corresponding to the minimum number of rings needed in a mesogen of that chemical type (e.g. 2 for ester, 3 for amide, 2.5 for a 1:1 esteramide, etc.).

The Mesogenic Index has been applied to a substantial body of data on polyesters and polyamides drawn from from published literature including 22 systems where the critical concentration for onset of liquid crystallinity has been established with reasonable accuracy (Tables 2 and 3). The characteristic value of the MI at the onset of a liquid crystalline phase is generally found to be remarkably close to 10. When applied to the more limited data available on polyurethanes, polyethers, polycarbonates, the rule holds up well.

The total score for an empirical formula unit [(A)-(B)]n, based on alternating acid and base monomers is first calculated according to whether the polymer is random or ordered.

For ordered copolymers, the constitutive rule is:

$$[MI]_{ab} = 2 \times \{\Sigma m_a \cdot [MI]_a + \Sigma m_b \cdot [MI]_b\} \tag{1}$$

where a and b refer to chemical units A and B, and m_a and m_b are the corresponding molar fractions such that $Em_a = Em_b = 0.5$. $[MI]_a$ and $[MI]_b$ are the respective contributions to the polymer Mesogenic Index $[MI]_{ab}$.

For ordered polymers of this type it is sufficient to add the MI value of the A and B groups together to obtain the polymer MI. The link group is regarded as part of the mesogen and no correction is necessary to account for the nature of the linking group.

For *random copolymers*, a knowledge of the minimum number of units along the chain to form a mesogen is also required (which is a function of the linking group) so that the average score for the polymer can be worked out. In the case of a polyester, it is well established that two units are sufficient to form a mesogen [47, 48, 62]. The MI value for the empirical formulation is therefore sufficient. In the case of multiple copolymers containing many monomer units, the average score per unit is first calculated and multiplied by the appropriate mesogenic length factor L, using Eq. (2):

$$[MI]_{ab}=L(\Sigma m_a[MI]_a+\Sigma m_b[MI]_b) \tag{2}$$

For example, if a polyester contains two acids and two diols each in equal concentrations of 0.25 moles, its $[MI]_{ab}$ is calculated from $L=2$, $m_a=m_b=0.25$. If the two acid units are terephthalic acid and iso-phthalic acid then they score $[MI]_a=6$ and 4 respectively and if the diol units are hydroquinone and methyl hydroquinone, then these score $[MI]_b=6$ and 5 respectively; Eq. (2) becomes

$$[MI]_{ab}=2(0.25\times6+0.25\times4+0.25\times6+0.25\times5)=10.5$$

The above polyester would therefore be expected to exhibit a liquid crystalline phase in the melt as it scores just above the minimum.

We have concluded from literature studies that there is a need for three units in polyamide mesogens, so that the empirical score per unit must be multiplied by 3. Thus, if the above example was a polyamide and not a polyester, then $L=3$, $m_a=m_b=0.25$, $[MI]_a=6$ and 4, but $[MI]_b=4$ and 3 (as the amide group scores 2 less than the ester); hence Eq. (2) yields

$$[MI]_{ab}=3\times(0.25\times6+0.25\times4+0.25\times4+0.25\times3)=12.75$$

a value well above the minimum for predicting a liquid crystalline polyamide. In this case we might expect from experience for the polyamide to be lyotropic.

A series of alternating polyamides synthesised by Aharoni [63] bear witness to the need for preformed polyamide units to contain at least three aromatic rings with amide links in order to exhibit liquid crystallinity. This does not apply if the rings are linked by, e.g., ether links (where no linking groups are present, the number of aromatic rings needed to achieve mesogenic properties is known to be four [64]). The author has discussed the case for polyamides in more detail in a previous [45] and a pending publication [46].

The appropriate length for other linking groups can be calculated from the MI approach. When this approach is applied to a wide range of published material, it is consistently found that LCPs have an MI score ≥ 10.

In order to discuss the application of the MI to polymers it was decided to classify them according to structural type, whether random, alternating, ordered, homo- or copolymer. All classifications can broadly be described by the general category [(A)-(B), (C)]; A and B refer to difunctional monomers such as diacids or diols, whereas type C refers to a hybrid monomer containing two different functional groups, e.g. hydroxy benzoic acid.

The calculation of the MI is still quite straightforward. Suppose a polymer comprises 75% of terephthalic acid/hydroquinone units (TPA/HQ) and 25% m-hydroxybenzoic acid (MHBA). As the corresponding MI scores for the units are 6 (TPA and HQ) and 4 (MHBA) the MI is calculated as follows:

$$MI=2(0.375\times(6+6)+0.125\times(4))=0.75\times(6+6)+0.25\times4=10$$

This predicts 75/25 ratio as the critical composition for this copolymer to be capable of exhibiting an LC phase; increasing the level of MHBA should destroy its ability to support a mesophase. In practice the prediction can only be tested if the polymer is either soluble or will form a stable melt. If the above polymer was fully ordered and always occurred in the sequence A-B-C, then the MI would be the sum of the monomer values, i.e. 17.

4.2
Individual Group Contributions

Individual group scores for over 50 different sub-units of polymers have been tabulated in Table 1, based on the following considerations. Each unsaturated unit is considered to have a positive effect in the chain direction, which may be enhanced by conjugation. Conjugation of linear structures increases rigidity and planarity of the mesogenic unit. A single C=C is the basic unit on which the MI is founded with a standard contribution of +1. On this basis, p-aromatic rings score +3 for having three double bonds contributing to the mesogenic unit along the chain.

Conjugation is generally considered to increase the overall score of a unit by 1. For example, two non-conjugated isolated aromatic rings score 6 by this method, but when linked together and sandwiched between polar groups, e.g. ether oxygen or ester groups, the effective score increases to 7. Similarly, a hydroquinone unit should score 4. In most cases, the MI contribution is consistent with the canonical form containing the most double bonds and reflects the contribution of resonance stabilisation energy.

Anisotropic double bonds have been scored on the basis of each C=N contributing 1.5, each C=O scoring +2 (and also N=N scoring +2). This reflects the heteroatom's increasing ability to polarise the double bonds.

Groups which do not contribute to rigidity, or do not act in the direction of the chain, have been assigned zero or negative contributions. Aliphatic units occurring in the backbone, which separate unsaturated rings or other conjugated systems from linking groups, score –1 per $(CH_2)n$ if n is odd or score 0 if n is even; this is justified by the fact that odd numbers of methylene groups give a gauche conformation which kinks the mesogenic unit. For example, p-Ar.(CH_2)-C(O)O scores (3–1+2+1)=5 compared to 6 without the CH_2 group. Flexible units

Table 1a. Contributions of functional groups to the Mesogenic Index

Chemical Formula	Chemical Structure	Mesogenic Index Group Contribution
1. *p*-phenylene		3
2. *m*-phenylene	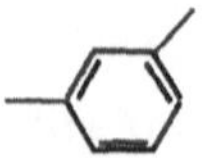	1
3. *p*-methylene phenylene	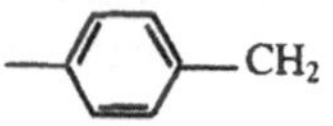	2
4. *p*-methylene benzyl *m*-methylene benzyl		1
5. *p*-(orthomethyl)phenylene *p*-(orthochloro)phenylene	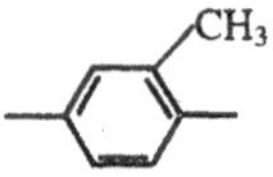 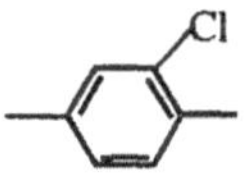	2
6. *p*-(ortho-alkoxy)phenylene	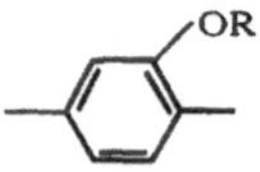	1
7. *p*-phenoxyether		4

between link groups, which are not integral to a mesogenic unit, are not scored, e.g. $(CH_2)n$ spacer units in $[\{Ar.COO\}\{(CH_2)nCOO\}]_n$.

In the different classes of *ordered* polymers discussed above mesogenic behaviour is best explained on the basis that either the [A] or [B] repeat unit comprises the structure $[(CH_2)_m\text{-Ar-}(CH_2)_n]$ where the MI is the sum of the contributions of the aromatic and linking polar parts. The alkylene units are not scored, but the contributions of the apparently isolated conjugated units in the A and B units are summed to find the Mesogenic Index. In this respect they are treated just like an alternating AB polymer.

Groups which are pendant to either the rigid units or the linking groups have been assigned a negative contribution to the Mesogenic Index. (They may interfere with resonance stabilisation in the chain direction). Groups which can usually be tolerated as a substituent on a ring without destroying crystallinity in aromatic polymers are normally assigned a score of –1, e.g. methyl, phenyl, chlorine and *t*-butyl. Larger groups which disrupt symmetry so much that the result-

Table 1b. Polyaromatic units

Chemical Formula	Chemical Structure	Mesogenic Index Group Contribution
8. p-biphenyl	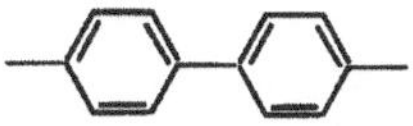	7
9. p-bis(orthomethyl phenyl)	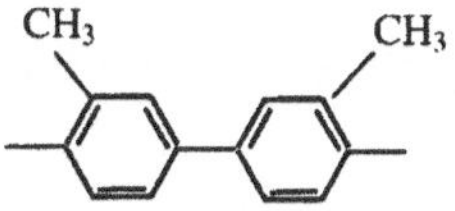 CH_3 ... CH_3	6
10. p-bis(ortho alkoxyphenyl)	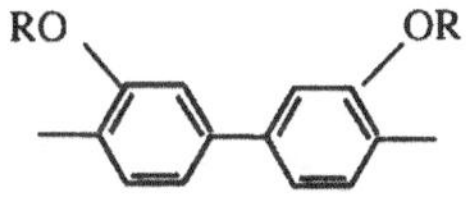 RO ... OR	5
11. stillbene	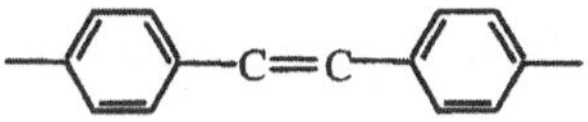 $-C=C-$	8
12. p-(diphenyl)ethylene	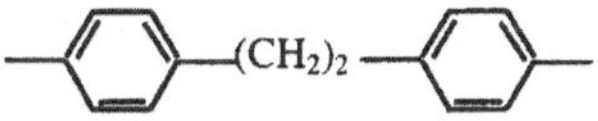 $-(CH_2)_2-$	6
13. p-(diphenyl)butylene	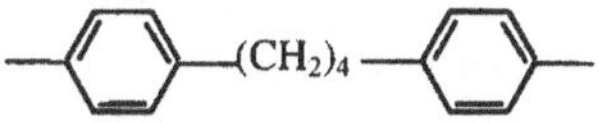 $-(CH_2)_4-$	6
14. p-(diphenyl)methane	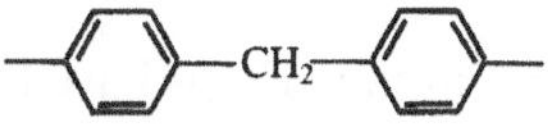 $-CH_2-$	1
15. p-(diphenol)	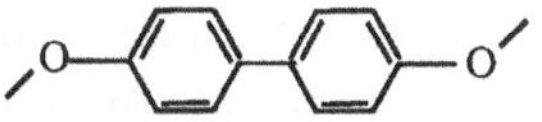 $O-\ \ -O$	7
16. p-(bisphenol A)	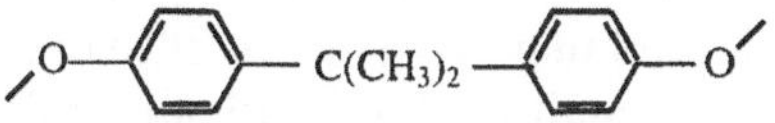 $O-\ \ -C(CH_3)_2-\ \ -O$	-1
17. p-(diphenyl)sulphone	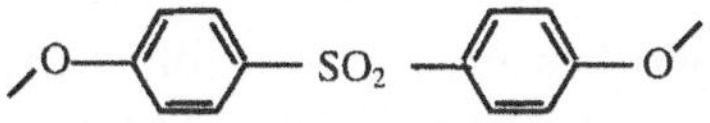 $O-\ \ -SO_2-\ \ -O$	-1

Table 1c. Multicyclic units

Chemical Formula	Chemical Structure	Mesogenic Index Group Contribution
18. terphenyl		10
19. p-(tris-benzamide)	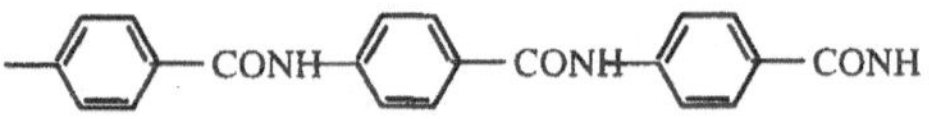 —⟨⟩—CONH—⟨⟩—CONH—⟨⟩—CONH	12
20. p-(bis phenoxy) phenyl	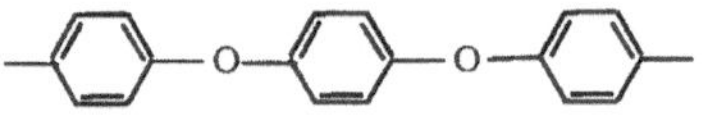 —⟨⟩—O—⟨⟩—O—⟨⟩—	2
21. p-(bis phenoxy phenyl) methane	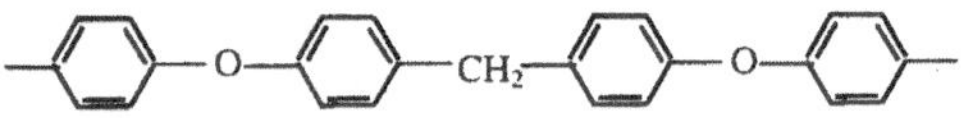 —⟨⟩—O—⟨⟩—CH₂—⟨⟩—O—⟨⟩—	3
22. p-(bis phenoxy phenyl) 2,2' propane	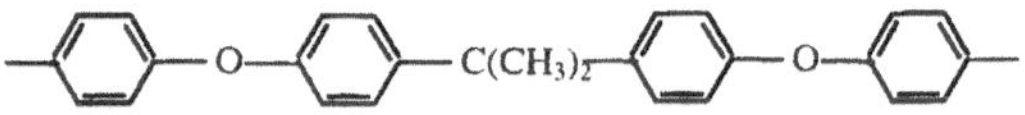 —⟨⟩—O—⟨⟩—C(CH₃)₂—⟨⟩—O—⟨⟩—	2
23. p-(bis phenoxy) biphenyl	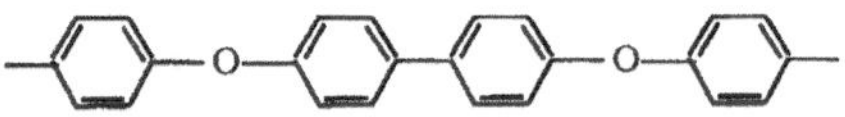 —⟨⟩—O—⟨⟩—⟨⟩—O—⟨⟩—	3

ant polymer is normally non-crystalline have been scored at –2, e.g. alkyl, alkoxy. If there are two substituents on opposite sides of the ring, relative to chain axis, then the score of each is deducted from the total. If the substituents are on the same side of the ring relative to the axis, the score is deducted for only one substituent.

Methyl groups pendant to linking groups have been scored in a similar way. An example is -CON(Me)- which scores 0, compared to +1 for -CONH-. Sterically, this assignment can be justified on the basis that the methyl substituent increases flexibility in the polymer chain because of the steric interaction between the N-Me and the ortho- H atoms on the phenyl ring-which reduces the double bond character of the C-N bond [65].

Linking groups between aromatic or other conjugated systems are similarly scored. NH is considered negative (–1) because of strong hydrogen bonding interactions between the chains, which mitigate against the formation of aligned nematic phases. On the other hand, ether oxygen acts only in the chain direction and is scored as 0 if isolated or +1 if it forms part of a linking group, e.g. [C(O)O] scores 3 and [OC(O)O] is scored at 4. It follows from these assignments that amide [CONH] scores 1, urethane [NHCO₂] scores 2 and imide scores 2.

Cyclo- aliphatic rings make some additional contribution to stiffness compared to aliphatic units. *trans-p*-Cyclohexylene has been estimated to contribute

Table 1d. Non-aromatic ring units

Chemical Formula	Chemical Structure	Mesogenic Index Group Contribution
24. *trans* p-cyclohexylene		2
25. *cis* p-cyclohexylene		1
26. *m*-cyclohexylene(*cis* or *trans*)		0
27. *p*-methylene cyclohexylene(*trans*) —CH₂—		1
28. *p*-dimethylene cyclohexane —CH₂— —CH₂ —		0
29. *p*-dicyclohexylmethylene	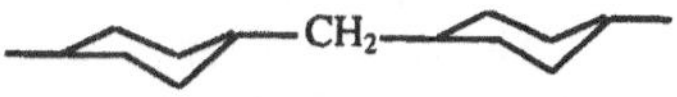	0
30. norbornene imide	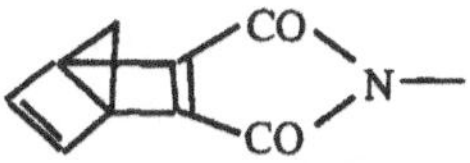	4
31. maleimide	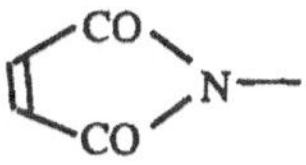	3

about +2, whereas *cis-p-*cyclohexylene is not considered to contribute more than the equivalent of a single rigid bond (+1). Norbornene rings and imide rings are also believed to contribute +2. These assignments have been made only on the basis of limited published data and may need revision.

Units which destroy the linearity of the chain must be treated as separate entities. These include *meta*-phenylene, 2,6-naphthalene, ether or methylene diphenylene and *meta*-cyclohexylene. The rule which is most generally applicable in this case is to subtract 2 units from the in-chain score. Hence *meta*-phenylene and *m*-cyclohexylene score +1 and 0 respectively.

The best fit for 2,6 or 2,7 substituted naphthalene is +3. This happens to be 2 less than the number of conjugate bonds and reflects the distortion from linearity caused by these naphthalene units. In the case of preformed units containing two aromatic rings which are joined together by a linking group, e.g. -CH₂-, -O-, -CH(CH₃)-, -CO-, -C(CH₃)₂-, -S- or -SO₂-, the contribution assigned reflects the

Table 1e. Spacer and linking groups

Chemical Structure	Mesogenic Index Group Contribution
SPACERS	
32. polymethylene	0
33. alkenyl [C=C]	1
34. azomethenyl [C=N]	1.5
35. carbonyl [C=O]	2
36. diazo [N=N]	2
37. conjugated butadiene unit [–C=C–C=C–]	3
LINKING GROUPS	
38. CONH (amide)	1
39. CON(Me) (*N*-methyl amide)	0
40. CON(*R*) (*N*-alkyl amid)e: CON<(imide)	–1
41. CO.O (ester)	3
42. OCO.O (carbonate)	4
43. OCO.NH (urethane)	2

effect of the linking group on chain linearity, i.e. -CH$_2$-, -O- or -S-<-CH(CH$_3$)- <-SO$_2$- or -C(CH$_3$)$_2$-. The overall best fit score for the preformed unit for each category has been set at 1, 0 and –1 respectively. Thus an ether group is considered to distort linearity to a similar extent to a meta-link in an aromatic ring. When two ether groups join three rings, the effect is considered to be that of two meta- links (i.e. +2).

An even number of CH$_2$ spacers between benzene rings can adopt a trans configuration and should not kink the chain. In the system Ar-(CH$_2$)n-Ar, an odd number for n is likely to give a gauche configuration leading to non-linearity. On this basis, the score is –1 for n≥3 (odd values). (n=1 is a special case).

A *p*-terphenyl group should contribute +10 on the basis of a conjugated unit. However a single methylene interposed between two rings would be expected to cause kinking and reduce this value to +5. A single spacer between each pair of rings should further reduce the score to +2. If two pairs of methylene linked rings are joined together with a central biphenyl ring, the logical contribution is expected to be +3.

By employing the rules outlined above, it is possible to score most condensation polymers on the Mesogenic Index. The contributions of the most common structures and linking groups are listed in Table 1a–e. In addition a supplementary series of values have been added in Table 1f, which have been derived to apply the MI to a wide range of polyimides. In general aromatic imide units are not as mesogenic as para- aromatic ester or amide units. The unit is not quite linear, and for practical purposes a best fit has been obtained by scoring the unit midway between a para- and a meta-oriented benzene ring, giving it a score of 2. This applies to both trimellitimide and pyromellitimide units. When two aromatic imides are linked the resultant biphenylene di-imide units scores just 4, 3

Table 1f. Functional group assignments based on polymides

Structure No.	Structure Group		Mesogenic Index Group Contribution
44.	$-S(CH_2)_n(CH_3)$		−1
45.	$-C_6H_5$		−1
46.	$- CH_3$		−1
47	$-Cl, -F$		0
48.	$-S(C_6H_5)$		−1
49.	$-O(C_6H_5)$		−1
50.	$-OR$		−2
51.	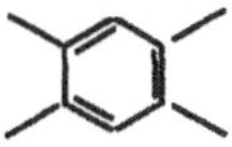	(diimide)	2
52.	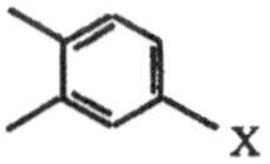	imide, where X=CH$_3$ or NO$_2$	2
53.	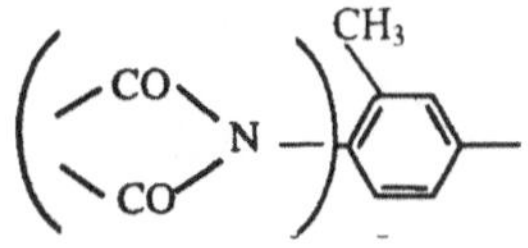	CH$_3$ group reduces ring score by 2 when adjacent	1
54.	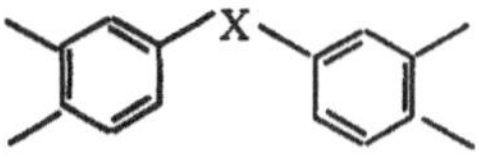	when X=O or S when X=CO, CF$_3$ or SO$_2$	2 0
55.	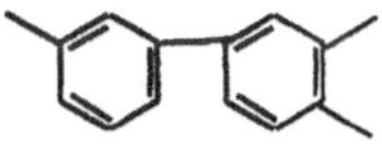		4
56.	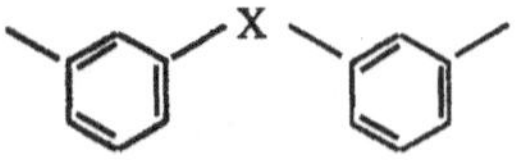		0

less than the *p*-biphenyl unit. Hence it is much less mesogenic than the *p*-biphenylenes but slightly better than a single *p*-phenylene unit. However, when a group such as ether or thioether interposes between two aromatic imide rings, the overall structure is stiffer and less bent than the corresponding diphenyl ether. Hence a score of 2 best fits its behaviour compared to 1 for the diphenyl ether unit. Another anomaly of the imide group is that it appears to cause more steric interference when a methyl group is adjacent to it in an ortho position on a neighbouring ring. A more accurate result is obtained by scoring −2 for the methyl group in this situation rather than −1 when it is ortho to an ester or amide group.

4.3
Determination of Critical MI Values

4.3.1
Polyesters, Polyamides, Poly(esteramide)s

Table 2 lists eleven *polyester* copolymer systems in which the compositional relationship to the presence of anisotropy has been established by the authors of

Table 2. Application of MI to critical compositions for LC phase formation in random polyesters

Polymer Code	Monomer Units	Critical Monomer conc. for LC phase (Mc %)	Polymer Type	Mesogenic Index
PE1	IPA/TPA/HQ	IPA<33	A_1A_2B	10.6
PE2	PHBA/(MeO)$_2$· PHBA	PHBA>50	C_1C_2	10.0
PE3	TPA/MeHQ/MHBA	MHBA>=20	ABC	10.4
PE4	TPA/t–CHD/c–CHD	t–CHD>25	AB_1B_2	10.5
PE5	TPA/MeHQ/CHDM	MeHQ>=7.5	AB_1B_2	9.4
PE6	IPA/HQ/PHBA	PHBA>20	ABC	10.4
PE7	TPA/HQ/MHBA	MHBA<=50	ABC	10.0
PE8	PET/PHBA	PHBA>35	ABC	10.0
PE9	TPA/ClHQ/Res	ClHQ>=20	AB_1B_2	10.4
PE10	TPA/HQ/ODBA	ODBA<30	A_1A_2B	10.8
PE11	IPA/ST/MeHQ	ST>=20	A_1A_2B	10.2

Table 3. Application of MI to critical compositions for LC phase formation in random polyamides and polyesteramides

Polymer Code	Monomer Units	Critical Monomer conc. for LC phase (Mc %)	Polymer Type	Mesogenic Index
PA1	PABA/PAMeBA	PABA>50	$C'_1C'_2$	10.5
PA2	PABA/PAPhAA	PABA>50	$C'_1C'_2$	10.5
PA3	PABA/11–AUnA	PABA>70	$C'_1C'_2$	9.3
PA4	DAB/TPA/AA	TPA>25	A_1A_2B (B=ordered)	10.5
PEA1	IPA/TPA HQ/PAPh	IPA=33 PAPh>0	A_1A_2BB'	10.6
PEA2	IPA/PHBA/PMeAPh	PMeAPh<30	AB'C	10.4
PEA2A	IPA/PHBA/PAPh	PAPh<30	AB'C	11.0
PEA3	PAPh/MeOTPA/IPA	IPA<33.5	A_1A_2B'	10.0
PEA4	PAPh/MeOTPA/PHBA	PHBA<37.5	AB'C	11.0
PEA5	PABA/MHBA/(MeO)$_2$·PHBA	(MeO)$_2$·PHBA<60	C_1C_2·C'	9.6
PEA6	PABA/PHBA/(MeO)$_2$·PHBA	(MeO)$_2$·PHBA=50 PABA<=40	C_1C_2C'	10.0

the papers[48, 50, 66–71,]. These include examples of [(A)-(B)], [(A)-(B),(C)] and [(C₁),(C₂)] random copolymer types. The critical Mesogenic Index value for the onset of liquid crystallinity lies in the range 10.2±0.4, i.e. less than one-half double bond unit difference between mesogens. Similar data published on random *polyamides* and *polyesteramides* is presented in Table 3 [49, 67, 72, 73]. Systems covered include [(A)-(B)], [(A)-(B'), (C)], [(A)-(B')], [(A)-(B), (A)-(B')], [(C), (C')], [(C'₁), (C'₂)]. Three lyotropic polyamide systems listed have a critical MI value of ≥10.5. All the polyesteramide systems, with one exception, have critical MI values below 11. These values are weighted to account for the proportion of 3-ring amide mesogens in the copolymers. The mean value for the 11 amide copolymers is 10.3± 0.5, i.e. very similar to the mean value for polyesters.

4.3.2
Polycarbonates

An excellent fit with published data is once more obtained by calculating the mean MI per mole of *polycarbonate* and multiplying this score by an appropriate "length factor". In this case 1.5 times the average monomer score enables a consistent prediction to be made.

Mahabadi and Alexandru [74] have published data on thermotropic copolycarbonates of biphenol (BP) and diethylene glycol (DEG) with hydroquinone (HQ), methyl hydroquinone (MHQ), *t*-butyl HQ, oxydiphenol (ODP) or bis(4-hydroxyphenyl)methane (DPM) or bisphenol A. The tabulated compositions and calculated MI values are recorded with the published data on mesophase transitions in Table 4. Mesophases appear only to be present when the MI score ≥10.1. However, the gap in values between mesogenic materials and non-mesogenic materials in the three series reported is about +1 and therefore it cannot be claimed on this evidence alone that polycarbonates have a similar critical value to polyesters and polyamides.

Further evidence to support the case for polycarbonates comes from the work of Kricheldorf and Lubbers [75]. They prepared several series of copolycarbonates of 4,4'-dihydroxybiphenyl and various other aromatic diols (Table 5). It is noteworthy that all their series of copolymers cease to exhibit a nematic phase when the MI£10.5. Hence by combining both sets of results it can be said that the critical value for polycarbonates lies between 10.1 and 10.5.

The 4,4'-dihydroxy biphenyl is essential to the formation of a mesophase because it contributes the largest score, i.e. 7, to the copolymers. It was present in all the mesogenic polycarbonates cited as mesogenic. The Mesogenic Index predicts that mono-ring aromatic polycarbonates probably cannot form a mesogenic phase because the maximum score which can be reached is 10.5 using hydroquinone as monomer (cf. Table 5). Liquid crystallinity is difficult to establish in this "borderline" polymer because it is highly crystalline, insoluble and infusible.

Table 4. Application of mesogenic index to copolycarbonates of biphenol and Deg

Reference Code	Copolycarbonate Composition (moles)								Tn (°C)	Melting Transition Tc (°C)	Mesogenic Index
	BP	BPA	DPM	ODP	HQ	MHQ	TBHQ	DEG			
PC11	100							94	–	181	11.3*
CRC1	89				11			94	142	172	11.0*
CPC2	79				21			94	115	154	10.7*
CPC3	69				31			94	–	137	10.4*
CPC4	50				50			94	–	–	8.3
PC12				100				94	–	–	8.3
CPC8	79					21		100	105	167	10.5*
CPC9	68					32		100	100	160	10.1*
CPC10	80						20	100	105	177	10.4*
CPC13	71	29						94	102	134	10.1*
CPC14	90	10						94	112	145	10.9*
CPC18	75		25					100	115	145	10.1*
CPC19	50		50					100	–	–	9.0
CPC20	–		100					100	–	–	6.8
CPC24	75			25				100	–	134	10.1*
CPC23	50			50				100	–	–	9.0
CPC22	25			75				100	–	–	7.9
CPC21	0			100				100	–	–	6.5

Table 5. Thermotropic polycarbonates derived from 4,4'-dihydroxybiphenyl and various diphenols

Polymer Code	Composition	Mesogenix Index
2a	BP/MeHQ 1:1	12.75[a]
2b	BP/MeHQ 1:2	11.5[a]
2c	BP/MeHQ 1:3	10.9[a]
2d	BP/MeHQ 1:4	10.5
3a	BP/ClHQ 1:1	12.7[a]
3b	BP/ClHQ 1:2	11.5[a]
5a	BP/ODP 1:1	12.0[a]
5b	BP/CDP 1:1	11.25[a]
5c	BP/SDP 1:1	10.5
5d	BP/SO2DP 1:1	10.5
5e	BP/BPA 1:1	10.5
6a	BP/ODP/MeHQ 1:1:1	11.25[a]
6b	BP/CDP/MeHQ 1:1:1	10.5[a]
6c	BP/SDP/MeHQ 1:1:1	10.0
7a	BP/BPA/HQ 1:1:1	10.5
7b	BP/BPA/HQ 1:1:2	10.5

a Liquid crystal phases were present in copolycarbonate

4.3.3
Polyimides and Polyetherimides

The values for polyimide groups and the corresponding mesogenic length were derived from work carried out by the author [43] on the synthesis and characterisation of liquid crystalline polyimides. Imide groups contribute +2 and have a mesogenic length of 2.

When the MI was first formulated, there was no useful published information upon which to assess the relationship between composition and mesophase in imide-containing polymers. Effort was therefore concentrated initially upon synthesising thermotropic low molecular weight oligomeric imides and applying the Mesogenic Index to these. The oligoimides listed in Table 6 were synthesised from 2 mol of nadic anhydride and various multicyclic diamines. All the diamines contained ether groups, except the first one which possessed an ester link. A good fit was obtained with the rule that the MI≥10, by postulating that the norbornene ring scored 2, similar to a cyclohexane ring. The imide group was taken to score 2 on the basis of consistency with other assigned values $NX_2=-2$, $CO=+2$). When the norbornene ring was replaced by maleimide with a single double bond, mesophases were not seen. This may indicate that the presence of an additional unsaturated ring structure such as is present in the nadic end unit is needed to ensure the occurrence of a mesophase in the di-imides of these multi-ringed diamines. Accordingly, maleimide was assigned a value of only one to

Table 6. Oligomeric imides

Polymer No.	Book No.	Chemical Structure	Melting Point (°C)	Mesogen Index	Comments
1	2627/88	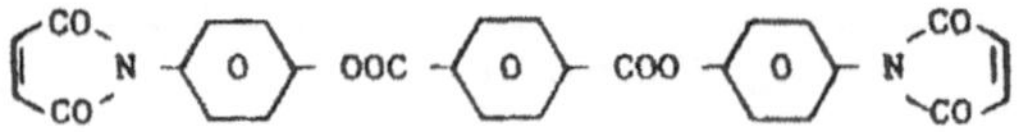	22[a] (melt)		
2	3040/73	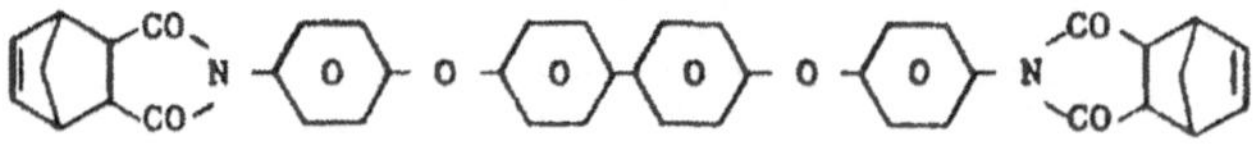		11[a]	
3	2103/73	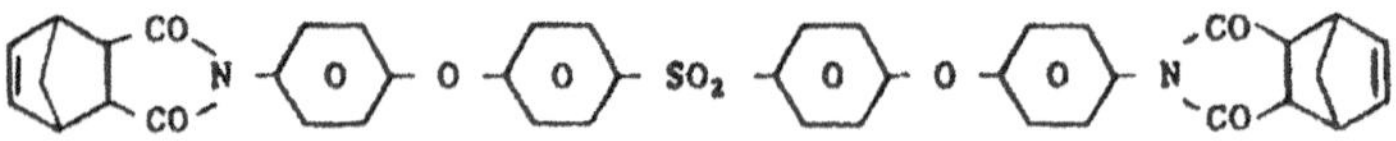	190	11[a]	Monotropic
4	2103/63	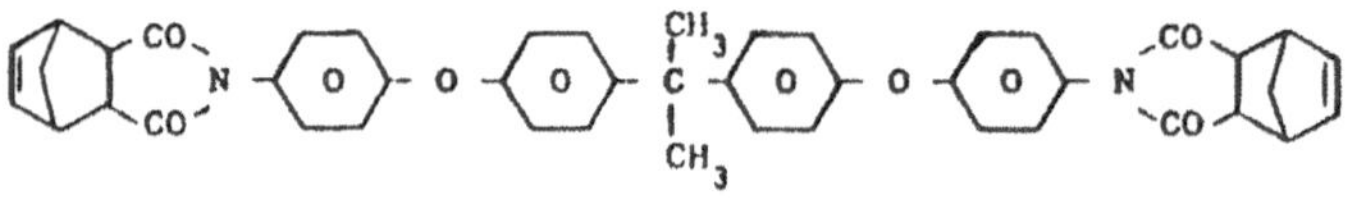		10[a]	Monotropic
5	2103/75	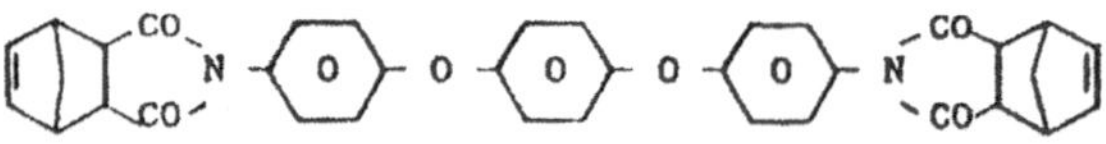	285	10[a]	Monotropic
6	3785/001	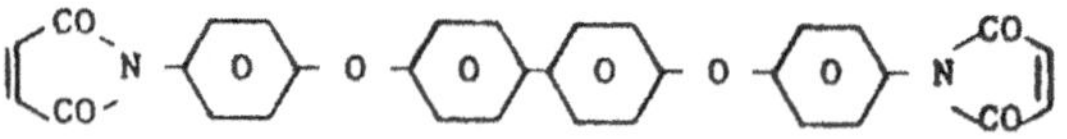		9	
7	2627/81	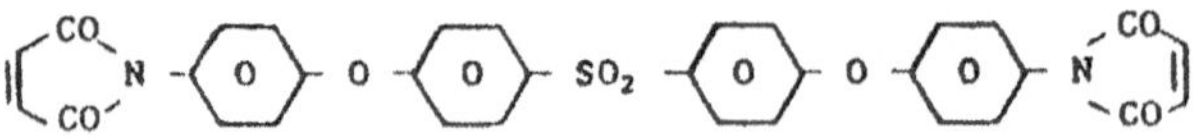		9	
8	2103/76	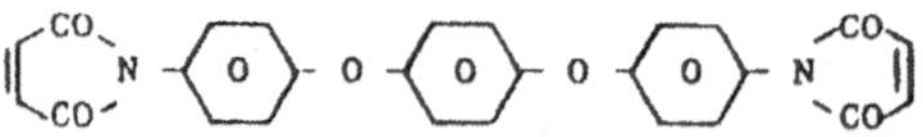		8	
9	2627/82	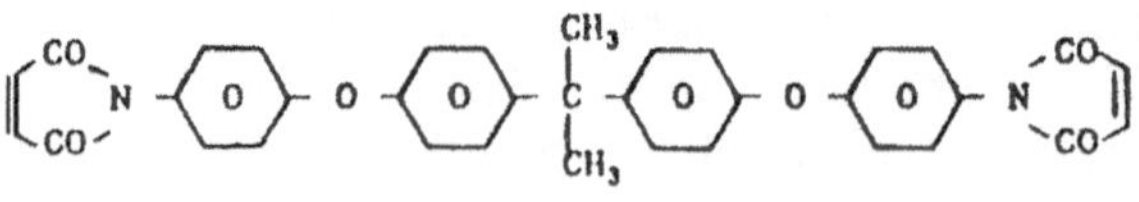		8	
10	2103/42	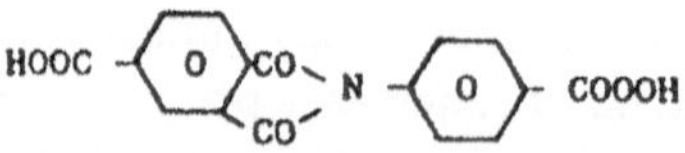		8	

a lc phase

Source: High Performance Polymers (1995) 7:421-431

reflect its single double bond, and the di-imides scored MI£9 on this basis, in accordance with their isotropic nature.

A series of much higher molecular weight oligoimides was next synthesised, in which neither ether nor ester groups were present [42]. The diamine contained two p-aminophenyl units linked by an aliphatic group $(CH_2)n$. Polyimides containing diamines with a biphenyl unit either linked directly or with two or four methylene groups interposed between the rings gave evidence of a transient mesophase in the melt upon rapid heating [43], which tended to disappear quickly once the nadic end groups started to decompose. The strongest phase was observed with the biphenyl unit. Mesophases were not observed with odd numbers of methylene groups and it is suggested that these create kinks in the polymer backbone which are not conducive to the rigid rod configuration normally required for mesophase formation.

The structures and corresponding calculated MI values are given in Table 7 (nos. 1–5). Mesophases were observed for scores of MI=12 and 13 but not for the odd structures (MI range 9.5–11). In addition, a series of imide-amide oligomers were synthesised with molecular weights above 1000, in which the terminal groups were nadic anhydride. MIs ranged from 8.3 to 10.7 but no positive identification of mesophases were made in any of these (Table 7, nos. 6–9)

A number of polyimides were synthesised by Liverpool University in 1989 [87, 88] and were tested at BP Sunbury for liquid crystal phases. The results are

Table 7. Oligomeric polyimides

Polymer No.	Book No.	Chemical Structure	Melting Point (°C)	Mesogenic Index
		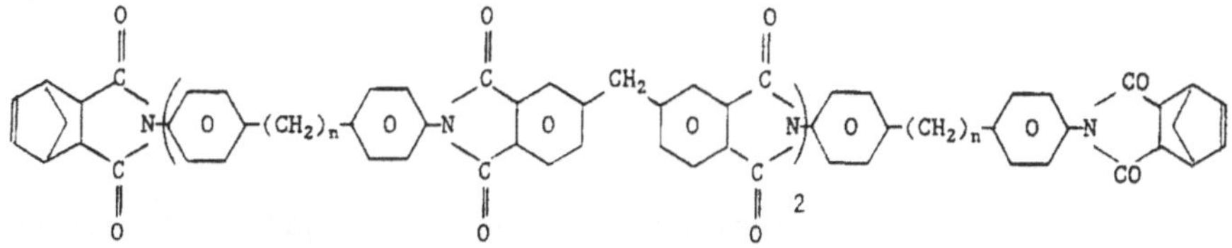		
1		n=0		13.0[a]
2		n=1		9.5
3		n=2		12.0[a]
4		n=3		11.0
5		n=4		12.0[a] (very weak birefringence)
6		NA-[BAPS-BTDE]₂ BAPS-NA		9.8
7	2103/70	NA-[BAPS-TPA-BAPS-TPA-BAPS]-NA	(Infusible)	10.7
8	2103/78	NA-[BAPS-IPA-BAPS]-NA		9.7
9	2103/76	NA-[BAPP-IPA-BAPP]-NA	255	8.3

NA=nadic anhydride; BAPS=bisaminophenoxy phenylsulphone; BDTE=benzene tetradicarboxylic diester; TPA=terephthalic acid; IPA=isophthalic acid; BAPP=bisaminophenoxy phenyl 2-2 propane
a An LC phase

Table 8. Thermotropic poly(etherimide)s prepared at Liverpool University

Polymer Ref. No.	Structure	Mesogenic Index	Nature of melt
PE1	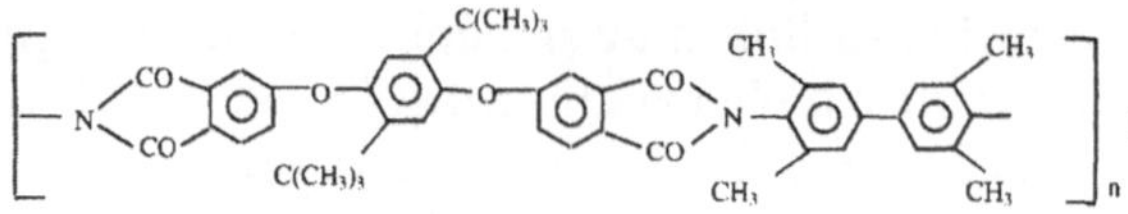	10	Transient LC phase
	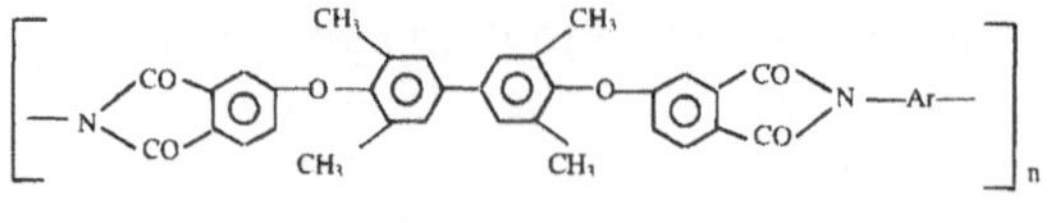		
PE2	where Ar= 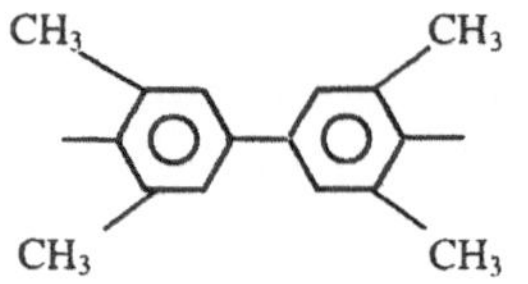	11	Transient LC phase
PE3	where Ar= 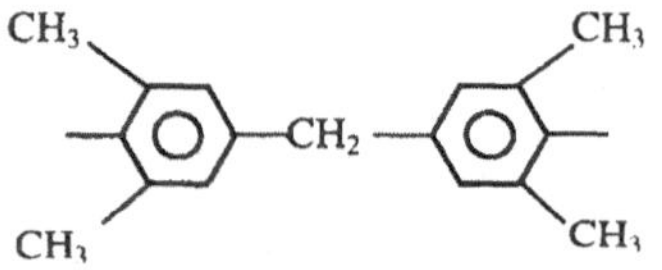	6	Isotropic
PE4	where Ar= 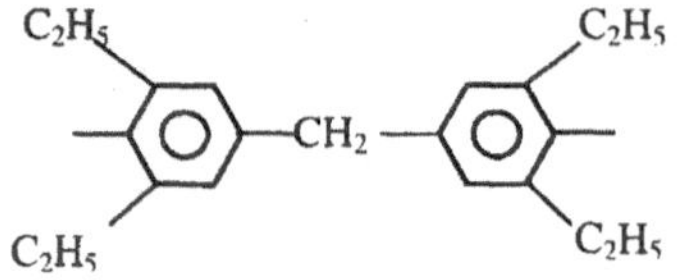	6	Isotropic

Source: Polymer (1993) 34:2865; Polymer (1994) 35:4215

Table 9. Determination of critical MI values in polyesterimides. The following table applies to the two varying polyesterimide compositions (25): (1) Copolyesterimide of m-amino phenol trimellitimide (A) and 2,6 hydroxynaphthoic acid (B); (2) copolyesterimide of m-amino phenol trimellitimide (A) and 1,4 hydroxybenzoic acid (B)

Ratio A:B (moles)	Mesogenic Index	Description of melt
100:0	8.0	isotropic
80:20	8.8	isotropic
60:40	9.6	isotropic
50:50	10.0	thermotropic
40:60	10.4	thermotropic
30:70	10.8	thermotropic
20:80	11.2	thermotropic

Source: Eur Poly J (1994) 30(8):549–556

given in Table 8. Although high melting, all but one showed a degree of mobility at high temperatures and transient liquid crystal phases were observed in two cases where the MI≥10. The mesophases gradually disappeared due to decomposition.

Direct evidence for these assignments was obtained from critical composition values obtained for two series of polyesterimides synthesised by Kricheldorf and co-workers [76] (Table 9). In both series, a preformed unit comprising the trimellitimide of *meta*-aminophenol is copolymerised with either 2-hydroxy naphthalene 6-carboxylic acid or 4-hydroxybenzoic acid. In both series, the critical value at which a mesogenic phase is first observed is 10.0, at a 1:1 composition. An isotropic phase was observed at 9.6 and below. Hence the critical value must lie between the limits of 9.6 and 10.

Hence the weight of evidence for polyimides points once again to a critical value of around 10 for the Mesogenic Index, on the basis of an imide group score of 2 and a Mesogenic Index length of 2. In the remaining section of this chapter we discuss the results of applying the MI approach to a large number of co-polyimides with esters, amides, ethers etc., much of this work being published by Kricheldorf and co-workers at Hamburg University in the past few years.

5
Application of MI to Copolymers Containing Imide Groups

The author's work on polyimide synthesis and its use to derive MI values for polyimides has been described in the previous section [43, 45–46]. These are virtually the only strictly true mesogenic polyimides which have been reported in the literature to date. However, in addition, there have been many reports of modified copolyimides such as copoly(etherimide)s, copoly(amide-imide)s, copoly(esterimide)s or copoly(imide-carbonate)s which are discussed below.

5.1
Copoly(etherimide)s

The structures given in Table 10 belong to a series based on biphenyl-3,3',4,4'-tetracarboxylic anhydride [BPTA] as the mesogen forming unit. Some of these have been reported to be thermotropic poly(etherimide)s [44]. Series 7 was based on BPTA alternating with aliphatic units derived from α,ω-diamino alkanes. Isotropic polyimides were formed with corresponding MI values of 7–8. Series 8 poly(etherimide)s were constructed from BPTA and α,ω-bis(4-aminophenoxy)alkanes. These were thermotropic with corresponding MI values 13–14 (i.e. above the critical value). Thermotropic poly(etherimide)s of this type were reported independently by Dolden [43]. Series 9 polyimides were constructed from the BPTA and the meta- or ortho- linked α,ω-bis(aminophenoxy)alkanes. The effect of the kink reduces MI to 10 for the meta linkage (borderline case-isotropic polyimide observed) and MI to 8 for the ortho- linked diamine (isotropic). Polyimide series 10 was formed from methyl substituted α,ω-

Table 10. Thermotropic poly(etherimide)s based on biphenyl tetracarboxylic imide

Polymer ref. no.	Structure	Mesogenic Index	Nature of melt
7.	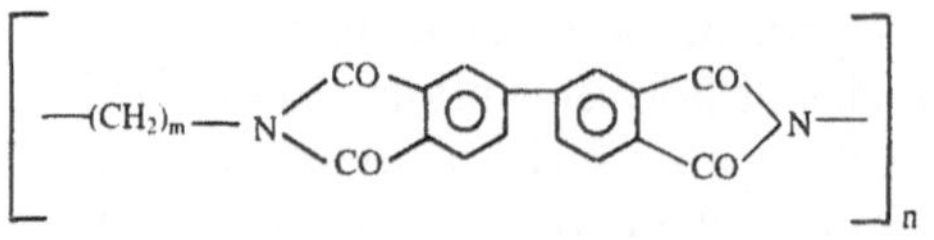		
	m even	8	Isotropic
	m odd	7	Isotropic
8.	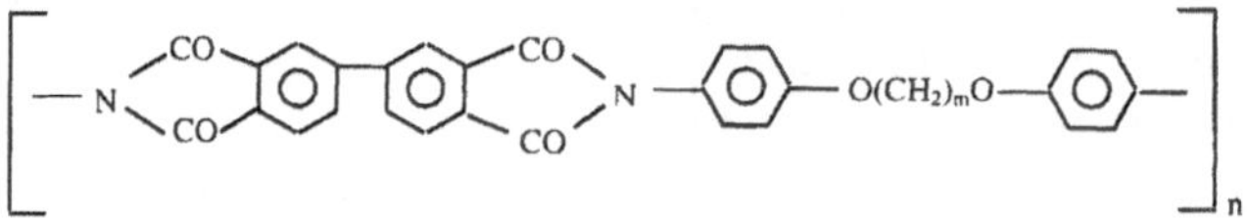		
	m even	14	LC phase
	m odd	13	LC phase
9.	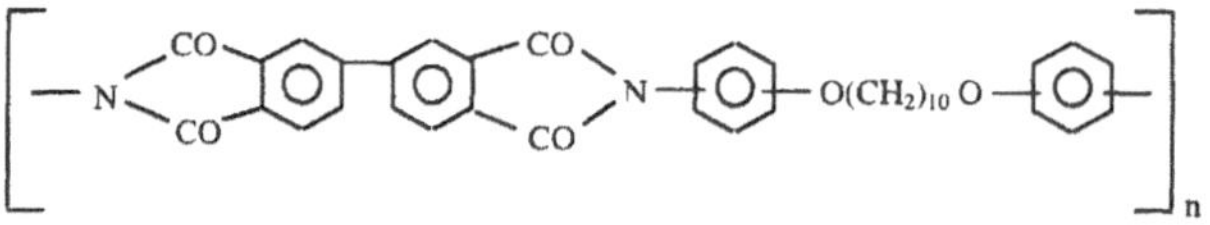		
	a. meta-	10	Isotropic
	b. ortho-	8	Isotropic
10.	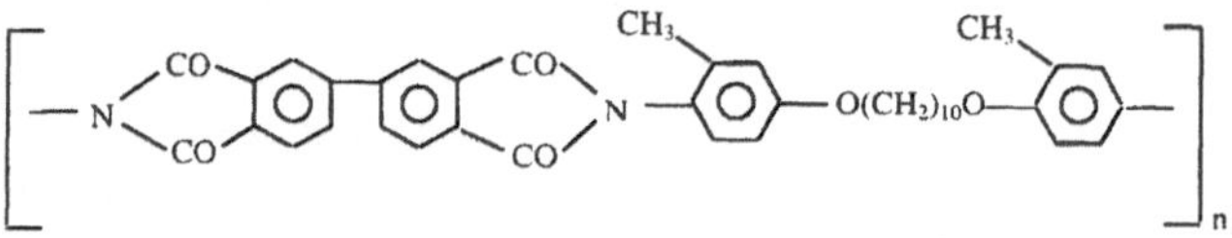		
	(counted MI score for $CH_3=-2$ because it is adjacent to imide group)	10	Isotropic
11.	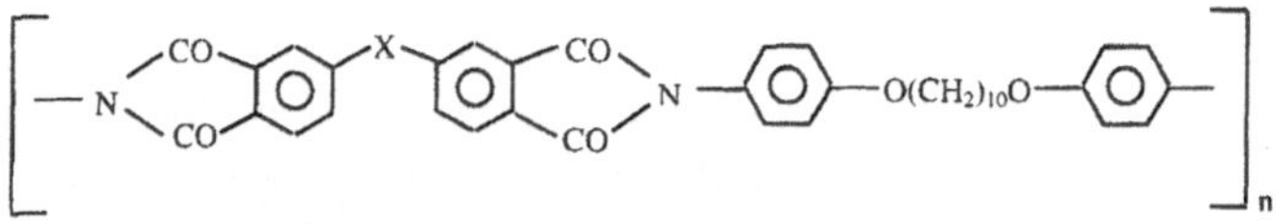		
	where $X=SO_2$, CO, $C(CH_3)_2$	10	Isotropic

Source: Polymer (1995) 36(9):1893–1902

bis(4-aminophenoxy)alkanes; the effect of the methyl substituent was to reduce the MI to 10 and an isotropic polyimide was obtained.

The synthesis of a poly(etherimide) from BPTA was conveniently carried out in dimethylacetamide by condensing equimolar amounts (0.01 mol) of BPTA and 1,12-bis(aminophenoxy)dodecane at 100 °C for 4 h. Imidisation was completed by adding 1:1 mix of acetic anhydride in pyridine as dehydrating agent and heating for a further hour. The product melted at ca. 300 °C to give a highly anisotropic melt but started to decompose before reaching its clearing temperature [43].

The application of the MI to these polymers depends upon choosing an appropriate value for the biphenyl group attached to imide groups. The normal value for a p,p'-biphenyl unit is 7, composed of three for each ring and one for conjugation. The view was taken here that when sandwiched between imide groups the biphenyl makes a comparatively poor mesogen which is not quite planar, and we have ascribed a value mid-way between the meta and para value for each ring (i.e. 2). Hence the full unit was considered to contribute 4. The MIs were calculated for the poly(etherimides) on this basis and accord with the rule that an MI>10 is needed for mesophase occurrence.

5.2
Copoly(amide-imide)s

A series of thermotropic poly(amide-imide)s were synthesised [38] from diamines produced by alkylation of silylated 4-nitrophenol with α,ω-dibromoalkanes, followed by hydrogenation of the resulting α,ω-bis(nitrophenoxy)alkanes. The diamines were reacted with trimellitic anhydride chloride in boiling m-cresol to produce polymer structure 21.

$$\tag{21}$$

For m=9–12, it was established that the co(polyamide-imide)s melted to a smectic A phase in the range 250–313 °C and cleared below 350 °C to an isotropic state. In Table 11, the structures and corresponding MI scores for polymer series 4–10 from this paper are given. Series 4 are all thermotropic and score between 12.5 and 13.75 depending on whether or not chain spacer is odd or even. Series 5 contained pendant methyl groups on the phenoxy rings adjacent to the imide groups which cause steric interference and substantially reduce the MI to 8.5, a level which correctly predicts isotropic behaviour. Series 6 contains metalinks which disrupt linearity and hence the polymer is isotropic with a similar MI score. The copolymer 8 has an average MI=14.5 and is reported to be thermotropic, whereas copolymer 9 has an MI=12 and is reported to be birefringent, although a mesophase was not definitely detected. Copolymer 10 is random and isotropic despite a borderline score.

A new class of diamine spacer was synthesised from α,ω-diaminoalkanes and 4-nitrophthalic anhydride; the resulting α,ω-bis(4-aminophthalimido)alkanes were polycondensed with one of the following acid chlorides: terephthaloylchloride, 2-phenylthioterephthaloylchloride, naphthalene-2,6-dicarboxylic acid or 4,4'-biphenyldicarboxylic acid chloride [40]. Thermotropic behaviour was con-

Table 11. Poly(amide-imide)s derived from trimellitic anhydride

Polymer ref. no.	Structure	Mesogenic Index	Nature of melt
4.	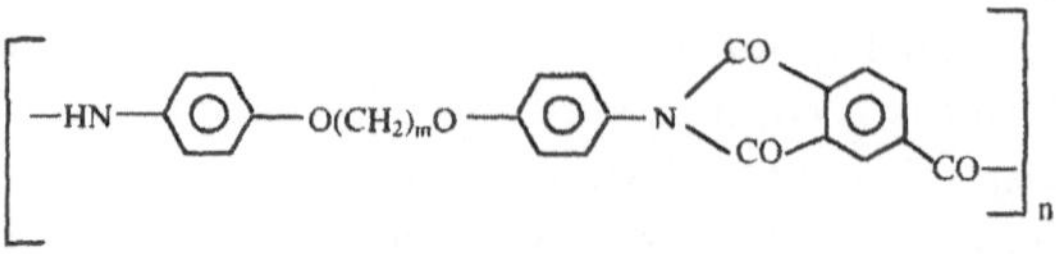		
	a, c – m odd	12.5	LC phase
	b, d – m even	13.75	LC phase
5.	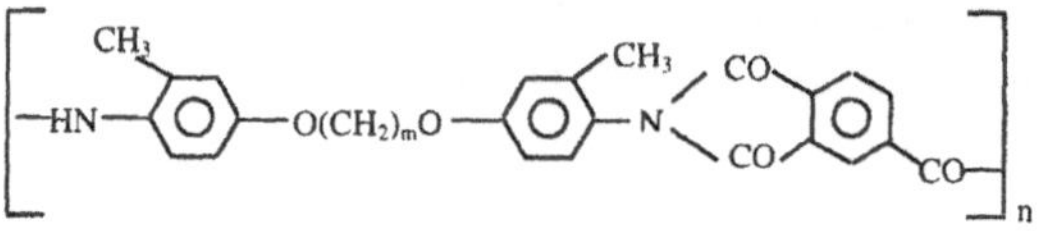	8.5	Isotropic
6.	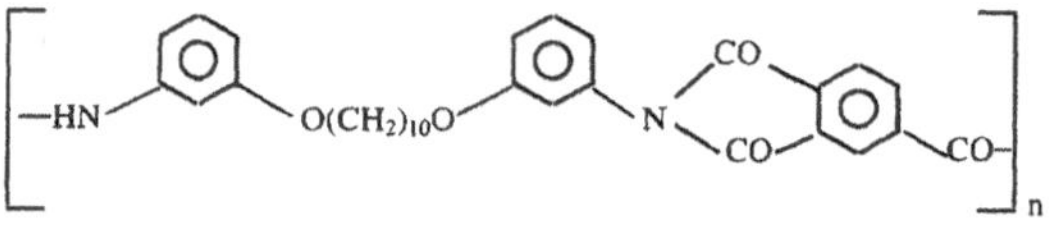	8.5	Isotropic
8.	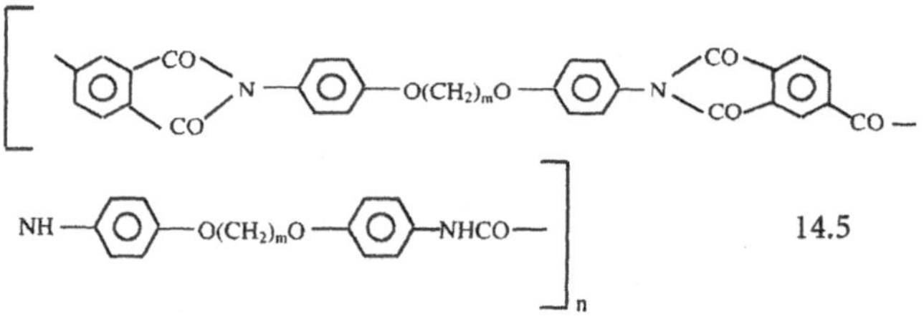	14.5	LC phase
9.	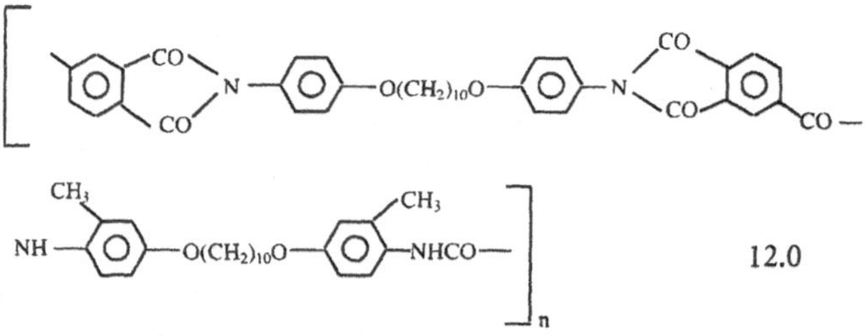	12.0	Birefringent
10.	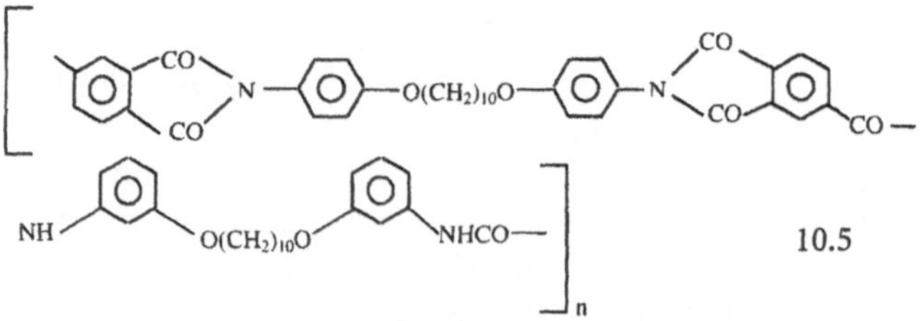	10.5	Isotropic

Source: J Poly Sci Pt A Polymer Chemistry (1995) 33:2241–2250

firmed in just one series containing the biphenyl dicarboxylic acid unit, in which both smectic and nematic phases were observed in the melt. The corresponding MIs are given with the polymer structures in Table 12 and confirm that only the biphenyl unit was capable of producing a mesophase.

A series of lyotropic poly(imide-amide)s were synthesised [39], in which an ordered unit comprising trimellitimide condensed with p-aminobenzoic acid and p-phenylene diamine (PPD) was copolymerised in varying ratios (0, 50 and 80%) with a terephthalic acid/PPD polyamide. The MI predicts all three polymers to be LC, but lyotropic behaviour was observed in only the first one (Table 13). This is particularly surprising because incorporating increasing quantities of the known lyotropic and more linear TPA/PPD units would be expected to produce definite lyotropic behaviour. Similarly, a copoly(amide-imide) of 3,3'-dimethyl benzidene and 4-amino benzoic trimellitimide was predicted by the MI to possess a mesophase but none was detected. These apparent failures may be due to the fact that the polymers which did show lyotropic behav-

Table 12. Thermotropic poly(amide-imide)s

Polymer ref. no.	Structure	Mesogenic Index	Nature of melt
5.	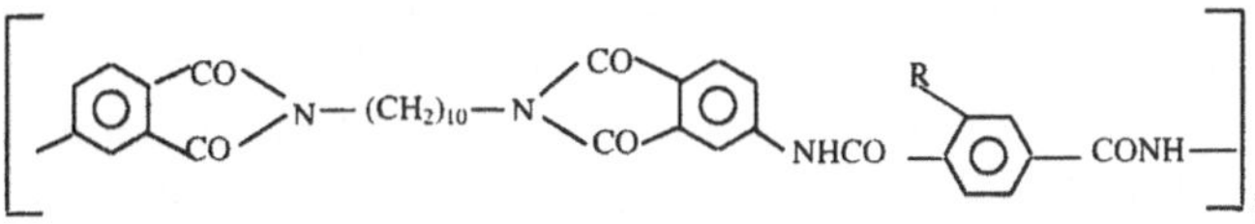		
	a. R=H	8.3	Isotropic
	b. R=S-C$_6$H$_5$	7.5	Isotropic
6.	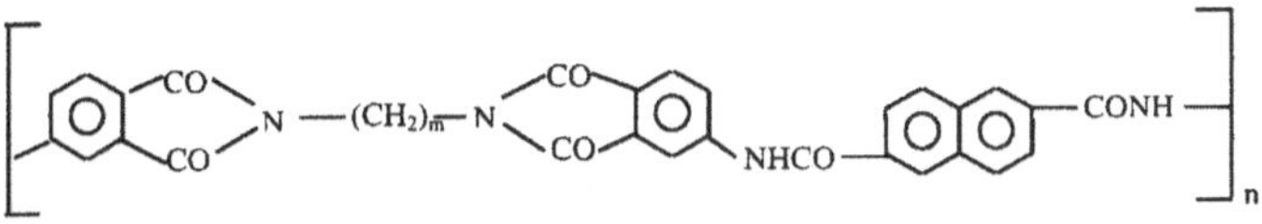		
	m=9	7.8	Isotropic
	m=10	8.3	Isotropic
7.	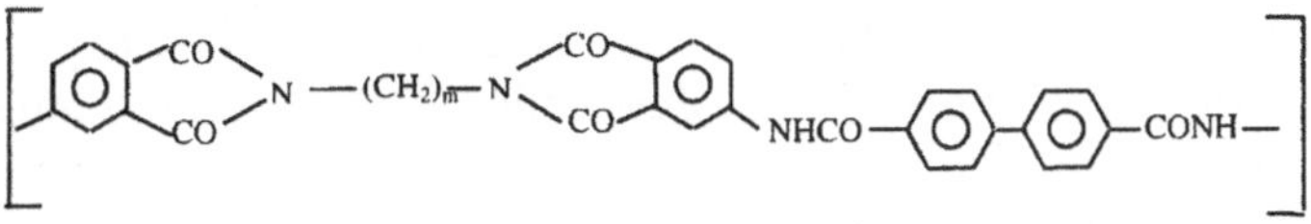		
	a. c. m=9	10.8	LC phase
	m=10, 12	11.3	LC phase

Source: J M S Pure Appl Chem (1995) A32(11):1831–46

Table 13. Lyotropic poly(amide-imide)s

Polymer ref. no.	Structure	Mesogenic Index	Nature of melt

a.	a:b=10:0	10.7	lyotropic
b.	a:b=5:5	11.3	Isotropic
c.	a:b=2:8	11.7	Isotropic

| 4. | 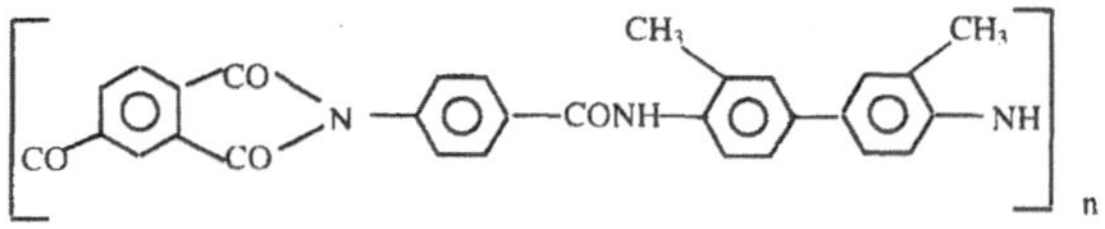 | 12.7 | Isotropic |

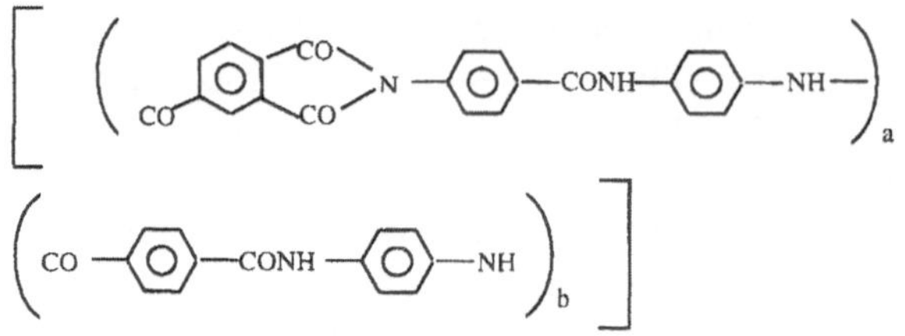

Source: Makromol Chem (1993) 14:395–400

iour could not be tested above 10% concentration in sulphuric acid. In most cases lyotropic behaviour was only observed at above 15% concentration.

A very recent synthesis of a thermotropic copolyimide contains an N-(carboxyphenyl) trimellitimide ester unit linked to a p-aminobenzoic acid via an amide linkage, which in turn is linked by an alkane diol unit back to the trimellitimide unit [86]. This is probably the first example of a copoly(ester-amide-imide) with mesogenic properties identified. Its MI score is 9.6.

5.3
Copoly(esterimide)s (PEIs)

5.3.1
PEIs Based on Trimellitic Anhydride

Table 9 lists two series of PEIs based on copolymers of precondensed units of trimellitic anhydride and m-aminophenol, reacted with varying molar fractions (20–100%) of 2-hydroxy naphthalene 6-carboxylic acid or 4-hydroxybenzoic acid. In both series an MI score of 10 corresponded to the onset of liquid crystallinity, and an isotropic copolymer was obtained for MI $\leq$ 9.6. A series of PEIs is given in Table 25, in which varying ratios of N-carboxyphenyl trimellitimide and isophthalic acid are co-reacted with phenylhydroquinone [82]. The MI values vary by only 0.1 units between compositions and more accurately pinpoint the transition between the mesogenic and isotropic state to the range 9.5–9.6.

Chang and Hung [83] have synthesised **22** below, in which an extra double bond is inserted into the N-(4-carboxyphenyl trimellitimide) unit. When copolymerised with an even chain alkane diol, an enantiotropic PEI is formed (MI=10); when the spacer is odd, only a transient monotropic phase is observed (MI=9.4). This is further evidence that the critical value for mesogenicity in PEIs is close to 9.5. It is of interest to note that the diacid employed in the copolymer, N-(carboxyphenylacrylic) trimellitic anhydride acts as the complete mesogenic unit for the copolymer.

$$\left[\ OC{-}\underset{CO}{\overset{CO}{\diamond}}{-}N{-}\bigcirc{-}CH{=}CH{-}OCO{-}(CH_2)m.CO\ \right]_n \qquad (22)$$

Chang and Hung have also synthesised a PEI containing N-(carboxyphenyl) trimellitimide units linked via alkylene spacers to a pyromellitic diimide unit [83] (**23**).

$$\left[\ OC{-}\underset{CO}{\overset{CO}{\diamond}}{-}N{-}\bigcirc{-}COO{-}(CH_2)m{-}N\underset{CO}{\overset{CO}{\diamond}}N{-}(CH_2)O{-}\ \right]_n \qquad (23)$$

When the number of alkylene spacers m was even (4 or 6) an enantiotropic smectic LC-phase was observed, whereas when m was odd only an isotropic melt was observed. According to the MI, an even value of m yields MI=9.5, while an odd value gives MI=9.

The five series of PEIs based on N-(4-carboxyphenyl) trimellitimide, cited above, clearly demonstrate that there is a critical value of the MI around 9.5 above which mesogenic properties are reported. This value is about half to one unit lower than observed with polyesters and polyamides. It is open to speculation whether this a real difference, possibly reflecting an inductive effect due to π overlap of neighbouring chains, or whether it is just within the bounds of normal variation.

Further structures of copolymers containing the trimellitic anhydride unit are given in Table 14 together with MI score and the nature of the melt. The structures are numbered as they appear in Kricheldorf's review article [41]. Ten PEI structures are given and once again all those with MI score ≥9.4 (eight copolymers) exhibited thermotropic behaviour in the melt. Eight of the ten PEIs contain long alkylene, alkylene ether or thioether segments in the main chain; one copolymer (no. 26) contained a mono-substituent in a ring ortho- to an ester group. and one (no. 35) contains an ortho- linkage. The latter possessed a mesophase despite the ortho- linkage scoring MI=10 (assumed ortho- link contributed zero). Five different substituents were employed (chlorine, phenyl, oxyphenyl, thiophenyl or iso-butyl. All five units have been assigned an MI score of –1 and all were thermotropic as predicted with a borderline score of 10. In the case of polymer no. 27, there are two ring substituents on opposite sides of the

Table 14. Poly(esterimide)s based on trimellitic anhydride

Polymer ref. no.	Structure	Mesogenic Index	Nature of melt
15.	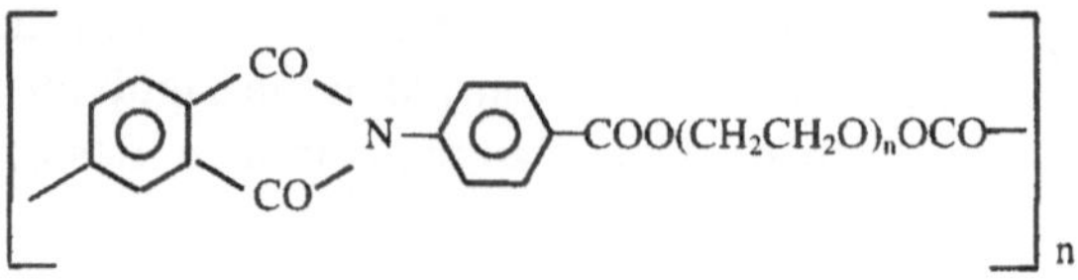	8.66	Isotropic
17.	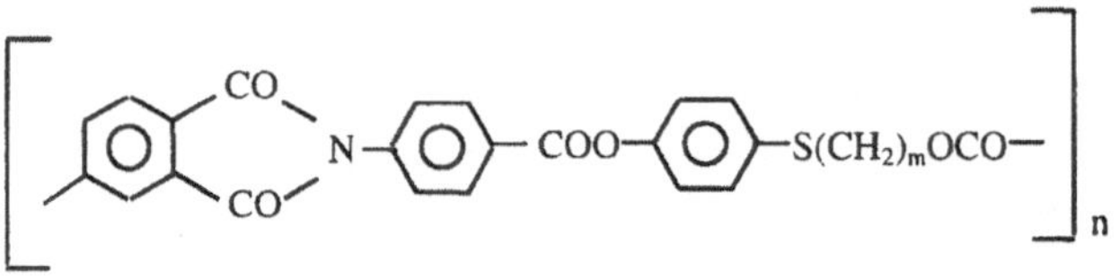		
	n even	10.7	LC phase
	n odd	10.0	LC phase
20.	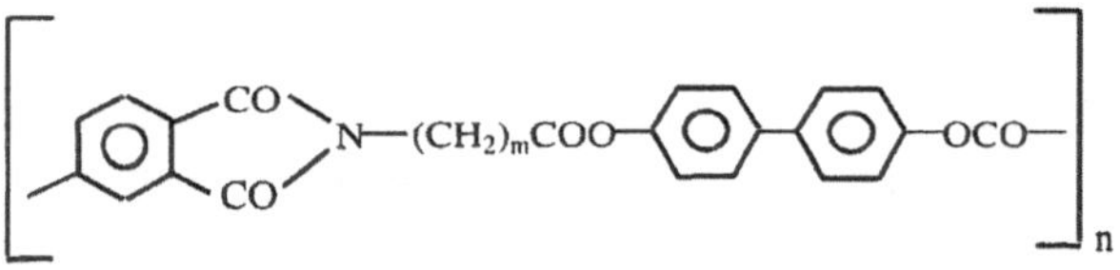		
	n even	11.3	LC phase
	n odd	10.7	LC phase
26.	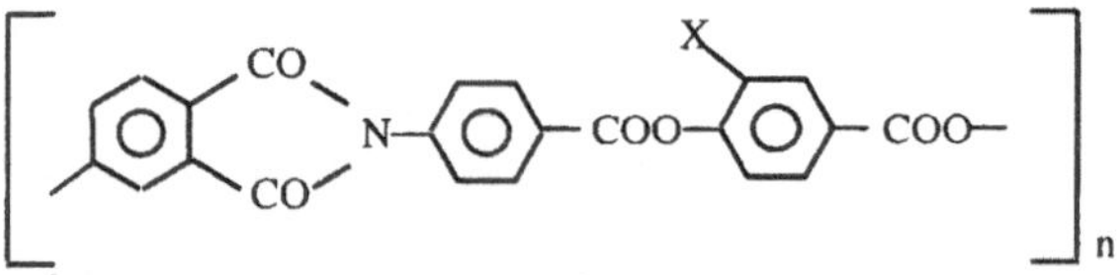		
	For X=C_6H_5, O C_6H_5, S C_6H_5, $C(Me)_3$	10.0	LC phase
	For X=Cl	10.3	LC phase
27.	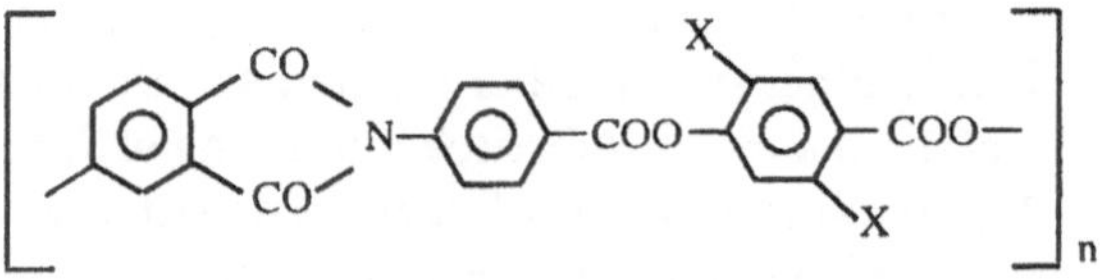		
	For X=OCH_3	9.4	LC phase
	For X=C_6H_5, O C_6H_5, S C_6H_5	9.4	Isotropic

Table 14. (continued)

Polymer ref. no.	Structure	Mesogenic Index	Nature of melt

18.

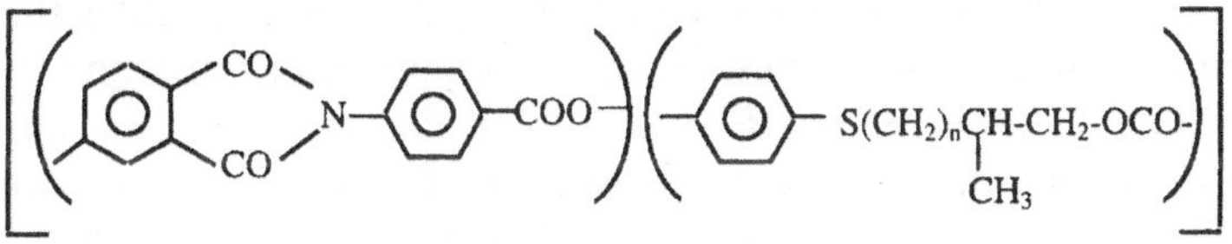

10.0   LC phase

9.

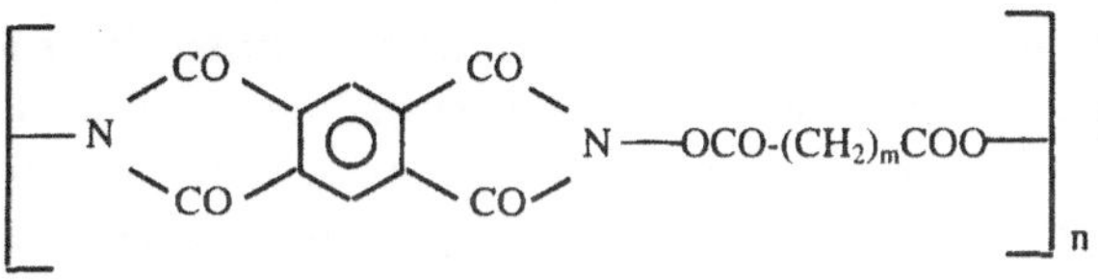

12.0   LC phase

11.

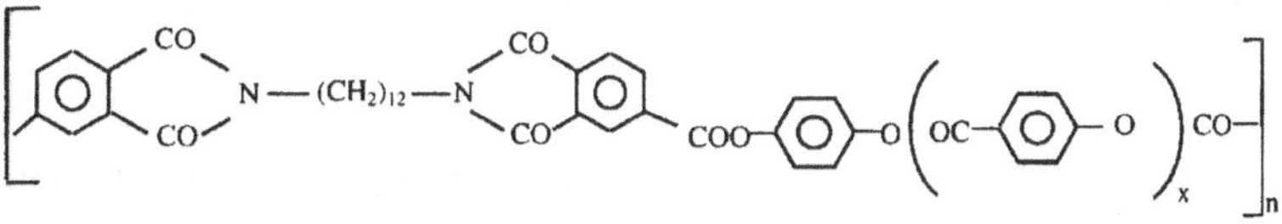

	Mesogenic Index	Nature of melt
For X=1	15.4	LC phase
For X=2	14.6	LC phase
For X=3	14.0	LC phase

13.

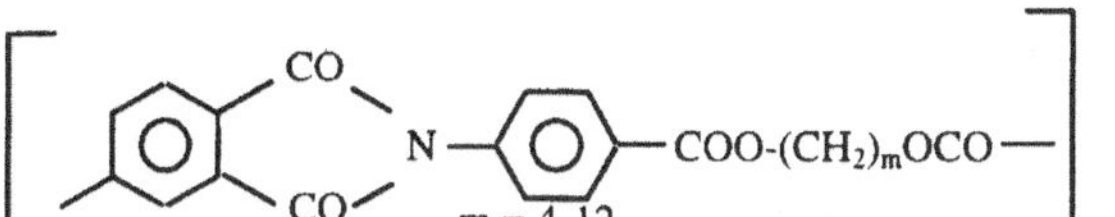

8.7 (m even) Isotropic
8.0 (m odd) Isotropic

35.

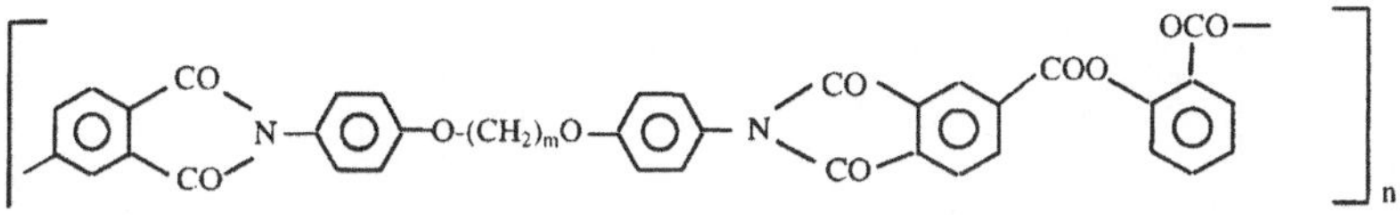

10.0   LC phase

Source: Mol Cryst Liq Cryst (1994) 254:87–108

ring. Under the MI rules they must be scored twice and this leads consequently to a borderline MI score of 9.4, with three out of four polymers being isotropic. A mesophase is observed when the ring substituents are methoxy.

Six out of ten of the above copolymers also contained units derived from tri-mellitimide and p-aminobenzoic acid. In the following five tables the PEIs are based on this unit. Table 15 contains four copolymers of this unit alternating with thioalkylene ester units [26]. It is evident that the ester unit must contain an additional aromatic ring in order to bring the MI above the critical value. Pendant methyl groups along the flexible chain have an adverse effect (scoring – 1) but this does not prevent polymer no.8 from achieving a thermotropic melt with a score of just 10.

Table 16 contains three series of copolymers in which the N-(4-carboxyphe-nyl) trimellitimide unit is co-reacted with various substituted hydroquinones [32]. In series no.1, all four polymers are liquid crystalline, scoring 10–10.3, despite a single ring substituent. In series 2, two ring substituents combine to drive the MI down to the critical value: four out of five polymers scoring 9.3 are isotropic; exceptionally, with two methoxy ring substituents a mesophase is observed. In the case of polymer no.3, the presence of three ring substituents lowers the MI even more and gives an isotropic copolymer.

Table 15. Poly(esterimide)s of trimellitimide of aminobenzoic acid and alkylene spacer

Polymer ref. no.	Structure	Mesogenic Index	Nature of melt
6.	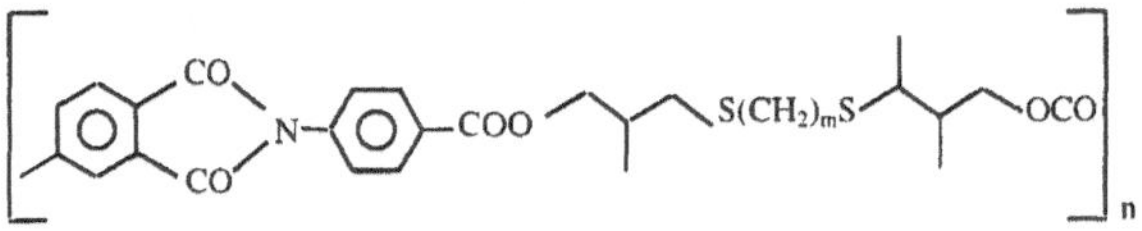	7.4	Isotropic
7.	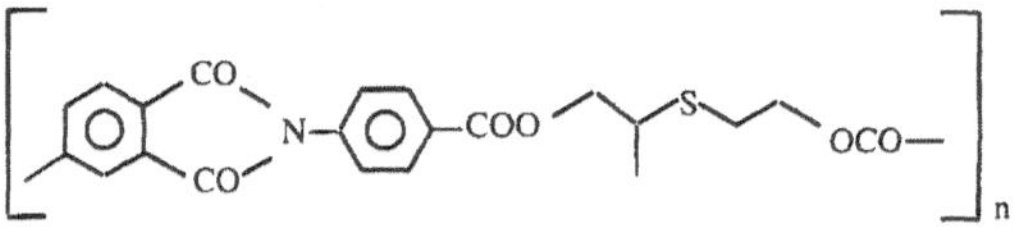	8.0	Isotropic
8.	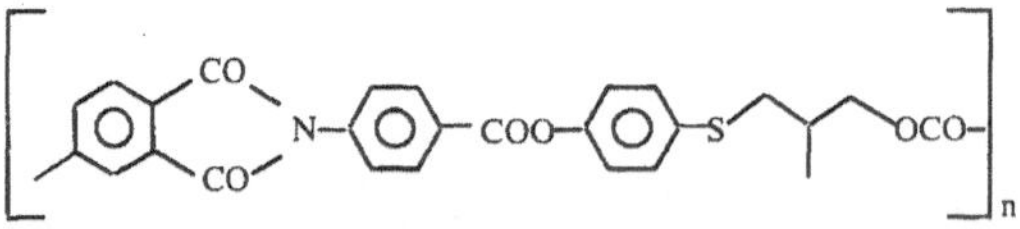	10.0	L.C. phase
9.	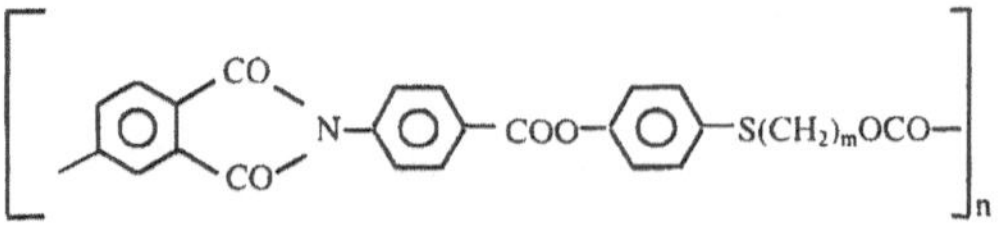	10.7	L.C. phase

Source: J Poly Sci Pt A Polymer Chem (1995) 33:427–439

Table 16. Poly(esterimide)s of carboxyphenyl trimellitimides and hydroquinones

Polymer ref. no.	Structure	Mesogenic Index	Nature of melt
1.	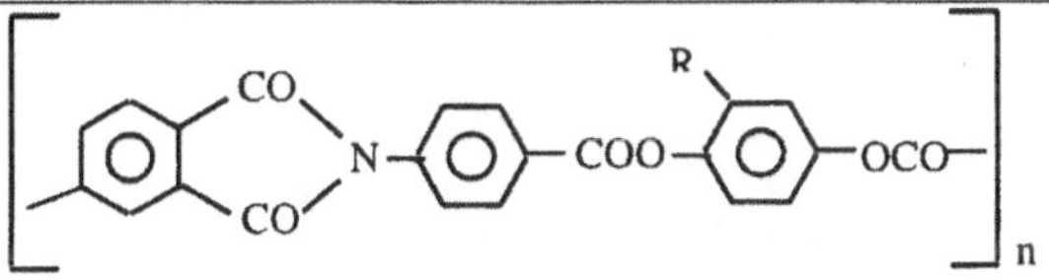		
	a. R=CH$_3$	10.0	LC phase
	b. R=Cl	10.3	LC phase
	c. R=C(CH$_3$)$_3$	10.0	LC phase
	d. R=C$_6$H$_5$	10.0	LC phase
2.	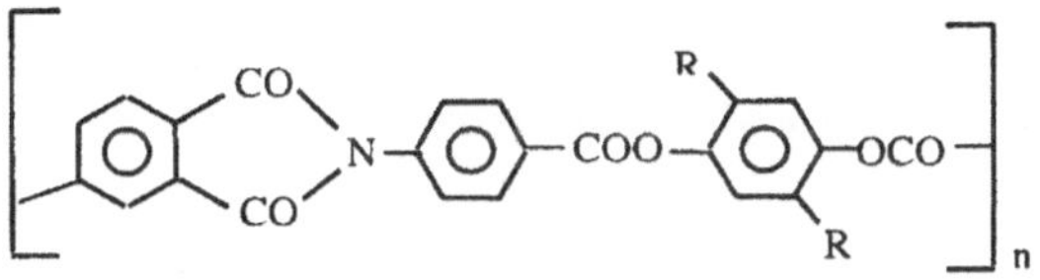		
	a. R=OCH$_3$	9.3	LC phase
	b. R=C(CH$_3$)$_3$	9.3	Isotropic
	c. R=C$_6$H$_5$, O C$_6$H$_5$, S C$_6$H$_5$	9.3	Isotropic
3.	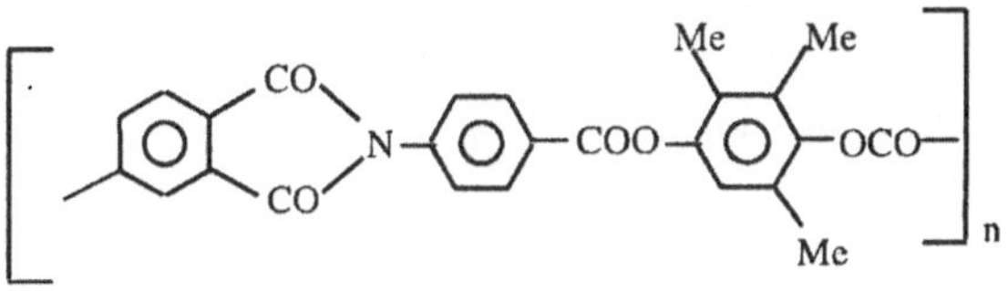	9.3	Isotropic

Source: Macromolecule (1993) 26(19):5161–5168

Table 17 comprises cholesteric PEIs derived from *N*-(4-carboxyphenyl) trimellitimide and phenylene diacrylic acid [77]. The phenylacrylic unit has been scored at +6 and reflects an additional unit for conjugation. Both units in polymer 7 score 14 and hence one might expect any combination of the two units to be mesogenic. The four ratios of the two units given for polymer no. 9 are both mesogenic as both constituent units are mesogenic in their own right. The methyl substituted thioalkene ester of *N*-(4-carboxyphenyl) trimellitimde is just mesogenic with a score of 10. All predictions of the MI are correct in this table.

Table 18 contains two copolymer series based on *N*-(4-carboxyphenyl) trimellitimide and various hinged diphenols [35]. In series 2, when X=O, S or C(CH$_3$)$_2$ the MI correctly predicts an isotropic polymer. For the biphenyl case, a mesophase is predicted but as the polymer is infusible the prediction cannot be tested. In one case, when the hinged group is carbonyl, an isotropic copolymer is incorrectly predicted. In order to make a correct prediction in this case, the

Table 17. Cholesteric poly(esterimide)s derived from phenylene diacrylic acid and 4-aminobenzoic trimellitimide

Polymer ref. no.	Structure	Mesogenic Index	Nature of melt
7.		/	
		14	LC phase
	(phenylacrylic acid is scored at MI=6 due to conjugation)		
8.	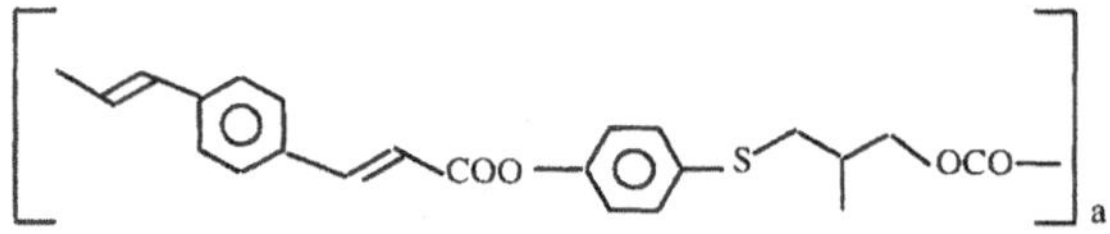 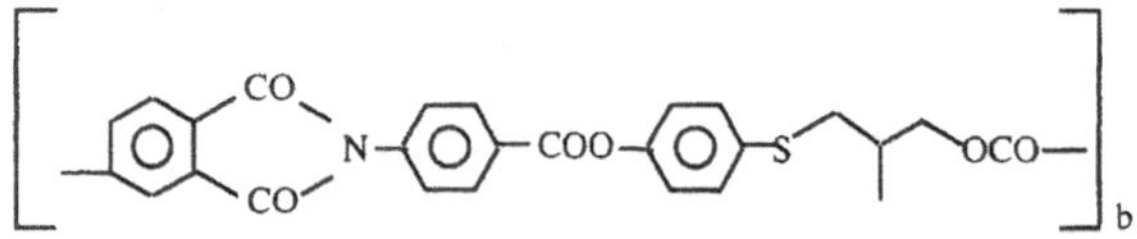	/	
	a:b::1:1	11.6	LC phase
	a:b::3:7	11.2	LC phase
	a:b::1:9	10.4	LC phase
	a:b::0:10	10.0	LC phase

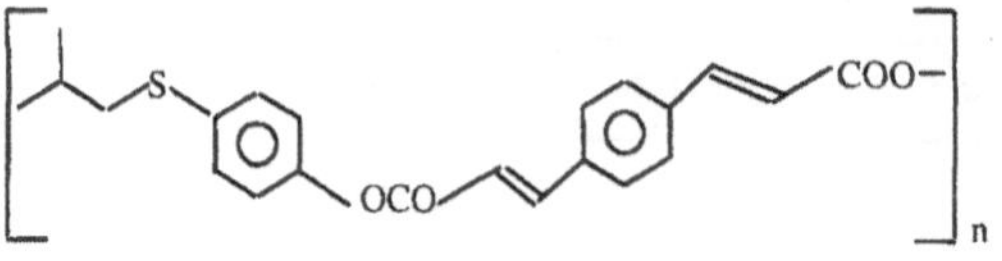

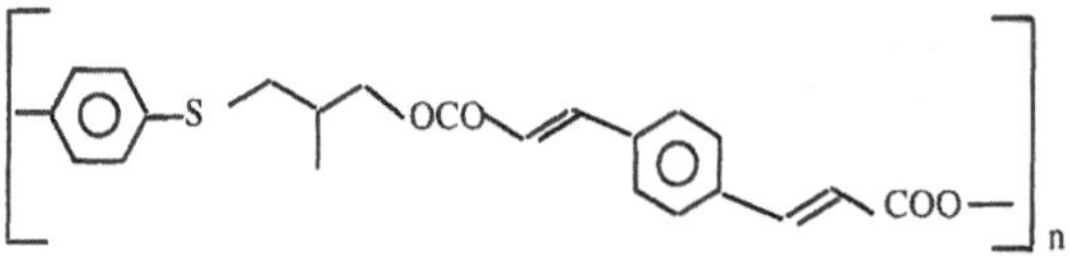

Reference: Polymer (1995) 36:1893

CO would need to make a contribution of at least +1 in conjunction with the two diphenyl groups, compared to MI=0 assigned. It might be expected that some conjugation acts in the main chain direction creating additional stiffness and mesogenicity. In which case one unit should be added for conjugation. However, we have too few examples with C=O groups to make a confident assignment: the value of zero was assigned originally for carbonyl diphenyl ester units. Two other PEI with this imide unit have been cited in the literature; in the first case a score of +1 would not alter the correct prediction of the MI for an isotropic polymer (see Table 22); however, if the two examples cited in Table 10 (polymer 11) were given a CO score of +1 then the MI would incorrectly predict a mesophase. It is therefore perfectly feasible that carbonyl diphenylenediimide makes a pos-

Table 18. Poly(esterimide)s from *N*-carboxyphenyl trimellitimide and diphenols

Polymer ref. no.	Structure	Mesogenic Index	Nature of melt
2. 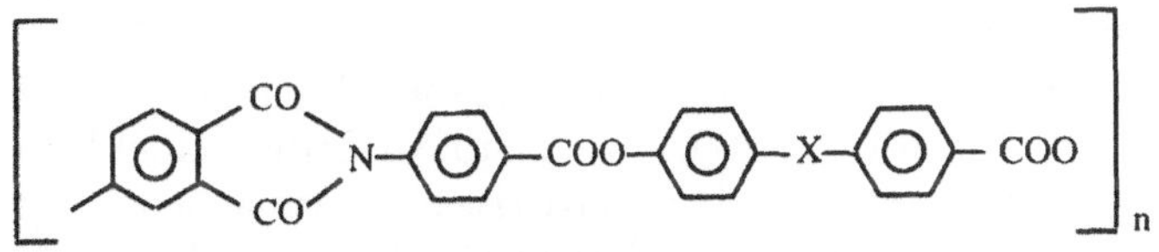	(polymer structure)		
a.	X=Nil (MI unit score=7)	13.3	Infusible
b.	X=O (MI unit score=1)	9.3	Isotropic
c.	X=CO (MI unit score=0)	8.7	LC phase
d.	X=S (MI unit score=1)	9.3	Isotropic
e.	X=C(CH$_3$)$_3$ (MI unit score=0)	8.7	Isotropic
3. 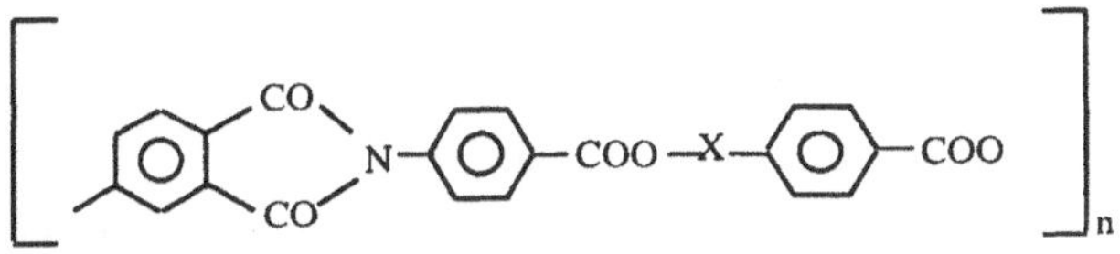	(polymer structure)		
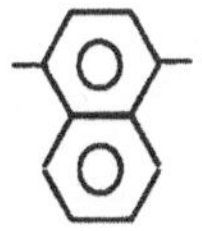 a.	X= (structure) (MI unit score=2)	10.0	Isotropic
b.	X= 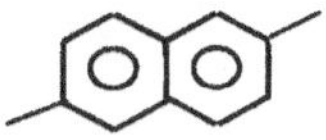 (structure) (MI unit score=3)	10.3	Infusible
c.	X= 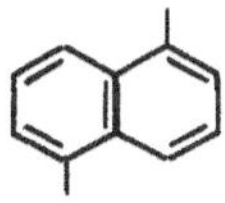 (structure) (MI unit score=3)	10.3	Infusible
d.	X= 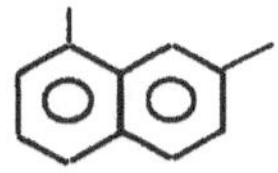 (structure) (MI unit score=1)	9.4	Isotropic

Source: Eur J (1992) 28(3):261–265

itive contribution to mesogenicity, but more examples of polyimides containing this unit are needed to judge the magnitude of the effect. It may be that in certain cases, π orbital overlaps occurs with neighbouring chains, and that the effective MI score of the polymer is enhanced by donor-acceptor interactions.

The second polymer series in Table 18 consists of a naphthalene comonomer reacted with the trimellitimide unit. In two cases a correct prediction of an isotropic copolymer is obtained. In two other cases the prediction of a mesophase is not tested as the polymers are infusible.

Table 19 lists two further series of polymers in which the *N*-(4-carboxyphenyl) trimellitimide unit is co-reacted with substituted and/or meta- oriented aromatic diols. Polymers 5a and 5b are correctly predicted to contain a mesophase, each having a single thioalkyl substituent on the hydroquinone co-monomer. However, when there are two substituents on the ring, the MI rule is to subtract –1 for each substituent-which reduces the MI score to 9.3 which is marginally below the critcial value of 9.5 needed to correctly predict LC phases for polymers 5c and 5d. The shorter chains give rise to a normal nematic phase even in the bisubstituted polymer, whereas it seems likely that the longer chains form a kind of layered structure in the melt which may be called a liquid sanidic phase or a biaxially oriented nematic phase. The inference of this is that there is likely a critical chain length at which alkyl substituents no longer have an antagonistic effect upon mesophase formation. The second series (no.6, Table 19) are correct-

Table 19. Polyesterimides derived from trimellitimide

Polymer ref. no.	Structure		Mesogenic Index	Nature of melt
5.				
a.	$X=H, Z=S(CH_2)_7CH_3$		10.0	LC phase
b.	$X=H, Z=S(CH_2)_{15}CH_3$		10.0	LC phase
c.	$X=Z=S(CH_2)_7CH_3$		9.3	LC phase
d.	$X=Z=S(CH_2)_{15}CH_3$		9.3	LC phase
6.				
a.	Ar= 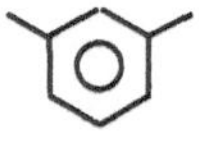	(MI unit score=1)	9.3	Isotropic
b.	Ar= 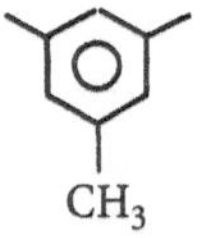	(MI unit score=0)	8.7	Isotropic
c.	Ar 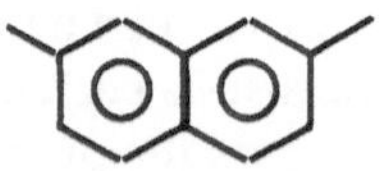 =	(MI unit score = 2)	10.0	Infusible

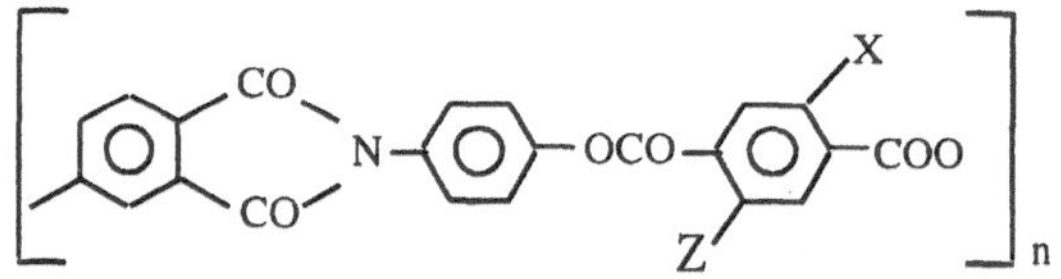

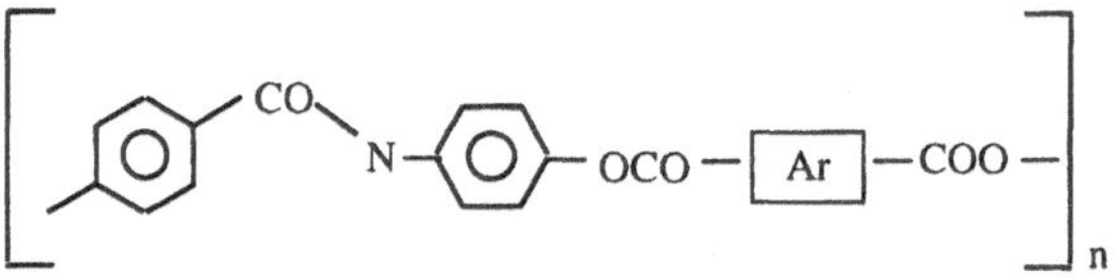

Source: Macromolecules (1993) 26(19):5161–68

ly predicted to be isotropic when resorcinol or methyl substituted resorcinol are reacted with the N-(4-carboxyphenyl) trimellitimide.

A series of ordered PEIs based on an alkylene diimide of trimellitic anhydride have been synthesised (24) by reacting with aromatic diphenols such as hydro-quinone, 2,6-dihydroxy naphthalene and 4,4'-dihydroxybiphenol [14]. All are reported to be mesogenic and contain a smectic A phase in the melt. The diacid unit has an MI score of 14, including the two ester groups formed from it, so that it might be expected to form a mesogenic polymer irrespective of whether the diol was aromatic or just a spacer. This applies because the diacid unit is precon-structed and should not rearrange in the melt. For these copolymers it is permis-sible to add the scores of the component units together.

$$\underset{HOOC}{\overset{CO}{\underset{CO}{\bigcirc}}}N-(CH_2)_m-N\overset{OC}{\underset{OC}{\bigcirc}}COOH \qquad (24)$$

Kricheldorf has synthesised some PEIs based on N-(4-carboxyphenyl) trim-ellitimide (NCPT) units coreacted with *ortho*-diphenols and, surprisingly, ob-tained thermostable enantiotropic nematic melts despite the fact that a linear conformation is energetically unfavourable [32]. On the basis of the Mesogenic Index, it would be anticipated that ortho- substitution would be even more un-favourable than meta-, and an MI score of zero would be appropriate. However, in a copolymer with the NPCT units this would lead to an overall MI score of 8.6-at least one unit below the critical score. One of the four PEIs prepared contained a methyl group on the ring which might be considered to lower the score by a further unit. Three of the four *ortho*-polymers therefore require a contribution of at least +1 for the ortho unit; the remaining polymer perhaps +2. The ques-tion arises as to where this contribution comes from. Kricheldorf offers an ex-planation based on donor-acceptor interactions between temporarily coplanar chain segments. Computer modelling clearly demonstrated that donor-acceptor (DA) interactions of nearly coplanar imide systems dominate the overall confor-mational energy situation.

Insufficient data exists otherwise to make an assessment of the MI contribu-tion of ortho- substituents with other linking groups, although logic dictates a destabilising effect. On the basis of the foregoing, it is suggested that a score of +1 is assigned to ortho- units in polyimides, due to DA enhancement, and zero in other systems.

5.3.2
Polyesterimides Based on Imide Diphenol

Four examples of this unit in PEIs have been reported in Kricheldorf's review [41] and are given in Table 20. Three of these gave a near borderline score of 10 and possessed mesophases, the other being isotropic with a score of 8.7.

Table 20. Poly(esterimide)s based on imide-diphenol units

Polymer ref. no.	Structure	Mesogenic Index	Nature of melt
12.	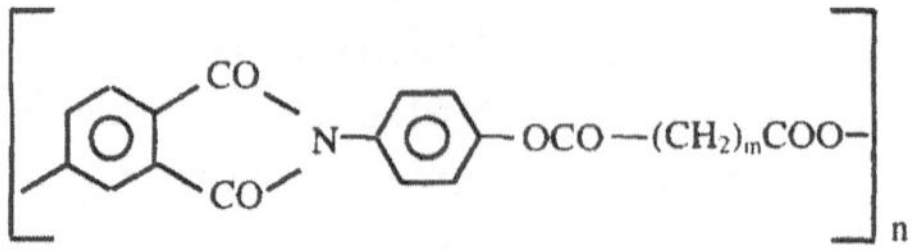	8.66	Isotropic
22, 23.	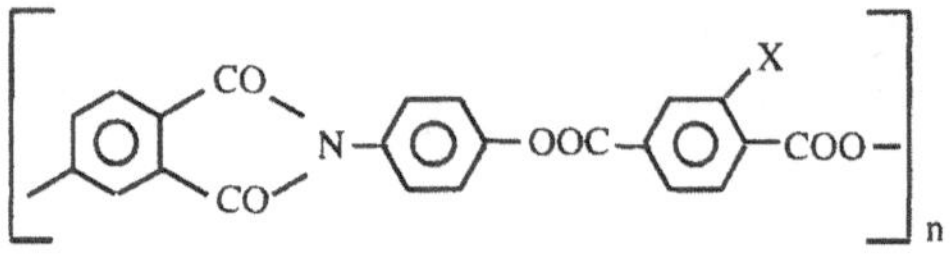 X=O or S	10.0	LC phase
24.	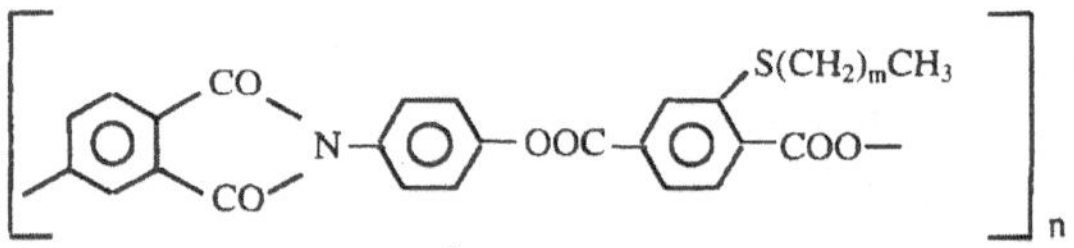 for m=7 or 15	10.0	LC phase

Source: Mol Cryst Liq Cryst (1994) 254:87–108

5.3.3
PEIs Derived from Pyromellitic Anhydride

None of the PEIs listed in Table 21 [41, 78] which contain a pyromellitic anhydride unit, either in conjunction with an aliphatic spacer or with a second aromatic unit, have been reported to contain a mesophase. On the basis of the assignment for the pyromellitic diimide unit of a MI score of 2 (midway between para- and meta-), all of the polymers are predicted to be isotropic. The closest score is obtained with a biphenyl unit as co-monomer (9.5) (which might be expected to be mesogenic with a NCPT unit) but this falls short by 0.5 on the MI scale. It is evident that, due to the lower mesogenicity of imides, at least three co-monomers with benzene rings are needed to boost the MI score to 10. The same comment applied to trimellitimide-based PEIs. On the basis of these results it does appear that the pyromellitic imide copolymers are rather less mesogenic than those containing the NPCT unit.

Table 21. Poly(esterimide)s derived from pyromellitic anhydride

Polymer ref. no.	Structure	Mesogenic Index	Nature of melt
8.	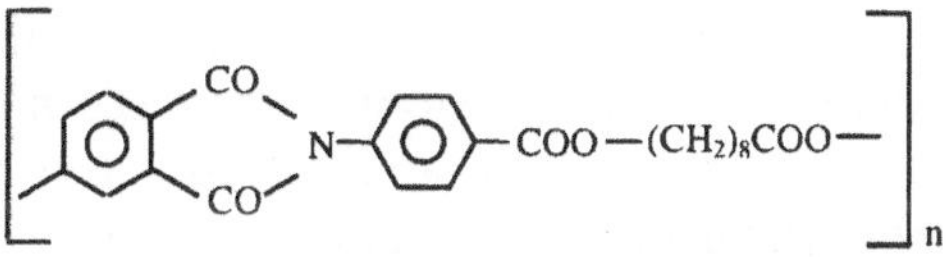	8.7	Isotropic
5.	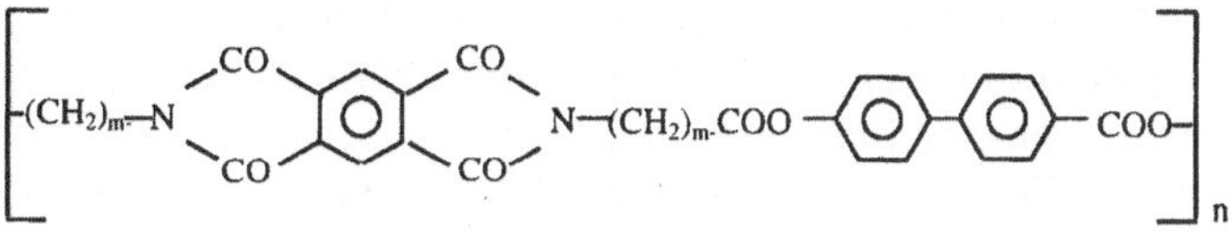		
	when m is even	9.5	Isotropic
	when m is odd	8.5	Isotropic
6.	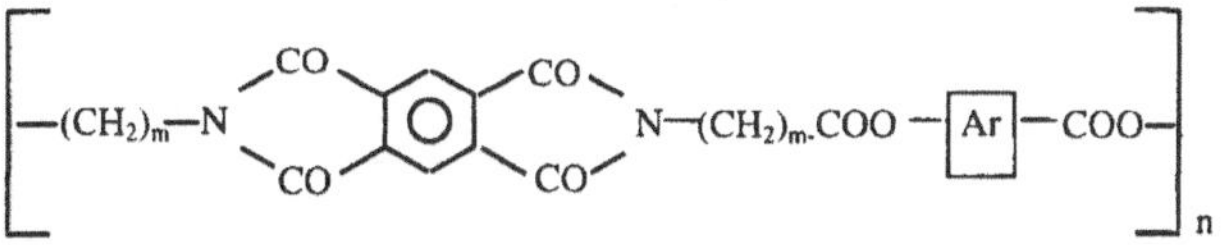		
a, c.	m even Ar= 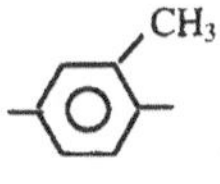 or 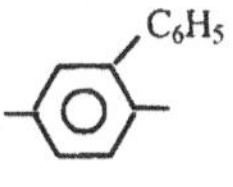	7.0	Isotropic
	m odd	6.0	
b.	m even Ar= 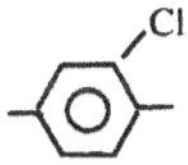 or 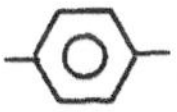	7.5	Isotropic
	m odd	6.5	Isotropic
d.	m even Ar= 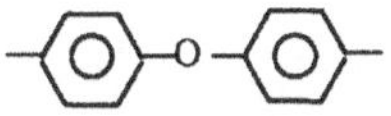	6.5	Isotropic
	m odd	5.5	
	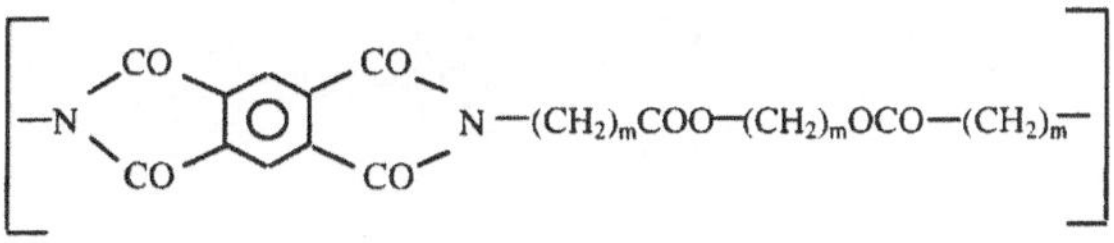	6.0	Isotropic
	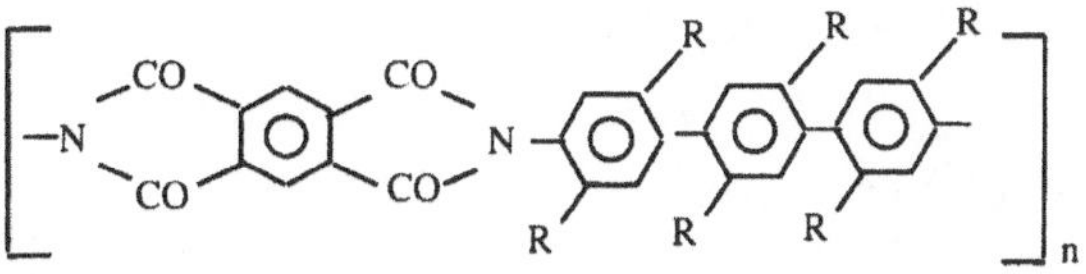	9.0 (R=alkyl)	Isotropic

Source: Makromol Chem (1993) 194:1209–1224

5.3.4
PEIs Based on Linked Biphenyltetracarboxylimide

Table 22 lists four polymer series of PEIs which contain an ether, sulphone or carbonyl linked diphenyl tetracarboxylimide unit [36]. Each of the four series contains two imide and two ester groups and have four aromatic co-monomers. Polymer series 7 and 9 also contain two units of p-aminophenol in conjunction with a bicyclic aromatic dicarboxylic acid, whereas polymer series 6 and 8 are based on m-aminophenol. The Mesogenic Index correctly predicts that series 6, 8 and 9 give rise to isotropic copolymers. However, in the case of series 7, in which the variable monomer was based on ether or thioether diphenol derivatives, thermotropic phases were observed in all three cases. When all four series are viewed as a whole it is clear that an MI value of 9.5 represents a borderline condition for a mesophase to occur. It also corresponds with the minimum value for a mesophase observed in NCPT-based PEIs.

Table 22. Poly(esterimide)s from diphenyl ether tetracarboxylic imide

Polymer ref. no.	Structure	Mesogenic Index	Nature of melt
6.	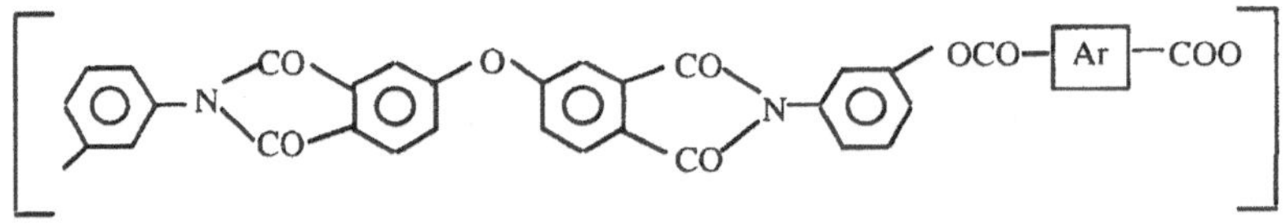		
a.	Ar= 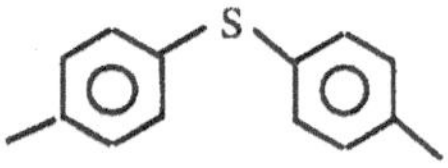	7.0	Isotropic
b.	Ar= 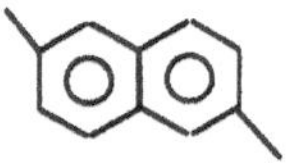	9.0	Isotropic
c.	Ar= 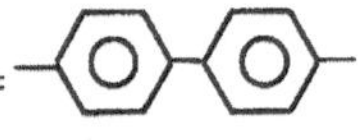	10.0	Isotropic
7.	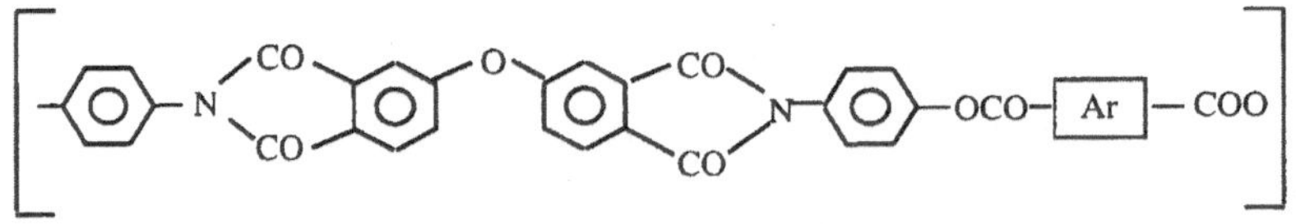		
a, b	Ar= 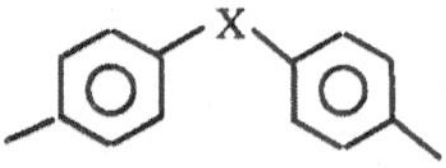 X=O or S	9.5	LC phase
c.	Ar= 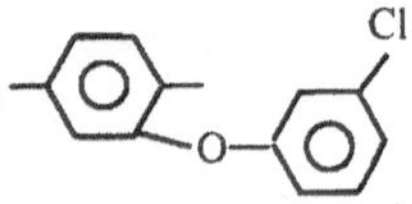	≥10	LC phase

(MI unit score ≥2)

Table 22. (continued)

Polymer ref. no.	Structure	Mesogenic Index	Nature of melt
d.	Ar= 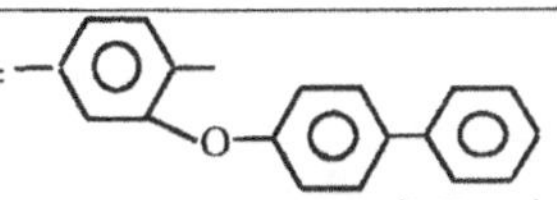 (MI unit score ≥3)	≥10	LC phase
8.	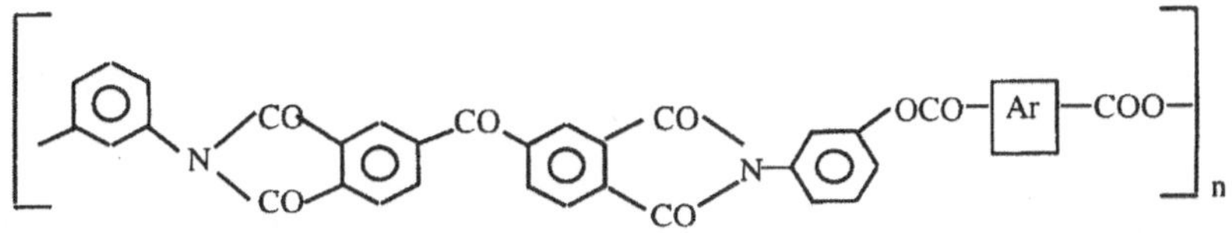		
a.	Ar= 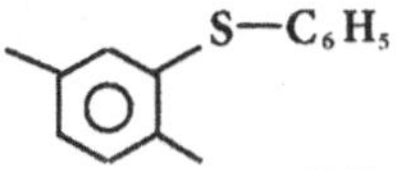 S—C$_6$H$_5$ (MI unit score =2)	7.5	Isotropic
b.	Ar= 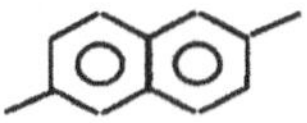 (MI unit score =3)	8.0	Isotropic
9.	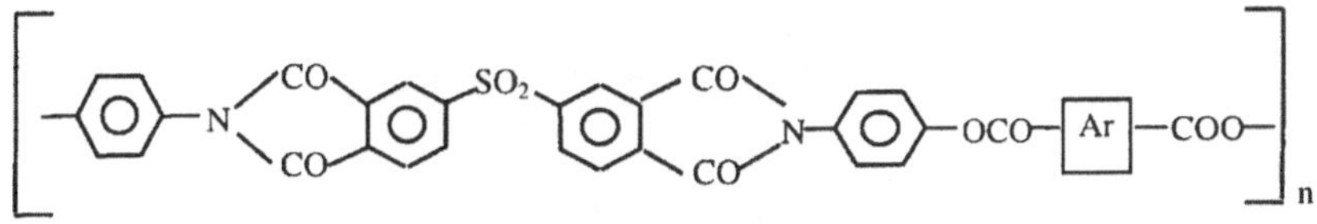		
a.	Ar= 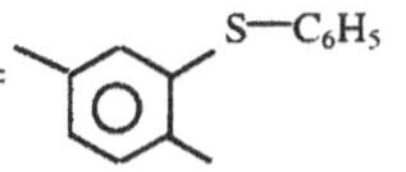 S—C$_6$H$_5$	9.0	Isotropic
b.	Ar= 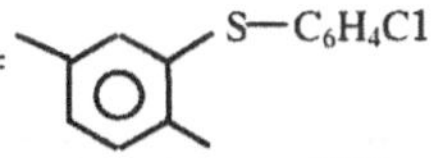 S—C$_6$H$_4$Cl (MI unit score =2–3)	9–9.5	Isotropic

Source: JMS Pure Appl Chem (1995) A32(2):311–330

5.3.5
PEIs Containing Bisphenyl Tetra-imide.(BPTA)

Table 23 demonstrates the mesogenicity of polymers containing biphenylene-3,3',4,4'-tetracarboxylic imide. A comparison of series 9 and 10 in the table shows that replacement of pyromellitic dianhydride with BPTA in a copolymer with two moles of *m*-aminophenol and an aromatic diacid does not lead to mesogenic polymers. The assignment of an MI score of 4 compared to 2 for pyromellitic anhydride does however raise the MI as far as the borderline condition of MI=9.5 when 2,6-naphthalene dicarboxylic acid is the co-monomer.

Table 23. Poly(esterimide)s based on bisphenyl tetracarboxylic acid

Polymer ref. no.	Structure	Mesogenic Index	Nature of melt
9.	OCO–Ar–COO (structure)		
a.	Ar=	7.5	Isotropic
b.	Ar=	8.5	Isotropic
10.	OCO–Ar–COO (structure)		
a, b.	Ar= where X=O or S	8.5	Isotropic
c.	Ar= (MI unit score=2)	9.0	Isotropic
d.	Ar=	9.5	Isotropic
11.	OCO–Ar–COO (structure)		
a, b.	Ar= (X=O or S)	10.5	LC phase
c.	Ar=	11.0	LC phase
d.	Ar= (MI unit score=4)	12.0	LC phase

Source: JMS Pure and Appl Chem (1994) A31(9):1315–1328

Mesogenic PEIs are obtained when *m*-aminophenol is replaced by *p*-aminophenol. A mesophase is correctly predicted in all four polymers of series 11 in Table 23. Each of the four different aromatic acids from ether diphenyl (MI score=1) to ether phenyl diphenyl (MI score=4) gives rise to a mesophase, unlike the series with pyromellitic. The Mesogenic Index approach confirms that BPTA is a much stronger mesogen than pyromellitic diimide. In fact, it predicts that if the fourth co-monomer in series 11 was an aliphatic spacer, a mesophase would still be obtained.

5.3.6
PEIs from N-(4-Carboxyphenyl) 4-Nitrophthalimide

Four polymer series are given in Table 24 [28]. Polymer no. 6 contains a 4-carboxy phenyletherphthalimide unit copolymerised with 4-aminobenzoic acid and a substituted hydroquinone. The MI correctly predicts that a chlorine substituent will lead to a mesophase but not a phenyl substituent.

Table 24. Poly(esterimides)s from N-4-(carboxyphenyl) 4-nitrophthalimide

Polymer ref. no.	Structure		Mesogenic Index	Nature of melt
6.				
	a.	X=Cl	10	LC phase
	b.	X=C$_6$H$_5$	9.7	Isotropic
7.				
	a.	X=Cl	10.3	LC phase
	b.	X=C$_6$H$_5$	10	LC phase
8.			8	Isotropic
9.			8	Isotropic

Source: Eur Poly J (1992) 28(10):1253–1258

Table 25. PEI copolymers of n-carboxyphenyl trimellitimide, isophthalic acid and phenyl-hydroquinone

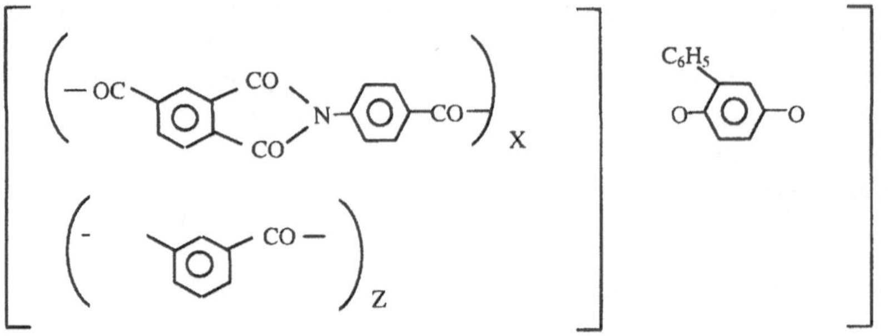

Polymer ref. no.	Structure	Mesogenic Index	Nature of melt
a.	X/Z=9:1	9.9	LC phase
b.	X/Z=8:2	9.8	LC phase
c.	X/Z=7:3	9.7	LC phase
d.	X/Z=6:4	9.6	LC phase
e.	X/Z=5:5	9.5	Isotropic
f.	X/Z=4:6	9.4	Isotropic

Source: J Poly Sci Pt A Chem (1993) 30:337

In series 7 both substituents form a mesophase (MI$\geq$10) when the hydroquinone is copolymerised with N-(4-carboxyphenyl) trimellitimide (Table 25). However, when the substituted hydroquinone in polymer 6 is replaced by aliphatic diols in polymers 8 und 9, then isotropic melt properties are obtained (MI=8). 4-Carboxy phenyletherphthalimide is a poor mesogen and relies on a predominance of ester mesogen in the PEI to achieve a mesophase. By comparison, N-4-carboxyphenyl) trimellitimide is a better mesogen. In PEIs with phenylhydroquinone and isophthalic acid (Table 25), it has a lower critical MI$\geq$9.5.

5.4
Copoly(imide-carbonates)

A series of copolymers containing both imide and carbonate units have been reported by Sato et al. [84, 85]. The structures are depicted in Table 26. Two copolymers have been synthesised and coreacted together in differing ratios; above a 1:1 ratio thermotropic behaviour was observed. The first copolymer was constructed from pyromellitic dianhydride separated from carbonate groups by aliphatic spacers. This polymer was isotropic on its own. The second polymer was a thermotropic polycarbonate containing a biphenyl unit and spacers. The MI score=9.1 at the onset of mesogenic properties. This value is surprisingly low and is about half a unit lower than observed with PEIs.

Hung and Chang [83] have synthesised a number of similar poly(imide-carbonate)s with an aromatic unit replacing one of the spacers, with structure 25.

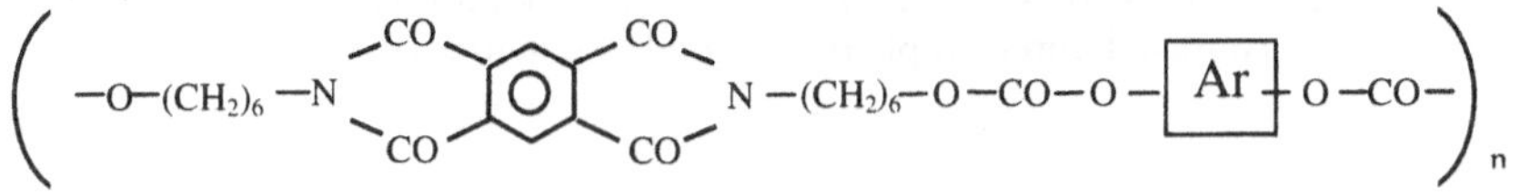

$$\left(-O-(CH_2)_6 -N \overset{CO}{\underset{CO}{\diagdown}} \underset{CO}{\overset{CO}{\diagup}} N -(CH_2)_6-O-CO-O- \boxed{Ar} O-CO- \right)_n \quad (25)$$

The aromatic units [Ar] included 1,3-phenyl; 1,4-phenyl; 2-chloro-, 1,4-phenyl; 2-methyl, 1,4-phenyl; 4,4'-biphenyl. In every case a thermotropic copolymer was obtained. On the basis of the MI system of scoring for random polymers, scores are in the range 7–9, i.e. well below the level anticipated for mesogenic copolymers. Yet this is the scoring method used for copolymers in Table 26. The calculation averages the mesogenic length L between imide and carbonate units (2 for imide and 1.5 for carbonate) giving L=1.75. The total score for four constituent units is a maximum of 2 with biphenyl units. The average score per unit is 5, which yields MI=5.25×1.75=9.2. It is evident that these polymers have significant built in order and that the mesogenicity of the pyrimellitic unit has been enhanced in some way. It is quite likely that there are strong DA-interactions operating between the pyromellitic unit and neighbouring carbonate groups which are electron rich. The overall contribution from this effect must be of the order of two units (effectively a double bond)! If we regard the copolymer as wholly ordered the MI in all cases is >>10. Both effects must play a part in making these copolymers mesogenic as the sequence of imide and carbonate groups are fixed.

Table 26. Copoly(imide-carbonate)s containing pyrromellitic dianhydride

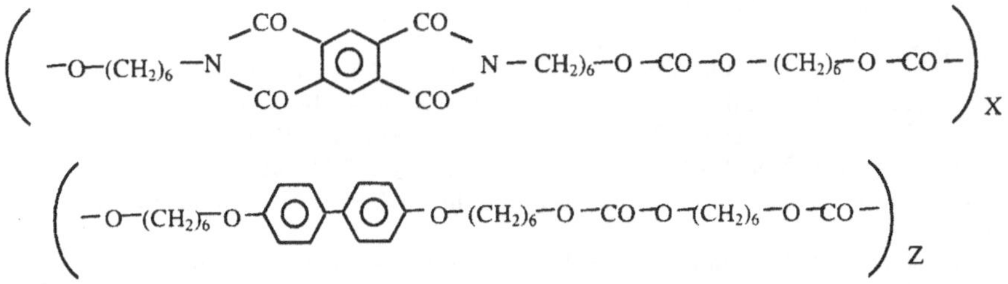

$$\left(-O-(CH_2)_6 -N \overset{CO}{\underset{CO}{\diagdown}} \underset{CO}{\overset{CO}{\diagup}} N - CH_2)_6-O-CO-O-(CH_2)_6-O-CO- \right)_X$$

$$\left(-O-(CH_2)_6-O-\bigcirc-\bigcirc-O-(CH_2)_6-O-CO-O-(CH_2)_6-O-CO- \right)_Z$$

Polymer ref. no.	Structure	Mesogenic Index	Nature of melt
a	X/Z=10:0	6.1	Isotropic
b	X/Z=8:2	7.3	Isotropic
c	X/Z=6:4	8.5	Isotropic
d	X/Z=5:5	9.1	LC phase
e	X/Z=4:6	9.7	LC phase
f	X/Z=2:8	10.8	LC phase

Source: Makromol Chem (1992) 193:1724; (1993) 194:2761

Acknowledgements. The author would like to extend his gratitude to BP Chemicals who originally gave him permission to publish his work on the Mesogenic Index, and to those members of staff at BP Sunbury, especially Paul Alder, who helped and inspired him in the evolution of the MI. He would also like to thank Professor Kricheldorf who made available his publications and encouraged him to apply the method to polyimides.

6
References

1. Kelker H, Hatz R (eds) (1980) Handbook of liquid crystals (1980). Verlag Chemie, Basle
2. Aharoni S (1981) J Poly Sci Phys Chem Ed 19:281
3. Whang W, Wu S (1988) J Poly Sci Part A Polym Chem 26:2749-61
4. Wenzel M, Ballauff M, Wegner G (1987) Makromol Chem 188:2865-2873
5. Dezern J (1988) J Polym Sci Part A Polym Chem 26:2157-2169
6. Irwin RW (1979) US Pat 4,176,223 assigned to Du Pont de Nemours
7. Makoto W, Fujiwara F, Hideo H (1989) Eur Patent 314,173 assigned to Idemitsu Petrochem Corp
8. Sillion B (1989) In: Aalen G, Bevington JC (eds) Comprehensive polymer science, vol 5, chap 30. Pergammon, Oxford
9. Alam S, Kandpal LD, Varma IK (1993) J Macromol Sci Rev Macromol Chem Phys 33: 291
10. de Abajo J (1992) In: Kricheldorf H (ed) Handbook of polymer synthesis, chap 15. Marcel Dekker, New York
11. Sroog CE (1991) Prog Polym Sci 16:561
12. Mustafa IF, Al-Dujaili AH, Alto AT (1990) Acta Polymerica 41:310
13. Kricheldorf HR, Pakull R (1987) Polymer 28:1772
14. Kricheldorf HR, Pakull R (1988) Macromolecules 21:551
15. Kricheldorf HR, Pakull R (1989) New Polymeric Mater 1:165
16. de Abajo J, de la Campa J, Kricheldorf HR, Schwarz G (1990) Makromol Chem 191:537
17. de Abajo J, de la Campa J, Kricheldorf HR, Schwarz G (1991) Polymer 32:942
18. Pardey R, Zhang A, Gabori PA, Harris FW, Cheng SZD, Adduci J, Facinelli JV, Lenz RW (1992) Macromolecules 25:5060
19. Berghahn M (1993) PhD thesis. Univ of Hamburg
20. Berghahn M (1993) PhD thesis. Univ of Hamburg
21. Kricheldorf HR, Pakull R, Buchner S (1989) J Polym Sci Part A Polym Chem 27:431
22. Kricheldorf HR, Pakull R, Buchner S (1988) Macromolecules 21:1929
23. Kricheldorf HR, Probst N (1995) Macromol Chem Phys 196:3511-3523
24. Kricheldorf HR, Probst N (1995) High Perform Polym 7:471-480
25. Kricheldorf HR, Probst N, Gurau M, Berghahn M (1995) Macromolecules 28:6565-6570
26. Kricheldorf HR, Bergmann M (1995) J Poly Sci Part A Polymer Chemistry 33:427-439
27. Kricheldorf HR, Brun C, Russanov A, Komarova L (1993) J Polym Sci Part A Polym Chem 31:279
28. Kricheldorf HR, Domschke A, Bohme S (1992) Eur Polym J 28(10):1253-1258
29. Kricheldorf HR, Schwarz G, Berghahn M (1994) Macromolecules 27:2540-2547
30. Kricheldorf H, Pakull R, Buchner S (1988) Macromolecules 21:1929
31. Kricheldorf H, Domschke A, Schwarz G (1991) Macromolecules 24:1011
32. Kricheldorf H, Schwarz G, Domscke A, Linzer V (1993) Macromolecules 26(19):5161-5168
33. Huner R, Kricheldorf HR (1990) Makromol Chem Rapid Commun 11:211
34. Kricheldorf HR, Linzer V, Bruhn C (1994) Eur Polym J 30(4):549-556
35. De Abajo J, De La Campa J, Kricheldorf HR, Schwarz G (1992) Eur Polym J 28(3):261-265

36. De Abajo J, De La Campa J, Kricheldorf HR, Schwarz G (1995) JMS Pure Appl Chem A32(2):311–330
37. Kricheldorf HR, Linzer V, Bruhn C (1994) JMS-Pure Appl Chem A31(9):1315–1328
38. Kricheldorf HR, Thomsen SA (1993) Makromol Chem Rapid Commun 14:395–400
39. Kricheldorf HR, Gurau M (1995) J Poly Sci Pt A Polymer Chemistry 33:2241–2250
40. Kricheldorf HR, Gurau M (1995) J MatSci Pure Applied Chem A32(11):1831–1846
41. Kricheldorf HR (1994) Mol Cryst Liq Cryst 254:87–108
42. Preston P, Soutar I, Woodfine B, Hay J, Stewart N (1990) High Perform Polym 2(1):47–56
43. Dolden JG, Alder P, Smith P (1995) High Perform Polym 7:421–431
44. Kricheldorf HR, Linzer V (1995) Polymer 36(9):1893–1902
45. Dolden JG, Alder P (1996) High Perf Polym 8:433–444
46. Dolden JG, Alder P (1998) High Perform Polym 10(3):249–272
47. Ciferri A (1982) In: Ciferri A, KrigbaumWR, Meyer RB (eds) Polymer liquid crystals, chap 3. Academic Press
48. Jackson WJ Jr, Morris JC (1985) J Appl Sci Applied Polymers Symposium 41:307
49. McIntyre JE, Milburn AH (1981) British Polymer Journal March:5 [PEA3–4, PEA7–12, PEA14, PEA18–19]
50. Osman MA (1986) Macromolecules 19:1827 [PE4, PE12, PE13]
51. Kricheldorf HR, Linzer V (1995) JMS Pure Applied Chem A 32(2):311–330
52. Kricheldorf HR (1993) Macromolecules 26(19):5161–5168
53. Flory PJ (1956) Proc R Soc London 234:73
54. Krigbaum WR, Brelsford E, Cifferi A (1989) Macromolecules 22:2487
55. Flory PJ, Ronca G (1979) Mol Cryst Liq Cryst 54:311
56. Ronca G, Yoon DY (1982) J Chem Phys 76:32
57. Windle A, Coulter P (1989) Presentation at Reading – Physical aspects of polymer science
58. Dowell F (1988) Mol Cryst Liq Cryst 155:457
59. Percec V, Yourd R (1989) Macromolecules 22:524
60. Antoun S, Lenz RW, Jin JL (1981) J Poly Sci Poly Chem Ed 19:1901–1920
61. Boeffel C, Kranig W, Speiss HW (1990) Phase structure of discotic liquid crystal polymers. 13th International Liquid Crystal Conference, July 22–27 Poster no POL-36P
62. Meurisse P, Noel C, Monnerie L, Fayolle B (1981) Br Polym J 13:55
63. Aharoni SM (1981) J Polym Sci Phys Chem Ed 19:281
64. Irvine PA, Flory PJ (1984) J Chem Soc-Faraday Transact 1 80 (7):1795–1806
65. Vitovskaya MG, Astapenko EP, Nikolaev VJ, Didenko SA, Tsvetkov VN (1976) Vysokomolek Soedin Set A 18:691
66. Kwolek SL (1988) Encyclopaedia of polymer science and technology, 3rd edn. Wiley, New York [PE1, PE10, PE27, PE28]
67. Chen WH, Chang TC (1988) J Poly Sci Polym Chem Ed 26:3269 [PE2, PEA5, PEA6, PEA13]
68. Chapoy LL (1985) In: Chapoy LL (ed) Recent advances in liquid crystalline polymers, chap 4. Elsevier, London [PE5, PE15]
69. Erdemir AA, Johnson DJ, Tomka G (1986) Polymer 27:441 [PE6]
70. Lenz RW, Jin JI (1980) British Polymer Journal Dec:132 [PE9]
71. Leblanc JP, Tessier M, Marachal E (1981) Macromolecular Prepats 89. Conference Proc, Oxford), p 61 [PE11]
72. Aharoni SM (1980) J Appl Poly Sci 25:2891 [PA1–3]
73. Preston J, Krigbaum WR (1982) J Polym Sci Polym Chem Ed 20:79 [PA4]
74. Mahabadi H, Alexandru L (1990) J Poly Sci Pt A Polymer Chemistry 28:231–243
75. Kricheldorf HR, Lubbers D (1990) Macromolecules 23:2656–2662
76. Kricheldorf HR et al. (1994) Eur Poly 30 (8):549–556
77. Kricheldorf HR, Linzer V (1995) Polymer 36:1893
78. Edwards WM, Robinson IM US Pat 2,710,853

79. Cor Koning E, Teuwen L, Meijer EW, Moonen I (1994) Polymer 35:4889
80. Asanuna T, Ockawa H, Ookawa Y, Yamashita W, Matsuo M, Yamaguchi A (1994) J Polym Sci Part A Polym Chem 32:2111
81. Schwarz G, Sun S, Kricheldorf HR, Ohta H, Olkawa H, Yanagudi A (1997) Macromol ChemPhys198:3123
82. Kricheldorf HR, Huner R (1992) J Poly Sci Pt A Polymer Chem 30:337
83. Hung TCh, Chang TCh (1996) J Poly Sci Pt A Poly Chem 34:2465
84. Sato T, Hirata T, Mikaida K (1992) Makromol Chem 193:1724
85. Sato T, Hirata T, Mikaida K (1992) Makromol Chem 194:2861
86. Reinecke H, De la Campa J, De Abajo J, Kricheldorf HR, Schwarz G (1996) Polymer 37:3101
87. Eastman GC, Paprotny J, Webster I (1993) Polymer 34 (13):2865–2874
88. Eastman GC, Paprotny J, Page PCB (1994) Polymer 35 (19):4215–4228 [PE3, PE7, PE8]

Received: March 1998

Author Index Volumes 101–141

Author Index Volumes 1–100 see Volume 100

Subject Index

Springer
Verlag
and the
environment

We at the Springer-Verlag firmly believe that an international science publisher has a special obligation to the environment, and our corporate policies consistently reflect this conviction. We also expect our business partners – paper mills, printers, packaging manufacturers, etc. – to commit themselves to using environmentally friendly materials and production processes. The paper in this book is made from low- or no-chlorine pulp and is acid free, in conformance with international standards for paper permanency.

 Springer